IET ENERGY ENGINEERING SERIES 97

Synchronized Phasor Measurements for Smart Grids

Other volumes in this series:

Volume 1 **Power Circuit Breaker Theory and Design** C.H. Flurscheim (Editor)
Volume 4 **Industrial Microwave Heating** A.C. Metaxas and R.J. Meredith
Volume 7 **Insulators for High Voltages** J.S.T. Looms
Volume 8 **Variable Frequency AC Motor Drive Systems** D. Finney
Volume 10 **SF_6 Switchgear** H.M. Ryan and G.R. Jones
Volume 11 **Conduction and Induction Heating** E.J. Davies
Volume 13 **Statistical Techniques for High Voltage Engineering** W. Hauschild and W. Mosch
Volume 14 **Uninterruptible Power Supplies** J. Platts and J.D. St Aubyn (Editors)
Volume 15 **Digital Protection for Power Systems** A.T. Johns and S.K. Salman
Volume 16 **Electricity Economics and Planning** T.W. Berrie
Volume 18 **Vacuum Switchgear** A. Greenwood
Volume 19 **Electrical Safety: A guide to causes and prevention of hazards** J. Maxwell Adams
Volume 21 **Electricity Distribution Network Design, 2nd Edition** E. Lakervi and E.J. Holmes
Volume 22 **Artificial Intelligence Techniques in Power Systems** K. Warwick, A.O. Ekwue and R. Aggarwal (Editors)
Volume 24 **Power System Commissioning and Maintenance Practice** K. Harker
Volume 25 **Engineers' Handbook of Industrial Microwave Heating** R.J. Meredith
Volume 26 **Small Electric Motors** H. Moczala *et al.*
Volume 27 **AC–DC Power System Analysis** J. Arrillaga and B.C. Smith
Volume 29 **High Voltage Direct Current Transmission, 2nd Edition** J. Arrillaga
Volume 30 **Flexible AC Transmission Systems (FACTS)** Y.-H. Song (Editor)
Volume 31 **Embedded Generation** N. Jenkins *et al.*
Volume 32 **High Voltage Engineering and Testing, 2nd Edition** H.M. Ryan (Editor)
Volume 33 **Overvoltage Protection of Low-Voltage Systems, Revised Edition** P. Hasse
Volume 36 **Voltage Quality in Electrical Power Systems** J. Schlabbach *et al.*
Volume 37 **Electrical Steels for Rotating Machines** P. Beckley
Volume 38 **The Electric Car: Development and future of battery, hybrid and fuel-cell cars** M. Westbrook
Volume 39 **Power Systems Electromagnetic Transients Simulation** J. Arrillaga and N. Watson
Volume 40 **Advances in High Voltage Engineering** M. Haddad and D. Warne
Volume 41 **Electrical Operation of Electrostatic Precipitators** K. Parker
Volume 43 **Thermal Power Plant Simulation and Control** D. Flynn
Volume 44 **Economic Evaluation of Projects in the Electricity Supply Industry** H. Khatib
Volume 45 **Propulsion Systems for Hybrid Vehicles** J. Miller
Volume 46 **Distribution Switchgear** S. Stewart
Volume 47 **Protection of Electricity Distribution Networks, 2nd Edition** J. Gers and E. Holmes
Volume 48 **Wood Pole Overhead Lines** B. Wareing
Volume 49 **Electric Fuses, 3rd Edition** A. Wright and G. Newbery
Volume 50 **Wind Power Integration: Connection and system operational aspects** B. Fox *et al.*
Volume 51 **Short Circuit Currents** J. Schlabbach
Volume 52 **Nuclear Power** J. Wood
Volume 53 **Condition Assessment of High Voltage Insulation in Power System Equipment** R.E. James and Q. Su
Volume 55 **Local Energy: Distributed generation of heat and power** J. Wood
Volume 56 **Condition Monitoring of Rotating Electrical Machines** P. Tavner, L. Ran, J. Penman and H. Sedding
Volume 57 **The Control Techniques Drives and Controls Handbook, 2nd Edition** B. Drury
Volume 58 **Lightning Protection** V. Cooray (Editor)
Volume 59 **Ultracapacitor Applications** J.M. Miller
Volume 62 **Lightning Electromagnetics** V. Cooray
Volume 63 **Energy Storage for Power Systems, 2nd Edition** A. Ter-Gazarian
Volume 65 **Protection of Electricity Distribution Networks, 3rd Edition** J. Gers
Volume 66 **High Voltage Engineering Testing, 3rd Edition** H. Ryan (Editor)
Volume 67 **Multicore Simulation of Power System Transients** F.M. Uriate
Volume 68 **Distribution System Analysis and Automation** J. Gers
Volume 69 **The Lightening Flash, 2nd Edition** V. Cooray (Editor)
Volume 70 **Economic Evaluation of Projects in the Electricity Supply Industry, 3rd Edition** H. Khatib
Volume 72 **Control Circuits in Power Electronics: Practical issues in design and implementation** M. Castilla (Editor)
Volume 73 **Wide Area Monitoring, Protection and Control Systems: The enabler for Smarter Grids** A. Vaccaro and A. Zobaa (Editors)
Volume 74 **Power Electronic Converters and Systems: Frontiers and applications** A. M. Trzynadlowski (Editor)
Volume 75 **Power Distribution Automation** B. Das (Editor)
Volume 76 **Power System Stability: Modelling, analysis and control** B. Om P. Malik
Volume 78 **Numerical Analysis of Power System Transients and Dynamics** A. Ametani (Editor)
Volume 79 **Vehicle-to-Grid: Linking electric vehicles to the smart grid** J. Lu and J. Hossain (Editors)
Volume 81 **Cyber-Physical-Social Systems and Constructs in Electric Power Engineering** Siddharth Suryanarayanan, Robin Roche and Timothy M. Hansen (Editors)
Volume 82 **Periodic Control of Power Electronic Converters** F. Blaabjerg, K. Zhou, D. Wang and Y. Yang
Volume 86 **Advances in Power System Modelling, Control and Stability Analysis** F. Milano (Editor)
Volume 88 **Smarter Energy: From Smart Metering to the Smart Grid** H. Sun, N. Hatziargyriou, H.V. Poor, L. Carpanini and M.A. Sánchez Fornié (Editors)
Volume 89 **Hydrogen Production, Separation and Purification for Energy** A. Basile, F. Dalena, J. Tong, T.N. Veziroğlu (Editors)
Volume 93 **Cogeneration and District Energy Systems: Modelling, Analysis and Optimization** M. A. Rosen and S. Koohi-Fayegh
Volume 95 **Communication, Control and Security Challenges for the Smart Grid** S.M. Muyeen and S. Rahman (Editors)
Volume 101 **Methane and Hydrogen for Energy Storage** R. Carriveau and David S.-K. Ting
Volume 905 **Power system protection, 4 volumes**

Synchronized Phasor Measurements for Smart Grids

Edited by
D. K. Mohanta and M. Jaya Bharata Reddy

The Institution of Engineering and Technology

Published by The Institution of Engineering and Technology, London, United Kingdom

The Institution of Engineering and Technology is registered as a Charity in England & Wales (no. 211014) and Scotland (no. SC038698).

First published 2017

The Institution of Engineering and Technology
Michael Faraday House
Six Hills Way, Stevenage
Herts, SG1 2AY, United Kingdom

www.theiet.org

British Library Cataloguing in Publication Data
A catalogue record for this product is available from the British Library

ISBN 978-1-78561-011-0 (hardback)
ISBN 978-1-78561-012-7 (PDF)

Typeset in India by MPS Limited
Printed in the UK by CPI Group (UK) Ltd, Croydon

Contents

Contributors xi
Foreword xiii
Preface xv

1 Synchrophasors for improving the performance of power system **1**
P.K. Agarwal

1.1 Summary 1
1.2 Evolution of the power grid 2
1.3 Challenges of modern grid 3
1.4 SCADA system 4
1.5 Synchrophasor technology 6
1.5.1 Phasor measurement 7
1.5.2 Time synchronised measurements 7
1.5.3 Wide area measurement system 9
1.6 SCADA vs WAMS 10
1.7 Word-wide deployment of WAMS 11
1.7.1 WAMS in Americas 11
1.7.2 WAMS in Europe 12
1.7.3 WAMS in Asia 12
1.8 Applications of synchrophasors data 13
1.8.1 Fault detection, classification and analysis 14
1.8.2 Low-frequency oscillations 17
1.8.3 Insight to coherency of generators 21
1.8.4 Remote detection of system separation 21
1.8.5 Synchronising two grids from remote 23
1.8.6 Improving power plant performance 24
1.9 Way-forward 27
Acknowledgement 28
References 28

2 An optimal redundancy criterion index (ORC) for optimal placement of phasor measurement units (PMU) for full observability of power grid **29**
Ranganath Vallakati and Prakash Ranganathan

2.1 Introduction 29

2.2 Synchrophasors 30
2.2.1 Advantages of synchrophasors over SCADA measurements 32
2.2.2 Challenges with synchrophasor measurements 32
2.3 Optimal placement of phasor measurement units (PMU) 33
2.3.1 Need for OPP for synchrophasors 34
2.4 PMU placement problem formulation using linear programming (LP) 34
2.4.1 Problem formulation 36
2.4.2 System with no zero-injection buses 36
2.4.3 System with zero-injection measurements 38
2.5 System observability redundancy index (SORI) 39
2.6 Optimal redundancy criterion (ORC) 41
2.7 OPP on standard test systems 43
2.7.1 Advanced Modeling and Programming Language (AMPL) 43
2.7.2 Zero-injection buses 44
2.7.3 Enforcing optimal redundancy criterion (ORC) 46
2.7.4 SORI 46
2.7.5 PMU placement costs 48
2.7.6 Justification of ORC index 49
2.7.7 Time computation using LP 52
2.7.8 Forecasting the scalability of OPP using linear regression 52
2.8 Conclusion 55
Acknowledgments 56
References 56

3 Wide-area measurement-based power network protection 59
Pratim Kundu, Saumendra Sarangi and Ashok Kumar Pradhan

3.1 Introduction 59
3.1.1 WAMS architecture 60
3.1.2 WAMS communication 61
3.2 Improved network protection during stressed conditions 62
3.2.1 Protection support during power swing 63
3.2.2 Improved zone 3 operation 69
3.3 Online identification of protection element failure 75
3.3.1 Faulted line identification 75
3.3.2 Component failure indices 75
3.3.3 Signal communication 76
3.3.4 The method 77
3.3.5 Case study: relay underreaching 78
3.4 Adaptive protection 82
3.4.1 Series compensated line 82
3.4.2 Three terminal line 85
3.5 Conclusion 94
Nomenclature 94
References 95

4 Synchrophasor-assisted visualization and protection of power systems 99
Sukumar Brahma

4.1 Traditional visualization with SCADA 99
4.2 State monitoring with synchrophasors 100
4.3 Disturbance monitoring and identification using synchrophasors 102
4.3.1 Feature extraction 104
4.3.2 Classification 105
4.3.3 Case study: disturbance identification using PMU-generated disturbance files 107
4.4 Use of synchrophasors for power system protection 111
4.4.1 Real-time constraints of power system protection 111
4.4.2 Suitability of PMU measurements for protection 111
4.4.3 Potential improvements in existing protection applications 113
4.4.4 Supervisory protection—potential use of synchrophasors to supervise local protection 114
4.4.5 Case study: synchrophasor-assisted out-of-step protection for generators 115
4.5 Conclusion 121
References 121

5 Using PMU measurements for enhanced power grid monitoring and protection 123
Xiangqing Jiao and Yuan Liao

5.1 Introduction 123
5.2 Monitoring of power system operation: state estimation 123
5.2.1 Background 123
5.2.2 State estimation solution 124
5.2.3 Use of PMUs for state estimation 126
5.3 Monitoring of power system model 127
5.3.1 Introduction 127
5.3.2 Online transmission line parameter estimation 129
5.3.3 Load model monitoring 136
5.3.4 System-level model monitoring 142
5.4 Wide-area fault location based on PMU measurements 143
5.4.1 Introduction 143
5.4.2 Fault location algorithm 143
5.5 Conclusion 145
References 145

6 Fault monitoring, detection, and correction using synchrophasor measurements in modern power systems 147
Anupam Mukherjee and Prakash Ranganathan

6.1 Clustering methods and openPDC interface 147

6.1.1 openPDC setup and data extraction process 147
6.1.2 Application of k-means clustering 148
6.1.3 Application of density-based spatial clustering of applications with noise clustering 152
6.1.4 Development of a novel method, multitier k-means clustering method 154
6.2 Need for a hybrid method 155
6.2.1 Development of multitier k-means method 155
6.2.2 Advantages of using proposed multitier k-means method 160
6.3 Results and discussions 161
6.3.1 Results obtained from data clustering algorithms 161
6.4 Conclusion 175
Acknowledgments 175
References 175

7 Transmission line fault detection, classification, and localization in smart power grids using synchrophasor measurements 177
Pathirikkat Gopakumar, Maddikara Jaya Bharata Reddy and Dusmanta Kumar Mohanta

7.1 Introduction 177
7.2 Real-time transmission line fault monitoring in power grids 177
7.2.1 Emerging methodologies for fault monitoring 178
7.3 EVPA- and ECPA-based transmission line fault detection 179
7.3.1 Real-time estimation of EVPA and ECPA 179
7.3.2 Fault detection using EVPA and ECPA 182
7.4 EVPA- and ECPA-based transmission line fault classification 191
7.5 EVPA- and ECPA-based transmission line fault localization 194
7.5.1 Parent bus identification methodology (bus to which faulty branch is connected) 195
7.5.2 Faulty branch discrimination methodology 195
7.5.3 Fault localization using ECPA in faulty branch 195
7.6 Conclusion 209
References 209

8 PMU-based vulnerability assessment of power systems 211
Guanqun Wang, Chen-Ching Liu, Evangelos Farantatos and Navin Bhatt

8.1 Introduction 211
8.1.1 Cascaded events and vulnerability assessment 211
8.1.2 PMU-based monitoring of power systems 213
8.2 PMU-based monitoring and assessment of rotor-angle stability 214
8.2.1 Maximum Lyapunov exponent (MLE) method 214
8.2.2 MLE calculation algorithm 215

8.3 Observability for PMU-based monitoring of system dynamics 218
8.3.1 Static observability 218
8.3.2 Topology observability 219
8.3.3 Dynamic observability 220
8.3.4 PMU-based monitoring of power system dynamics 221
8.4 Example 222
8.5 Conclusion 225
Acknowledgment 226
References 226

9 Synchrophasor applications for load estimation and stability analysis 231
Tushar, H. Lee, P. Banerjee and A. K. Srivastava

9.1 Introduction 231
9.2 Voltage stability assessment using synchrophasor data 232
9.2.1 Review of existing PMU-based voltage stability 232
9.2.2 Industrial voltage stability application 233
9.2.3 Decentralized voltage stability assessment 240
9.3 PMU-based load modeling 243
9.3.1 Definition of load modeling 243
9.3.2 Classification of load models 244
9.3.3 Review of estimation of load model parameters 249
9.3.4 Advantages of using synchronized phasor measurements for load modeling 250
9.3.5 Parameter identification techniques 250
9.3.6 Estimation of static load parameters 252
9.4 Transient stability assessment using synchrophasor data 257
9.4.1 Review of existing techniques for transient stability 258
9.4.2 MLE-based transient stability assessment 262
9.5 Conclusion 269
References 270

10 State estimation in the presence of synchronized measurement 277
Sanjeev Kumar Mallik, Saikat Chakrabarti and Sri Niwas Singh

10.1 State estimation 277
10.1.1 SCADA measurements 278
10.1.2 Phasor measurement unit 278
10.1.3 State estimation functions 279
10.1.4 Measurement model for static state estimation 279
10.1.5 Assigning weights to the measurements 280
10.2 Conventional state estimation 282
10.3 Linear state estimation 284
10.4 Postprocessing sequential state estimation 286
10.5 Preprocessing hybrid state estimation 287
10.6 Regularization method 290

Chapter 2 discusses the optimal redundancy criterion for placement of PMUs, keeping in view of the fact that the up-gradation of monitoring of system for utilities is an expensive issue. Other challenges that arise during contingency situations such as failure of a PMU or a communication line, the observability of system is lost and thus situation awareness of the operator is decreased in terms of taking appropriate decisions. Thus, concept of redundancy in measurements has been incorporated while formulating optimal placement of PMUs.

Chapter 3 deals with the efficacy of using WAMS data to prevent mal-operation of distance relays during events such as power swings, load encroachment and voltage instability. Synchronized measurement–based adaptive distance relays setting of series compensated and three terminal lines to validate the methodology to discriminate faults from other events to facilitate Zone 3 operation of relays.

Chapter 4 explores the possibility of creating another visualization layer independent of supervisory control and data acquisition (SCADA) system, so as to provide more insight into real-time dynamic events occurring in power systems. A concept of supervisory protection is also explored to increase security of protection schemes by detecting misoperation. Published research findings facilitate critical evaluation regarding feasibility of these concepts.

Chapter 5 discusses the role of PMUs for dynamic model evaluation and correction for efficient state estimation as well as wide-area fault location.

Chapter 6 focuses on extracting useful information from a huge data (at least 1.5 TB in a month) to control smart grid operations such as fault detection and correction using hybrid multi-tier clustering approach.

Chapter 7 puts forth real-time self-healing protection methodologies through fault detection, classification and localization applicable to transmission lines having any configuration unlike the conventional algorithms applicable to specific configurations.

Chapter 8 deliberates on another aspect of self-healing of smart grid using maximal Lyapunov's exponent (MLE) to detect an unstable power swing and to predict rotor angle instability. Based on vulnerable patterns, commensurate portioning control can be initiated to reduce overall system vulnerability by guarding against cascading failures.

Chapter 9 presents the application of synchrophasor measurements for online assessment of voltage stability, real-time load parameter estimation and transient stability prediction. The existing techniques available in the literature for transient stability assessment have been reviewed. Application of MLE for PMU-based online transient stability detection is also discussed.

Chapter 10 explores the hybrid state estimation to integrate the advantages of synchronized phasor measurements with the conventional SCADA measurements. However, the presence of PMUs may cause ill-conditioning, leading to inaccurate estimates as well as to lack of convergence of estimates in extreme cases. In order to overcome such limitations in PMU, current phasors in different forms with uniform distribution of measurement errors are presented to overcome inaccuracy in estimates.

Chapter 11 explores the wide-area security employing Decision Tree-Fuzzy based data mining for catastrophic predictors. It has better transparency compared to black-box models such as support vector machine. The transparency and interpretability of predictor ease its implementation and maintenance.

D. K. Mohanta and
M. Jaya Bharata Reddy

Chapter 1

Synchrophasors for improving the performance of power system

P.K. Agarwal[1]

1.1 Summary

In 2003, the National Academy of Engineering described the vast networks of interconnected aluminium wires and power plants commonly known as the power grid as the greatest engineering achievement of the twentieth century. The electric power grid has expanded from the small cluster of customers served by localised power generators to a nationwide system of systems that delivers electricity to cities and towns of a country. Though the electric grid is like a single-giant machine but is owned and operated by many stakeholders, jointly. The system operator coordinates the joint efforts of the stakeholders from a central control room to keep the power grid operational and live all the time. Conventionally, the central control room has a Supervisory control and data acquisition (SCADA) system which acquires real-time values of electrical parameters by integrating the signals from current transformer (CT) and voltage transformer (PT) installed in the field. The integrated analogue values are converted to digital values through analogue-to-digital converter which is scanned every 2–4 s by the remote terminal unit. In SCADA, data is updated once in 4–10 s at the control centre. System operators, based on the values obtained by SCADA system, take necessary control action to maintain the system parameters within acceptable limits. The updating time of 4–10 s was sufficient for the job of network operators as the electric grid was small in capacity & size and consisting of isolated clusters confined to a small geographical area. However, the brisk pace of addition of generating capacity consequence to the rapid pace of industrialisation resulted in interconnections of geographically apart locations with the long transmission lines. In addition, continuous pressure for grid optimisation is transforming the grid into a complex network. To manage such a large and complex network, system operators worldwide are finding the conventional SCADA system limited in its capability.

Limitation of SCADA systems ability has motivated the system operators worldwide to find an alternative technology which is capable of handling complex

[1]Power System Operation Corporation Ltd, New Delhi 110016, India

and extensive interconnected network. Synchrophasor technology is the technology which has the capability to capture the dynamic behaviour of the grid, unlike the SCADA system which can capture only steady-state conditions of the grid. Synchrophasor technology measures the phasor quantities of the grid voltage and current at a very fast rate (25–30 times a second). Though the concept of phasor quantities existed since the advent of the alternating current (AC) system of electricity generation, its practical use could be possible only in early years of the twenty-first century consequent to the advancement of Information and Communication Technology. The invention of the phasor measurement unit (PMU) enabled the measurement of phasor values of voltage and current. In synchrophasor technology, the measurement of phasors is done in a synchronised manner at several locations with the help of global positioning system (GPS). The measurements are known as Synchrophasor measurements, and the technology is known as Synchrophasors technology. The network of PMUs, deployed all over the grid, is called wide area measurement system (WAMS) or wide area monitoring (WAM) system.

Countries, worldwide, started deploying PMUs in early 1995 which accelerated at the beginning of the twenty-first century. India began its journey to synchrophasor initiatives in the year 2010 with the implementation of a simple pilot project which has now grown to a national level project having 68 PMUs deployed at strategic locations in Indian grid. Further to this, a full-fledged scheme is also under implementation in which about 1,600 PMUs and 35 PDCs will be installed covering the whole of the Indian network. System operators, in India, are utilising synchrophasor data in many innovative ways, for improving the performance of power grid.

This chapter focuses on the introduction of synchrophasor technology and its practical use for enhancing real-time monitoring and control of the electric grid. Many real-life examples have been illustrated with the help of case studies. Real-time event analysis and low-frequency oscillation monitoring are the most basic applications of synchrophasor data which is possible with only a small number of PMUs deployed. Cause analysis of critical incidence and applying corrective action are some other areas where PMUs data is utilised for improvement of the grid performance.

1.2 Evolution of the power grid

Wikipedia defines the power grid as 'An electrical grid' that is an interconnected network for delivering electricity from suppliers to consumers. It consists of generating stations that produce electric power, high-voltage transmission lines that carry power from distant sources to demand centres, and distribution lines that connect individual customers [1].

Evolution of existing power grid started with the establishment of first generating station Pearl Street Station in New York City, the USA in 1882, which was the first complete electric power system connecting 100-V generator that burned coal to power a few hundred lamps in the neighbourhood. Subsequently, many similar self-contained, isolated grid systems were built across America. During this era, two broad types of systems developed: the AC and direct current (DC) networks. Thomas Edison, who designed Pearl Street, was a proponent of DC. In a DC, the electrons flow in a complete circuit, from the generator, through wires and devices and back to the generator. William Stanley, Jr. built the first generator that used AC. Instead of electricity flowing in one direction, the flow switches its direction, back and forth [2]. AC has a significant advantage in that it is possible to transmit AC power at high voltage and convert it into low voltage to serve individual users.

From the late 1800s onward, a patchwork of AC and DC grids cropped up across the country, in direct competition with one another. Small systems consolidated throughout the early 1900s, and local and state governments began – together regulations and regulatory groups. However, even with regulations, some business houses found ways to create elaborate and powerful monopolies. Public outrage at the subsequent costs came to a head during the Great Depression and sparked Federal regulations, as well as projects to provide electricity to rural areas, through the Tennessee Valley Authority and others [2].

AC had a decided advantage over DC for the transportation of electricity over long distances, as it was much easier and cheaper to step-up and step-down voltage. This is due to Ohms law: $I = V/R$, where I is the current, V is the voltage and R is the resistance. By increasing voltage, resistance is reduced. The more resistance exists, the more electricity is lost as heat. Also, the higher the voltage, the smaller the wire can be used. As a result, DC generators had to be located within a mile of the load, whereas AC generators could be constructed much further away [3].

Around the same time, the idea of economies of scale began to be applied to the electrical industry. It became more and more apparent that a large centralised power plant was much more efficient (and cheaper) at providing electricity over a wide area than a small power plant was at providing electricity over a small area. Just before the turn of the century, Westinghouse built a hydroelectric power plant at Niagara Falls, and using the AC technology sent the electricity to Buffalo, NY – 20 miles away! In this way, grid started expanding in the quest of reliability and availability forming today's vast grid and complex network of generators, transmission line and transformers. Evolution of electric grid took place in every country almost in a similar manner [3]. Today's power grid is big and complicated to operate.

1.3 Challenges of modern grid

The challenges in grid operation have increased manifold as a result of enlarged system size; the brisk pace of capacity addition; long distance power flows; multiple players; increasing competition in the electricity market; emphasis on optimisation; climate change; large-scale integration of renewable energy sources;

and increasing customer expectations [4]. The quest for optimisation and ever growing demand for electricity are pushing the operation of grid nearer to its limits. It is a very critical issue for power system operators to keep the electricity grid stable and reliable all the time. The ability of the system operators to take decisions in real-time is directly dependent on the level of situational awareness derived from the data/information available to them in real-time [4]. For this, the grid operators need reliable, fast and accurate information acquiring tools for interacting with the massively large and complex power system. Due to dynamic nature of the power system, the characteristics of the electrical system machine keep changing randomly in response to inherent variability in the grid due to change of load, generation, transmission configuration, variable generation, etc. These challenges have forced the power system fraternity to rethink the power system operation.

Historically, SCADA system has been used as decision support tool for maintaining the performance of electric grid. Inherently, SCADA is an after-the-fact grid parameter-monitoring tool. The operator monitors various parameters that is line/transformers loadings, bus voltages, grid frequency, etc. and takes appropriate action reactively to keep parameters within acceptable limits. However, there is always a time gap between the occurrence of grid events and availability of its details except some limit violation alarms or breaker opening alarms. This shortcoming limits the operators' ability to initiate appropriate action to prepare for the next immediate contingency, if any. Though energy management system (EMS) part of SCADA system provides the applications like contingency analysis, security assessment, etc., but their outputs rely heavily on the credible inputs from SCADA system state estimator. These limitations of SCADA system is emerging as a bottleneck in maintaining the required level of performance of the power grid since some years.

1.4 SCADA system

SCADA is the tool used by utilities worldwide for decades for operating and managing the electric grid. Figure 1.1 shows a typical architecture of a SCADA system.

In SCADA system, the values of grid parameters like voltage, frequency, MW, MVAR, etc. of a substation or generating station are obtained from CT and PTs with the help of transducers. The transducer is an electronic device which transforms electrical signals into equivalent current values generally ranging from 4 to 20 mA. The equivalent current signal is fed to an analogue-to-digital converter board installed in RTU for converting it into digital values suitable to transmit on communication lines. A process, running into the RTU, scans (typically 2–4 s) these analogue-to-digital converter boards periodically and acquires the values of the grid parameters and transmit them to SCADA computers of central control room. The complete process of obtaining values from CT and PT and getting them at control room takes about 4–6 s. Hence, we can say that in SCADA system data are updated every 4–6 s. However, any change in close/open status of circuit

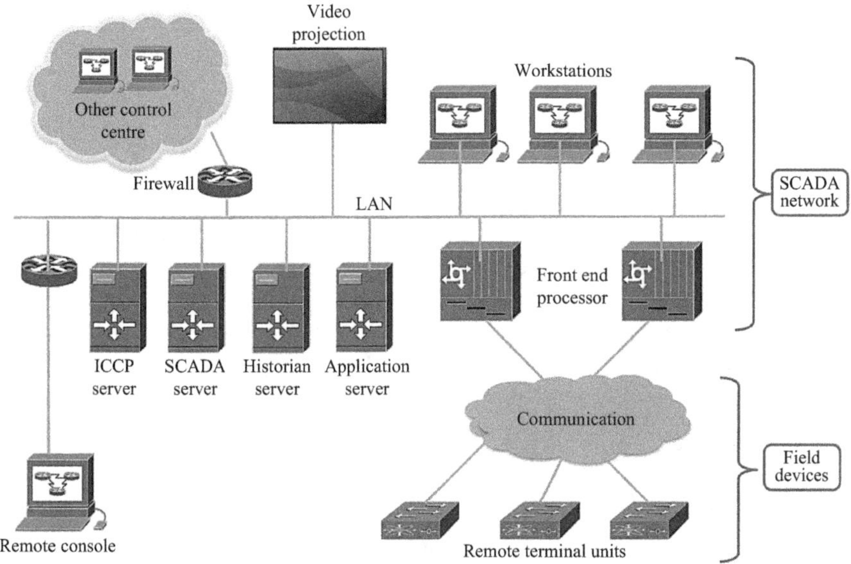

Figure 1.1 Typical architecture of SCADA system

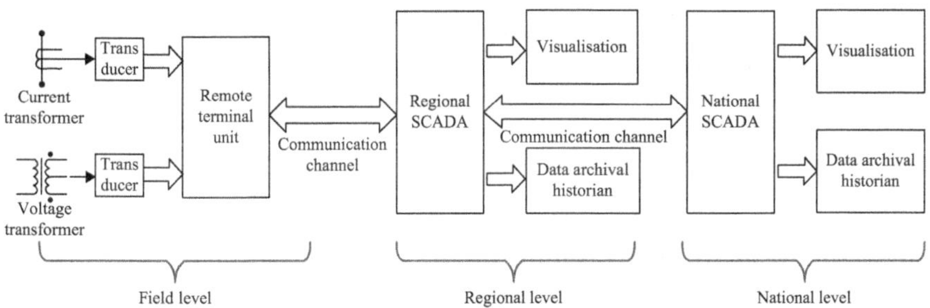

Figure 1.2 Data acquisition process in SCADA system

breaker/isolators is transmitted to control room immediately. In this way, SCADA system maintains the real-time updated view of entire power grid in the central control room and displays on screens helping the grid operators to manage the grid. Figure 1.2 shows the process of data acquisition in a SCADA system.

The SCADA system, in a large grid, has multiple layers of control centres, each managing a particular part of the power grid. Indian power system is the third largest power system in the world. Its transmission network is spread all over India and has many cross-border connections with neighbouring countries like Bhutan, Nepal, and Bangladesh. Indian power grid is demarcated into five regional grids, namely north, west, south, east, and North-east. All five regional grids are connected synchronously. Each region comprises many control areas defined by state boundary. Each state and each regional grid has its own SCADA system and

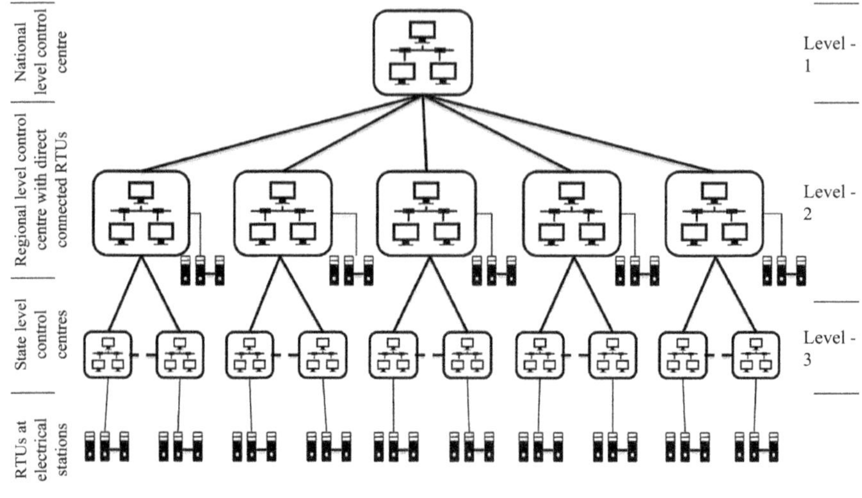

Figure 1.3 Hierarchical SCADA system for large grid

control centres. All these SCADA systems are ultimately connected to National Control Centre SCADA system in a hierarchical manner as shown in Figure 1.3.

Data acquired through RTU is transferred to direct connected SCADA system and again moved to SCADA system of next level in the hierarchy. Hence, typically, each data has to cross two to three levels of hierarchy before reaching the National Control Centre. Due to this total delay between obtaining values from CT/PT and display of data to grid operator may take 6–10 s.

1.5 Synchrophasor technology

Synchrophasor technology is the emerging new technology being used increasingly worldwide for managing power grid. Synchrophasor technology is a technology wherein phasor quantities of power system parameters like voltage and currents are measured, unlike SCADA system which measures only the magnitude. In Synchrophasor system, phasor measurements are done simultaneously by all the devices which are time synchronised with GPS. Synchrophasor technology has the potential to change the economics of power delivery by allowing increased power flow over existing lines. Synchrophasor data could be used to allow power flow-up to line's dynamic limit instead of to its worst-case limit. Synchrophasor technology will usher in a new process for establishing centralised and selective controls for the flow of electrical energy over the grid. These controls will affect both large-scale (multiple states) and individual transmission line sections at intersecting substations. Transmission line congestion (over-loading) and protection and controls will, therefore, be improved on a multiple region scale (US, Canada, Mexico) through interconnecting independent system operators (ISOs) [1].

1.5.1 Phasor measurement

In 1893, Charles Proteus Steinmetz presented a paper on simplified mathematical description of the waveforms of AC electricity. Steinmetz called his representation a phasor. With the invention of PMU in 1988 by Dr Arun G. Phadke and Dr James S. Thorp at Virginia Tech, Steinmetzs technique of phasor calculation evolved into the calculation of real-time phasor measurements that are synchronised to an absolute time reference provided by the GPS. We, therefore, refer to synchronised phasor measurements as synchrophasors. Early prototypes of the PMU were built at Virginia Tech, and Macrodyne built the first PMU (model 1690) in 1992 [1].

A phasor is a complex number that represents both the magnitude and phase angle of the sine waves found in electricity. Phasor measurements that occur at the same time are called 'Synchrophasors'. Although it is commonplace for the terms 'PMU' and 'synchrophasor' to be used interchangeably, they actually represent two separate technical meanings. A synchrophasor is the metered value, whereas the PMU is the metering device. In typical applications, PMUs are sampled from widely dispersed locations in the power system network and synchronised from the common time source of a GPS radio clock. Synchrophasor technology provides a tool for system operators and planners to measure the state of the electrical system and manage power quality [1].

PMUs measure voltages and currents at principal intersecting locations (critical substations) on a power grid and can output accurately time-stamped voltage and current phasors. Because these phasors are truly synchronised, accurate comparison of two quantities is possible in real time. These comparisons can be used to assess system conditions – such as frequency changes, MW, MVARs, kV. (clarification needed). The monitored points are preselected through various studies to make extremely accurate phase angle measurements to indicate shifts in system (grid) stability. The phasor data is collected either on-site or at centralised locations using PDC technologies. The data is then transmitted to a regional monitoring system which is maintained by the local ISO. These ISOs will monitor phasor data from individual PMU's or from as many as 150 PMUs – this monitoring provides an accurate means of establishing controls for power flow from multiple energy generation sources (nuclear, coal, wind, etc.) [1].

A PMU block diagram is shown in Figure 1.4. Each block is detailed below. A voltage or current level matched with the requirements of the converters are obtained by using appropriate measurement transformers. The analogue AC waveforms are digitised by an analogue-to-digital converter. The role of the anti-aliasing filter placed before the signal sampler is to remove the components of the signal whose frequency is equal or greater than one-half the Nyquiste (sample) rate [5].

1.5.2 Time synchronised measurements

As PMUs are located at different locations in the grid, time synchronisation is essential to provide the reference point for the synchronised measurements. Every PMU produces accurately time-stamped values that are sent to a server for further processing. The measurements from different PMUs are combined and provide a comprehensive overview of the entire interconnection and grid stress; alternatively, they are used to trigger corrective actions to maintain stability [6]. To accurately provide the wide view of geographically spread electric grid, the measurements at all the locations must be at the same instant, that is measurements must be time synchronised.

State estimation is one of the applications of power system management system, popularly known as State Estimator, which estimates the voltage and angle of each bus in the grid from various parameters measured by the SCADA system. State estimator output is the fundamental input data to various other applications in EMS. The unsynchronised measurement introduces the time skewed data leading to the inaccuracies. The time synchronised measurements provide accurate and true state of the system.

The analysis of any grid events requires mapping and relating of sequence of events captured from various devices installed in the system. Time synchronisation of all these devices is a critical requirement for accurate analysis of the event. Time-synchronised measurement provides the fast and accurate relation of the various sequence of events and hence in timely and accurately assessing the grid events.

The process of measurement of phasor quantities by PMU is shown in Figure 1.5. Here, unlike SCADA system measurement, PMU samples current and voltage

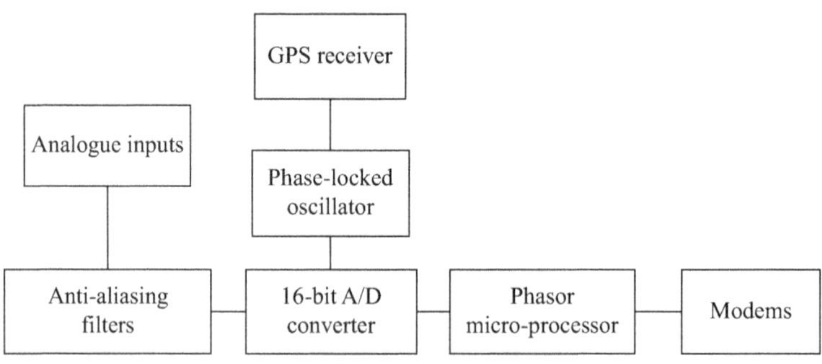

Figure 1.4 Block diagram of PMU

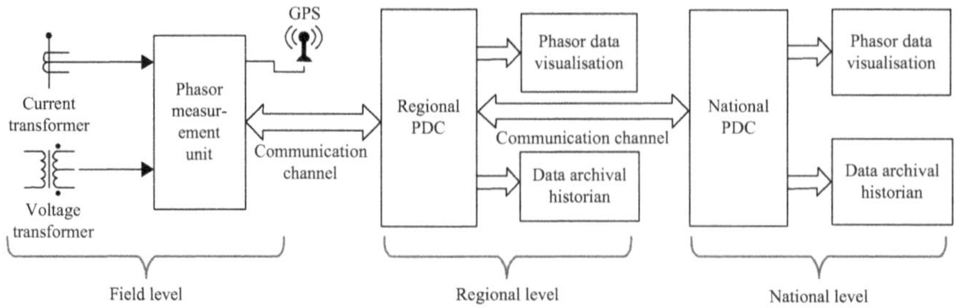

Figure 1.5 Data acquisition process in WAM system

directly from current and PTs. In addition, each PMU is synchronised with a GPS for its time reference providing accuracy better than 1 µs. For time-synchronised measurement, PMU divides each second into equal parts based on the configured sample rate. The first measurement is taken at the start of each second and repeated for each part of the second. All the PMUs follow the same philosophy and are time synchronised with GPS, they produce the measurements taken at the same instant of time from each location.

1.5.3 Wide area measurement system

Power grid consists of many electrical stations which should be measured for effective and efficient monitoring of the power system. Hence, PMUs are installed on nearly all stations or at least on the minimum number of buses so as the calculation of voltage and angle is possible also for the stations which are not measured with PMUs. Typical installations of synchrophasor measurement system have many numbers of PMUs connected to a central system known as PDC via various communication links. The PDC receives the time-stamped measurements from all PMUs and consolidates the same in a packet having the same time-stamped data and made available to the applications like visualisation, oscillation monitoring system, data historian, etc. This network of PMUs, PDCs, historian, visualisation and applications is known as WAMS. Figure 1.6 shows a typical architecture of WAM System.

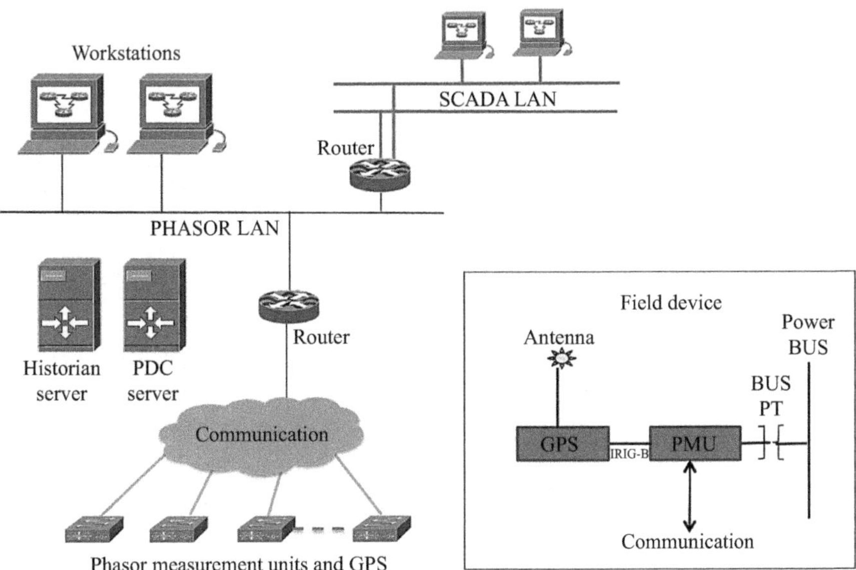

Figure 1.6 Architecture of typical WAM system

1.6 SCADA vs WAMS

Since decades, SCADA system has been used for operation and monitoring of power grid by the system operators worldwide. Historically, power grid was a simple network of generating stations and transmission lines (mostly radials) and did not require elaborate monitoring scheme. It was only necessary to evaluate the steady-state conditions on few seconds time scale rather than continually observe the operating state on millisecond time scale. But the growth of demand, the growth of distributed energy sources such as renewable generation introduces more short-term and unpredicted fluctuations and disturbances. The emerging challenges and the pressing need to ensure stable and reliable operation of the power system motivated the grid operators worldwide to implement emerging new technologies for grid monitoring. World-over the synchrophasor technology is increasingly being used for supplementing the conventional SCADA/EMS for providing refined and fast measurement capable of overcoming the challenges in managing increased variability and uncertainty in the grid.

The time resolution of SCADA measurements is generally of the order of 4–10 s, whereas for WAMS it is 20–40 ms. The USP of Synchrophasor technology or WAM System is the measurement of phasor quantities of voltage and current which gives the value of angular difference at any two points of the grid. The direct measurement of phasor values was not possible in SCADA system except some indirect calculation from the measurements of voltage and current done by transducers. The Indian system operators are utilising the calculation of the angular difference between two pockets of Northern Grid of India to avoid separation under skewed load despatch or power swing conditions between these two pockets (see Figure 1.7). As it was not possible to telemeter the phase angle separation between

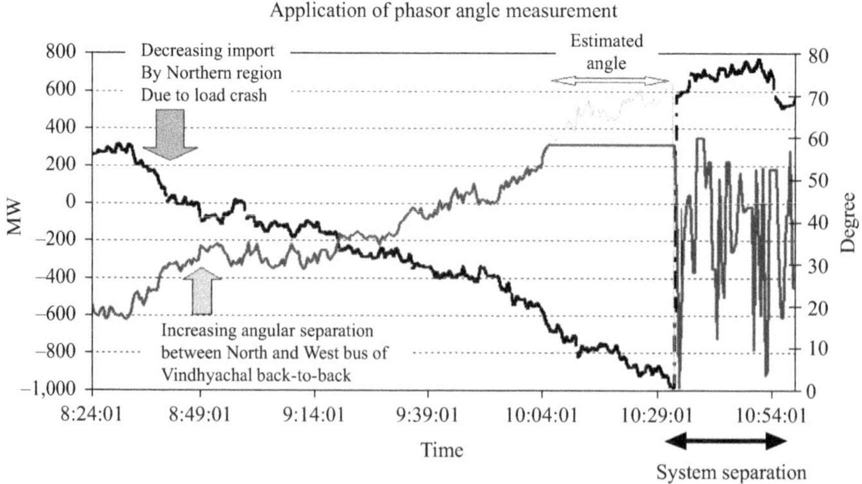

Figure 1.7 Phase angle measurement in SCADA system

Table 1.1 Comparison of SCADA and WAMS

Description	SCADA	WAMS
Data updating	Slow, once in 4–10 s	Fast, once in 20 ms
State measurement	Steady state measurement only	Dynamic state measurement also
Angle measurement	No direct measurement	Direct measurement
State estimation	Non-linear and complex calculation	Linear, easy to calculate

two distant nodes in the grid with conventional SCADA system, the angular separation between them was calculated at the control centre with the help of real-time power flow, bus voltages and network reactance using standard equation Sin $= P \times X/V1 \times V2$ [7]. A comparison of SCADA system and WAM system can be understood form Table 1.1.

1.7 Word-wide deployment of WAMS

Original use of PMUs began in parts of both the Western (such as Bonneville Power Administration (BPA) and Western Area Power Administration) and Eastern Interconnections (such as New York ISO) in pilot projects during the 1990s. Use of Synchrophasors has been increasing since 2004 when the United States–Canada blackout investigation report recognised that many of North Americas major blackouts have been caused by inadequate situational awareness for grid operators and recommended the use of synchrophasor technologies to provide this real-time wide-area grid visibility [8].

1.7.1 WAMS in Americas

The deployment of PMUs was first taken by The BPA of USA in its wide-area monitoring system. The role of PMUs data in uncovering the 2003 North American blackout renewed the DOE (Department of Energy) interest in PMUs deployment. According to North American Reliability Council (NERC), a valuable lesson from the August 14 blackout is the importance of having time-synchronised system data recorders. NERC investigators laboured over thousands of data items to synchronise the sequence of events. That process would have been significantly improved if there had been a sufficient number of synchronised data recording devices. Today there are over 1,700 production-grade PMUs deployed across the United States and Canada, streaming data and providing almost 100 per cent visibility into the bulk power system. In Brazil, a huge blackout in March, 1999 revived the interest in PMU applications, which was derailed due to difficulties faced by Brazilian economy during that decade and the restructuring of the electric energy sector delayed, mainly for dynamic performance analysis during disturbances.

1.7.2 WAMS in Europe

Switzerland and Italy, in 2003, started using WAM systems as an additional tool for power-system operation and analysis. The Croatian transmission system operator HEP-TSO commissioned the first phasor measurement installation in 2003. HEP-TSO acquired the first dedicated PDC as a part of a turn-key system that included five PMUs from the same manufacturer.

1.7.3 WAMS in Asia

China's power transmission grid is run by two major transmission system operators, the State Grid Corporation of China and the smaller China Southern Power Grid. Together, they deployed a number of PMUs as a part of nationwide mandate requiring that all sub-stations in 500-kV networks and above and all generators larger than 100 MW be monitored by PMUs. As in 2013, approximately 2,400 PMUs sets had been deployed in power grid in China.

India started its Synchrophasors initiative in 2010 by installing a pilot project consisting of eight PMUs and one PDC. As in January 2016, total 68 PMUs have been installed by the central utility as well as state utility. First, WAMS project was implemented in India in the year 2009 with a simple pilot project with technical requirements which can be met with readily available products. The project was located in Northern Region (NR). In this pilot project, four PMUs each with a GPS were installed at strategically selected locations, and one PDC was installed at regional control centre managing the northern power grid in India. Project became operational in 2010. Subsequently, five more PMUs were added to NR project. The inputs to PMUs were taken from three-phase voltages of the 400-kV buses and three-phase currents of the feeders of the substations.

In 2011, an Indian origin vendor came up with indigenously designed PDC and installed a demo project in Southern Region (SR) with four PMUs. In this PDC, an oscillation monitoring application was also installed providing small signals modal analysis of the grid on real-time basis. Subsequently, one more demo project was commissioned in Western Region by using open PDC and two PMUs. After experiencing the utility of Synchrophasor data in managing and improving the performance of the grid, similar pilot projects were taken up in other regions which are already operational.

Commissioning of Regional pilot projects were isolated from each other but the regional grids were operating synchronously (except SR grid), hence the need of a unified availability and visualisation of Synchrophasor data from all the PMUs was felt. Consequently, all regional pilot projects were integrated forming a National level WAM project. Further, to improve the observability, additional PMUs were installed.

In the national pilot project, a PDC was installed at National Load Despatch Center, and all regional pilot projects PDCs are integrated with this. At present data from about 68 PMUs located all over India is available at National PDC. The current deployment of PMUs in India is shown in Figure 1.8. Apart from this, some states of India has also started deploying PMUs. The state of Gujarat has planned

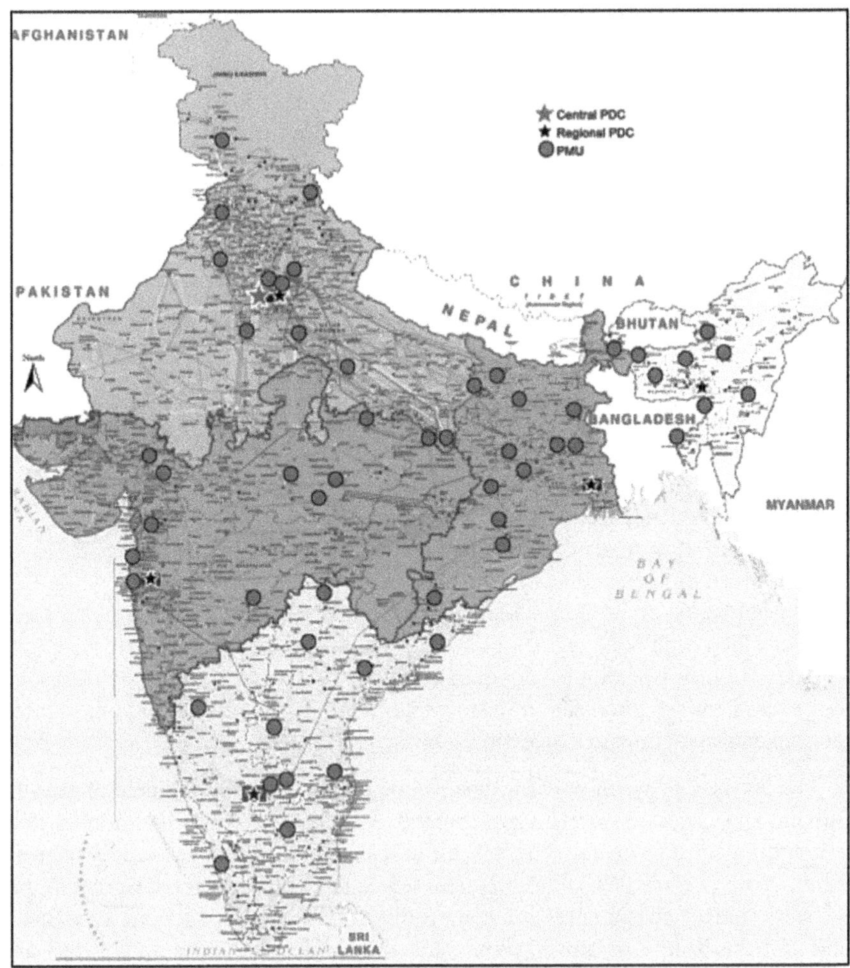

Figure 1.8 PMU deployment in India

for installation of 100 PMUs. Some PMUs have already been installed, and some of them have been integrated to national PDC.

1.8 Applications of synchrophasors data

Synchrophasors data is capable of providing wealth of operational information of electric grid which SCADA was not able to dig out. Synchrophasors data can capture network dynamics, unlike SCADA system which provides steady-state visibilities only. These include fast monitoring of angular separation, oscillatory stability, voltage stability and islanding detection. These solutions bring additional situational awareness for the grid operator [9]. Since implementation of first pilot project in India, grid operators have been utilising synchrophasor data in many ways for their real-time grid operation as mentioned below:

1. Improvement of situational awareness.
2. Detection of occurrence of transmission line tripping/revival within a flow-gate by observing:
   (i) Real-time monitoring of trends of critical grid parameters like frequency, d*f*/d*t*, angular separation.
   (ii) Step change in angular separation.
   (iii) Step change in line current (MW/MVAR).
3. Detection of occurrence of generator tripping by observing:
   (i) Sudden decline in frequency.
   (ii) Increase in d*f*/d*t* value.
   (iii) Change in angular separation.
   (iv) Decrease in voltage magnitude.
4. Detection of occurrence of line circuit breaker auto-reclose event indicated by kink in frequency trend.
5. Detection of occurrence of load crash/load throw off by observing.
6. Sustained high frequency.
7. Sustained abnormal phase angle separation.
8. Sustained high voltage.
9. Detection of islanding of any part of grid.
10. Subsystem synchronisation during restoration by using standing phase angle separation and phase sequence.

The various uses of synchrophasor data can be understood with the help of real-life examples and case studies. The synchrophasor systems are operational in India since 2010 and Indian grid operator have been utilising the synchrophasor data in many innovative ways since then. Power System Operation Corporation Ltd. has compiled such case studies and examples as a compendium and made these reports available in public for the benefit of power system fraternity at large. As the author was also involved in the preparation of these reports, many examples and case studies have been adopted from these reports and have been presented in a lucid manner. The reports are available at website http://www.posoco.in. Many examples have been presented here. The examples and case studies have been categorised in following groups:

1. Fault detection, classification and analysis
2. Low-frequency oscillation
3. Insight to coherency of generators
4. Electrical island detection and their resynchronisation with the grid
5. Model validation
6. Improving power plant performance
7. Monitoring of natural disasters.

1.8.1 Fault detection, classification and analysis

The availability of synchrophasors has considerably enhanced the wide area monitoring and situational awareness of power system behaviour under steady state as well in transient/dynamic conditions. The real-time trends of Synchrophasor data at control centre have become first-hand information for grid operator to view and analyse any transient phenomenon occurring in the grid. Various events that went un-noticed with present SCADA system can now be detected and analysed in almost real-time, opening up a whole new era in power system monitoring and control. With PMU data at the control centre, the grid operators can observe the signature of events as first-hand information and then do further analysis by SCADA alarm system.

Historically, analysis of any grid event is carried out with the help of relay protection flags, disturbance records (DR)/event logger (EL) outputs made available by the stakeholders. However, often challenges are faced on account of issues such as non-availability/healthiness/failure of recording instruments, utility's inhibitions in data sharing, control area jurisdiction, time synchronisation, portability/compatibility of DR/EL records, latency and skewness in SCADA data. The availability of synchrophasor data now has become an effective tool for analysis of grid events and facilitated preparation of an accurate 'first information report' of an event occurring in the grid.

Figure 1.9 shows an event of multiple times fault in HVDC (high voltage direct current) line due to storm in that area. A series of faults were observed from the display of trend data from PMU installed at HVDC terminal in the state of Gujarat, located in western India. The HVDC terminal is a part of HVDC line from Gujarat to Haryana (located in northern India). This indicated the possibility of multiple faults near HVDC location. HVDC station operator confirmed the same. The system control, room Operator, immediately directed HVDC station to shift to the Reduced Voltage Mode of Operation reducing the possibility of further faults.

Figure 1.10 the real time display of voltage signals received from a PMU located at that UMPP (Ultra Mega Power Project) in Gujarat. A voltage dip and simultaneous fall in frequency were observed from the display. Maximum dip in the voltage of this PMU was observed which indicates a fault Phase-Phase Fault. Operator identified that a transmission line from this UMPP has tripped on

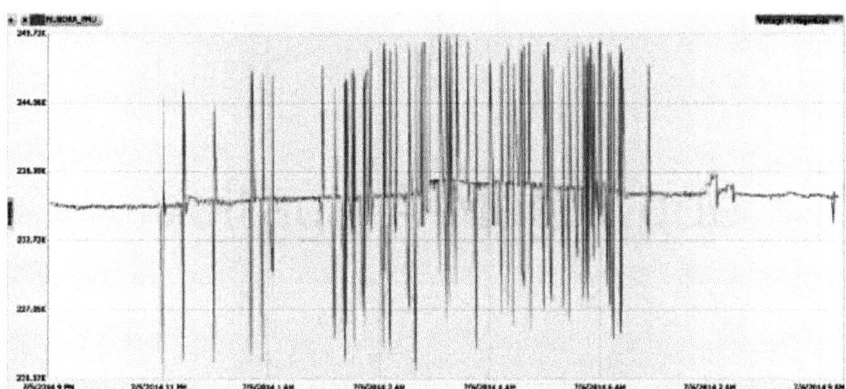

Figure 1.9 Detection of multiple fault in HVDC line

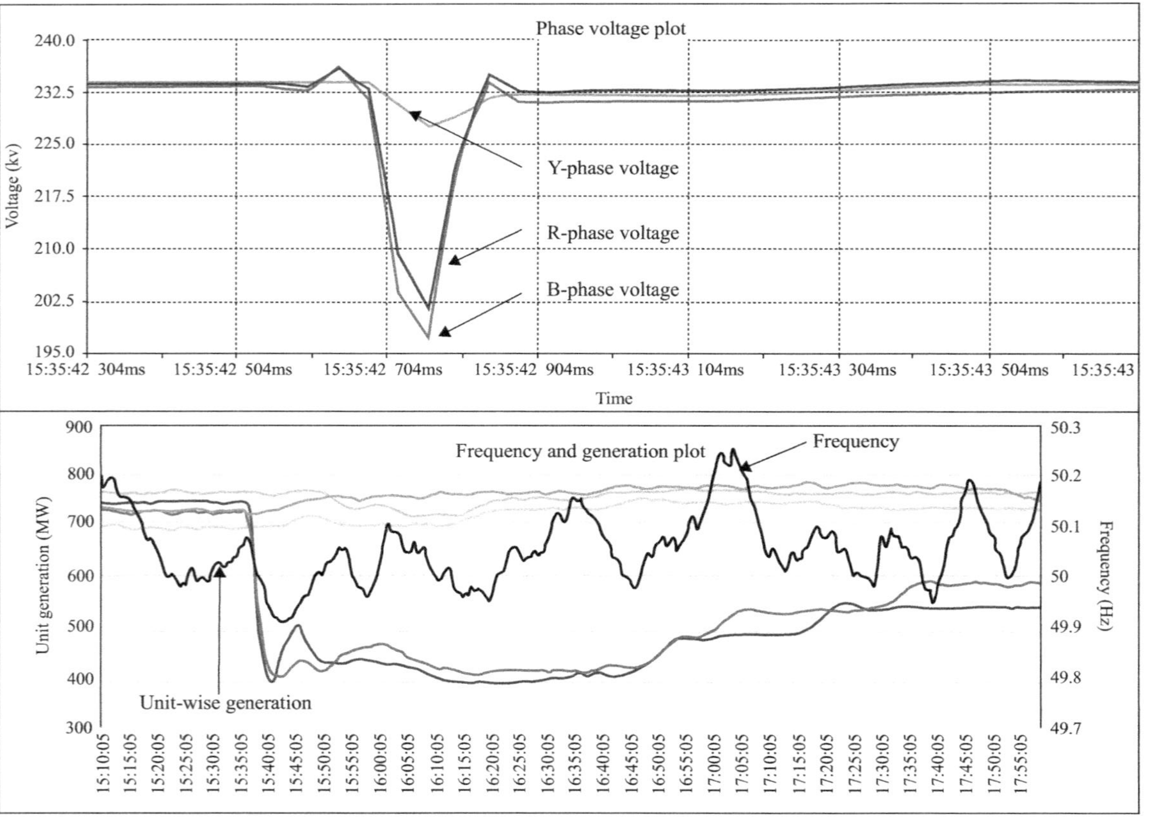

Figure 1.10 Real-time detection of fault and SPS action

phase-to-phase fault and consequently special protection scheme (SPS) operate which backed-down the generation. The operator informed the concerned constituent for necessary load balance and monitored the line loadings in a parallel network for any instance of overloading.

1.8.2 Low-frequency oscillations

The power system is a large non-linear system with lots of oscillation modes. The oscillations include electromechanical oscillations, control modes oscillations and sub-synchronous resonance. The change in the electromechanical torque of a synchronous machine following a perturbation has two components. One component is called the synchronising torque and determines the torque change in phase with rotor angle perturbation. The other component is called damping torque and determines the torque change in phase with speed variation. Rotor angle stability depends on both components of torque. Lack of synchronising torque causes non-oscillatory instability or monotonic instability in the system and lack of damping torque result in oscillatory instability in the system. Rotor angle stability is of two types which are small signal stability (small disturbance in the power system) and transient stability (large disturbance in the power system). Small signal stability is the ability of power system to be in steady state after a small disturbance. The instability due to this is mainly attributed to insufficient damping torque. Transient stability is associated with the ability of power system to maintain synchronism when subjected to large disturbances like line fault, bus fault, generator outage, etc. On the other hand, the instability arising due to this is the result of insufficient synchronising torque.

Small signal instability is due to insufficient damping torque leading to low-frequency electromechanical oscillations (LFO) in the system which is oscillatory in nature. If there are N generators in a system, then the total number of such LFO modes would be $N-1$. During low-frequency oscillations, mechanical kinetic energy is exchanged between synchronous generators of the inter-connected system through tie lines. Most of these oscillatory modes in normal power system state are well damped. However, they get excited during any small disturbance in the system and lead to oscillation in power system parameters like rotor velocity, rotor angle, voltage, currents power flow, etc. Due to oscillation in parameters, protection equipment may undesirably operate leading to cascade tripping in the electrical system. Therefore, it is necessary to detect such modes and initiate corrective actions to ensure system reliability and security. Among these parameters, the rotor velocity of the generators and the power flow in the network are most important. The rotor velocity variation causes fatigue to the mechanical parts of turbine generator system. The power-flow oscillations may amount to the entire rating of a power line when they are superimposed on the stationary line flow and would limit the transfer capability by requiring increased safety margins.

The low-frequency oscillation classification is in general dependent on the power system for which it has to be used on the basis of Eigenvalue analysis based on the system model. The classification helps in monitoring the variables which may be participating globally or alone. The participation factor of each variable in one of the oscillatory modes helps in deciding the classification of modes. In general, low-frequency oscillation can be basically of four different types as given below:

- Inter-area mode (0.1–0.7 Hz): Inter-area modes are associated with swinging of a group of generators in one part of the system with a group of generators in other parts due to weak interconnecting lines between two power systems. Those are also referred to as global mode.
- Local mode (0.7–2.5 Hz): It is either due to oscillation of one generator against the remaining of the system (very similar to the one-machine infinite-bus system) or oscillation of one generator against another, both located close to each other (two generators in the same power plant).
- Control mode: These are in system due to poor design of controllers of AVR, HVDC, SVC, AGC, etc. These are also referred to as regulating mode.
- Torsional mode (10–40 Hz): These modes are associated with the turbine-generator shaft system and associated rotational components.

The ranges defined above may overlap on other depending on power system but in general, this is accepted in power system fraternity.

With the current SCADA system, electricity system operators are not able to identify LFOs in the system due to slow updating rate of data which is once in every 4–10 s (analogue values). The oscillations at generator level, that is intra-plant oscillations or local mode oscillations were assumed as it appeared as hunting in the generators, whereas the inter-area modes were not visible to system operators by any means apart from simulation studies. The SCADA data reporting rate is comparatively slow which are not useful in detecting the oscillation or the changes going in the system in sub-seconds.

With synchronised phasor measurements availability at a reporting rate of 25–50 frames/second from PMU, now operator is able to visualise such oscillations in the system. Tools and techniques are also in development to detect the source of such oscillations and to analyse them in real time and take corrective action before they create further complexities in the system. The detection of LFOs and their history is of great help in planning and implementation of damping controllers of HVDC, TCSC, etc. At present, PMUs have enabled the operator to visualise such LFOs whose source can be tracked with the availability of the optimum number of PMUs giving complete observability of system.

Figure 1.11 shows the oscillations capture during tripping of an extra high voltage tie line. It can be seen that oscillations remained quite some time but died down subsequently because of positive damping of the same. Bottom graph shows real-time state of positive damping and insufficient damping with different coloured dots. The vertical axis shows the mode, that is frequency of oscillation. The red dot (indicated in figure with a label) indicates very low value of damping and all other dots indicate some value of positive damping. As there was only one such dot which got converted into normal dot (positive damped) subsequently, these oscillations were not dangerous to the grid.

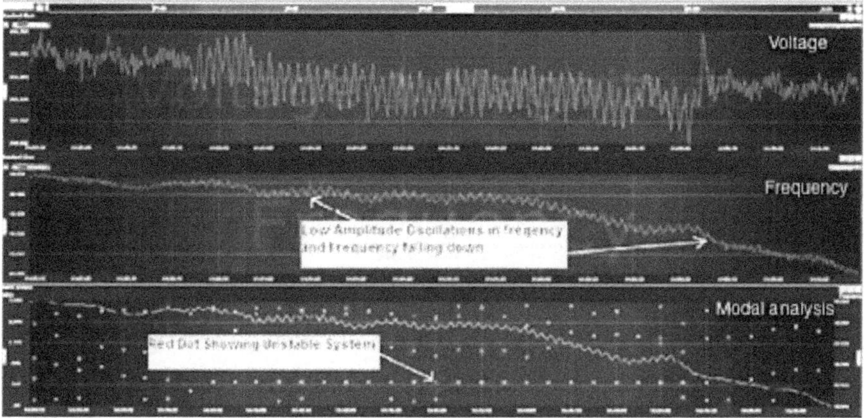

Figure 1.11 Low-frequency oscillation due to tripping of tie line

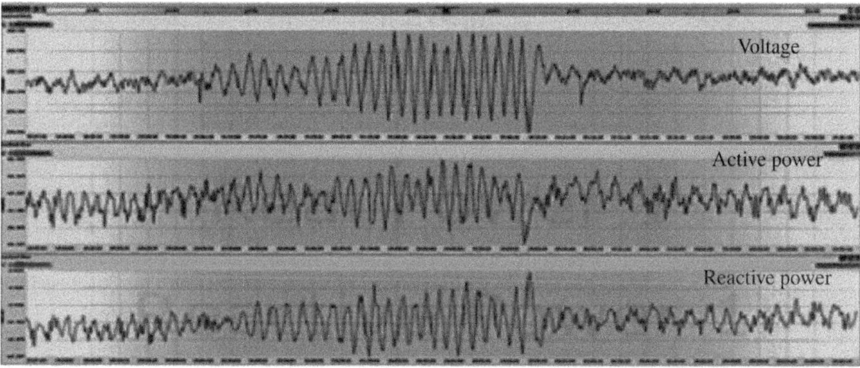

Figure 1.12 Low-frequency oscillation during synchronisation of thermal unit

Similarly, Figure 1.12 presents a case of situations when low-frequency oscillations were observed from the nearby PMU while synchronising the 500-MW thermal unit in one of the regional grid of India. The figure shows the oscillations observed in voltage, active power and reactive power signals. As oscillation were for very small period (few milliseconds) and were with negative damping, those were harmless to the grid.

Oscillations monitoring system (OMS) is a fundamental application which is possible with even small numbers of PMUs installed. Since oscillations are like perturbations in a pond which starts from a point and propagate all around, they are sensed by any PMUs present in the grid. The small frequency oscillations are always present in the system due to continuous changing condition of the grid; however, they are not harmful as they die down in a short time, that is they are not negatively damped, that is not of increasing amplitude. OMS provide all this

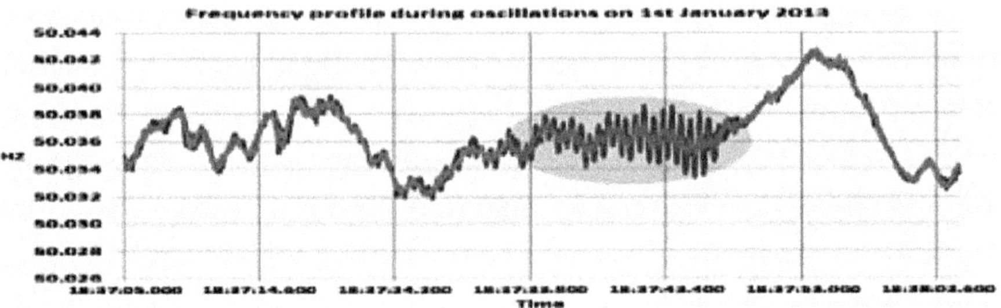

Figure 1.13 Low-frequency oscillations in thermal power plant frequency

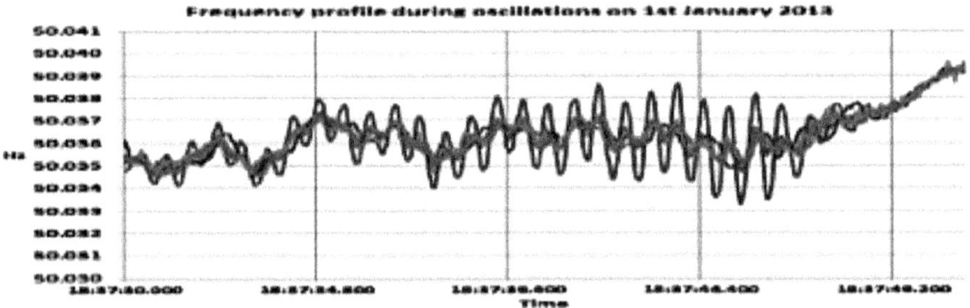

Figure 1.14 Zoomed view of oscillations

information to the grid operator. OMS do the modal analysis of signals captured by the PMUs and provide damping and other information to the operators.

Low-frequency oscillation has been observed in Indian system on several occasions. This case study presents low-frequency oscillation observed at only one location. No oscillation was observed in rest of the system as observed in the PMU installed across the different region of NEW grid. This indicated that this is a case of local mode of oscillation.

Figure 1.13 shows the frequency plot which shows that low-frequency oscillations observed only in the PMU installed at a Thermal Power Station near New Delhi and not in other nearby PMUs. The mode that was observed is 1.67 Hz, which is in the range of Intra-plant oscillation, which was present only in that thermal power plant frequency signal. Figure 1.14 shows the zoomed view of the oscillation which is of growing nature. During this case, all over India data was collected for oscillation and once again it was found that it was localised only as no oscillations were reported from any other generators. During investigation, it was found that there was an event of malfunctioning of EHC (electro hydraulic controller) of governor of 490 MW Unit of thermal power plant. It was further clarified that low-frequency oscillations were on account of testing of the valve control system on 490 MW Unit of thermal power plant, and the same unit was hunting from 350 to 470 MW. The oscillation frequency was 1.66 Hz.

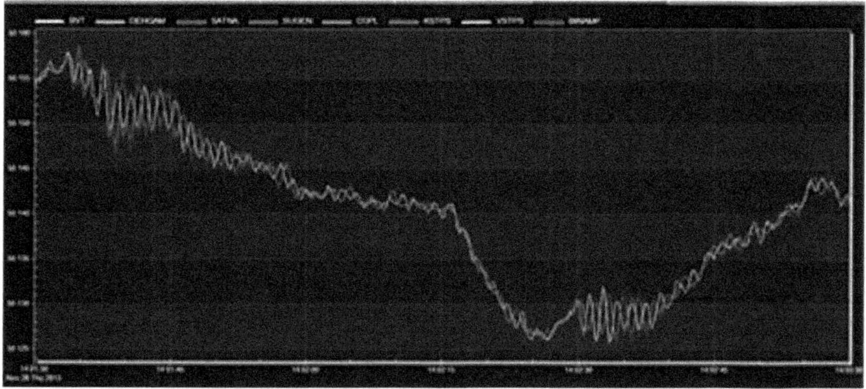

Figure 1.15 Zoomed view of oscillations

1.8.3 Insight to coherency of generators

Oscillation captured by PMUs are very useful for giving the insight of coherency of generators in the system. Figure 1.15 shows the event occurred in coherent group of generators located in the eastern and western part of Western Grid. The event which triggered the swinging of generators in the two areas is yet to be identified but for the system operator having such a plot in real time gives enough warning that the system is being exposed to inherent stress. This helps one to avoid any nasty 'surprises'. The plot in Figure 1.15 reflects the stressed condition of lines connecting the eastern and western part of the Western Grid.

1.8.4 Remote detection of system separation

System separation is the phenomena, in the power grid, in which some part of the grid gets disconnected due to tripping of connecting tie lines. Timely and early detection of this event is of utmost important for grid operator. Since separated part of the grid is in general always unstable due to distorted load generation balance, early detection and timely action may reduce the risk of further damage/tripping to separated part of the grid. Measurement of angular separation with the help of synchrophasor data is very useful for the detection of system separation condition in the grid. Angular separation is the measure of phase angle difference between any two locations. WAMS always need a reference point for display of the angular difference at various parts of the grid. If any of the parts loses synchronisation with the rest of the grid due to tripping of tie lines, WAMS displays a defined pattern of angular difference varying between $-180°$ and $+180°$. The frequency of variation depends on the quantum of difference between the frequencies of two systems. Figure 1.16 shows an event in which system of the south of Odisha (one of the Eastern states of India) separated from the rest of the gird due to tripping of all tie lines connected to that area. The display shows a real-time angle of four locations with respect to the PMU located in Northern part of India. At one instant, angle of one of the PMUs installed at sub-station of south Odisha, a state in eastern part of

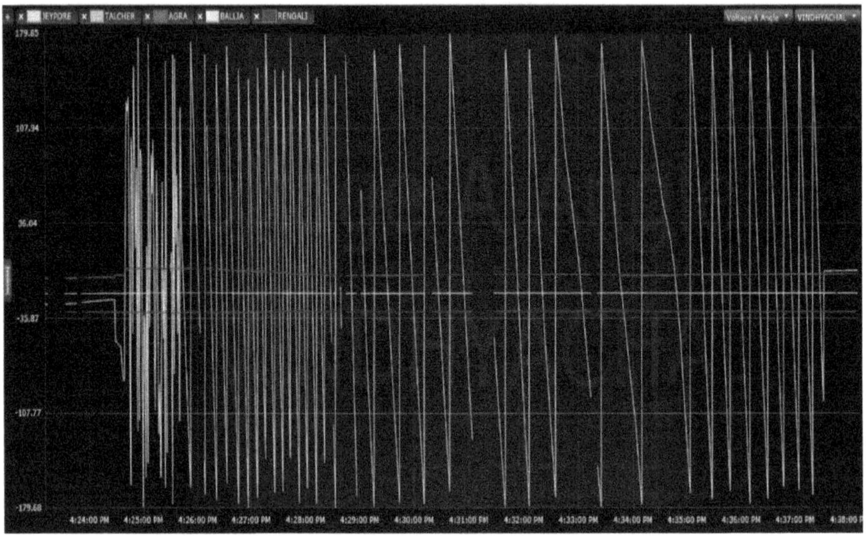

Figure 1.16 Angle trend after system separation

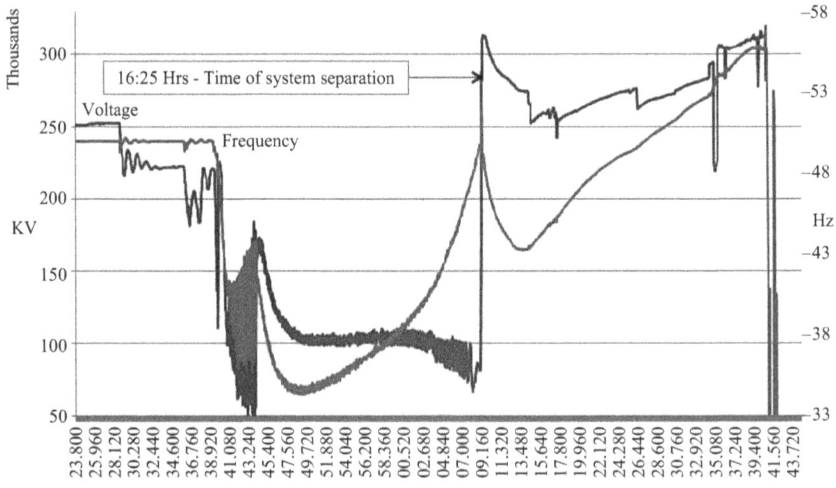

Figure 1.17 Plot of voltage and frequency during the event

India, started displaying the values between +179.85 and 179.68. The angle from other three PMUs, installed in northern part of India, remained more or less the same. It depicted the system separation phenomena and prompted the operator for taking corrective action reducing the possibility further outages in the grid.

Figure 1.17 shows the plot of voltage and frequency from the PMU located at the sub-station inside the south Odisha. It gives the detailed account of the event

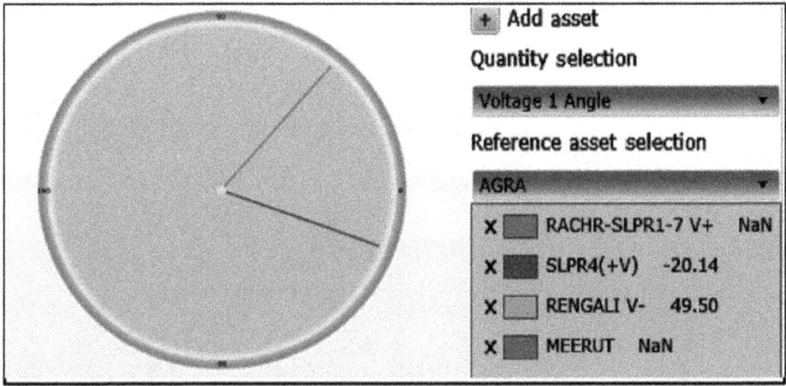

Figure 1.18 Synchronisation for two grids

occurred before separation of south Odisha from rest of the grid. It shows the fault in R-phase in which voltage values shoot up to the range of 100–300 kV at the time of separation.

1.8.5 Synchronising two grids from remote

Synchrophasor data found very helpful in monitoring and controlling of synchronisation of two grids from a remote control room. Generally, synchronisation of an element with the grid is done from the sub-station by observing synchroscope dial and closing the circuit breaker at appropriate time. Use of synchroscope is convenient as far as synchronisation of single element concerned. However, for synchronization of two grids require extensive monitoring and controlling of frequency difference and angular difference between two grids and connecting breaker should be close when both are with in required limits. Since grid conditions are always changing dynamically, the conventional method of using synchroscope may not be feasible in this case.

On 31 Dec 2013, India need to synchronise its Southern grid with the rest of the grid for realisation its goal of 'One Nation and One Grid'. For this, synchrophasor data was used and both grid were synchronised successfully. Figure 1.18 shows the WAMS display used during the event. A synchroscope like display was created for monitoring of phasors of both the grids for getting the moment when both are revolving with the nearly same speed with a minimum angle between them. Nearly the same speed signifies that difference in frequency between the two is minimum. The minimum angle difference between the two grids prevents the sudden large load flow between two grids to avoid unnecessary low-frequency oscillations in the system.

Figure 1.19 shows the real-time trend of frequency, voltage and power flow of two stations located in two grids being synchronised. These real-time trends were very useful for matching the frequency of both the grid and breaker was closed when the frequencies were quite close.

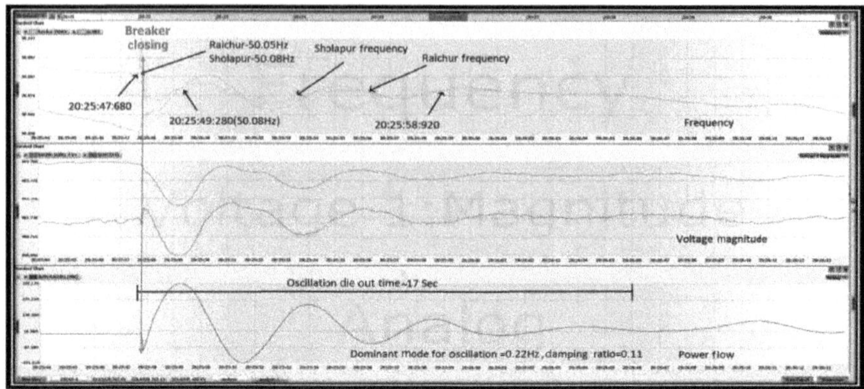

Figure 1.19 Synchronisation for two grids

1.8.6 Improving power plant performance

1.8.6.1 Ascertaining the performance of PSS

The power system stabiliser (PSS) is a supplementary control system of a generators excitation system. The output signal of the PSS is injected into the summing junction of the exciter block in the generator to damp out low-frequency oscillations of the power system. It is the most economical way to mitigate this kind of low-frequency oscillations. For the PSS to perform its role, it is crucial to tuning optimally the internal parameters of the PSS, which are composed of lead-lag time constants and the gain. The tunings of the PSS and its application have been studied and applied to power systems around the world since the 1960s. The underlying reasons for tuning PSS parameters are to compensate for the phase lag due to the electrical system, generator and excitation system and to provide electrical torque in step with speed via the excitation system and generator [10].

PSS performance is often evaluated from the damping of the local mode, the generator swinging against the rest of the power system. This mode is usually at frequencies between 1 and 3 Hz. Stronger network ties and lighter loading tend to give higher local mode frequencies, and weaker ties and heavier loading tend to give lower local mode frequencies. PSS performance must be designed to provide acceptable performance over a wide range of system conditions, which may result from different operating conditions (such as lines out-of-service and varying load levels). However, normally it is not straightforward to detect when the PSS needs tuning. But the synchrophasor measurement could help determining the status of PSS tuning.

Figure 1.20 shows the oscillation experienced by a hydro plant located in northern part of India due to PSS. Tehri Hydro Electric Power Plant having 1,000-MW capacity is located in the north of NR of India. There are four 765-kV transmission lines emanating from this plant for evacuation of its generation. These all four lines were charged at 400 kV at that time. Tehri power station has complained the experiences of oscillation in the system while increasing the loading of lines.

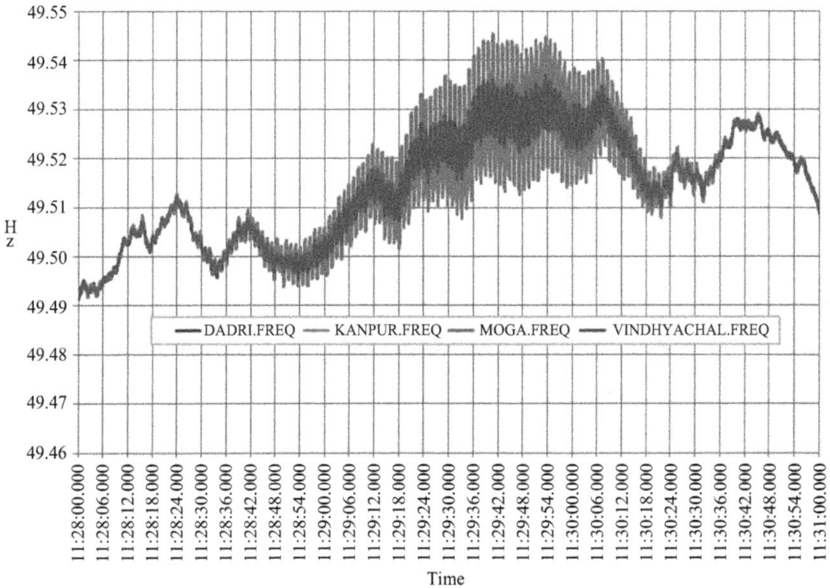

Figure 1.20 Oscillations due to need of PSS tuning

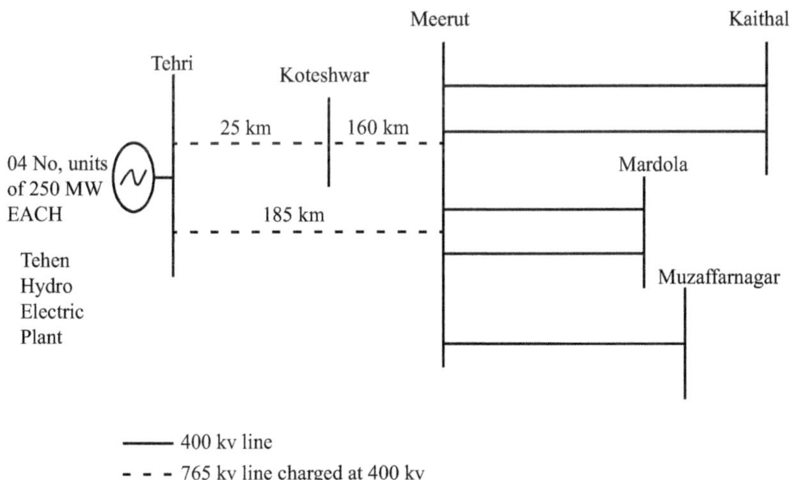

Figure 1.21 Evacuation from Tehri Hydro Plant

An exercise was carried out to examine the oscillation phenomena and to ascertain the load-ability of the line. The test was performed by keeping only two out of four lines in service (refer to Figure 1.21) and generation at Tehri HEP was increased in a controlled manner. At certain power flow, the machine at the hydro plant started experiencing the local mode of oscillations. Oscillation disappeared as soon as the power flow on the lines decreased. Since the oscillations were in local mode and were appearing without any disturbances, it was concluded that PSS of the machine needed tuning. After tuning the PSS, the oscillations were not experienced even with the increase of flow.

1.8.6.2 Validation of performance of special protection scheme (SPS)

A hydroelectric power station Karcham Wangtoo is located in the north of Northern India having the capacity of 500 MW (2 × 250 MW). Two dedicated transmission lines were planned for evacuation of its generation to nearby pooling station at Abduallapur. Due to some reason, the commissioning of these lines got delayed, but the plant was ready to generate to its full capacity. To facilitate the evacuation of power generation, studies were carried for exploring the possibility of the evacuation of the generation through existing transmission lines via nearby power station Naptha Jhakri. As the evacuation of more generation without grid security during any contingency may lead to cascade tripping, a special protection scheme was implemented. This SPS should trip the machines at Karcham Wangtoo in the case of the outage of any of the evacuation link, and line loading of any of the remaining lines increases more than 800-MW load. However, during the operation of SPS at an incidence, system-wide oscillations were observed as seen from the plot of PMU data. On investigation, it was found that there was an inadvertent delay of about 10 s in the tripping of a unit through the operation of SPS as shown in Figure 1.22. The delay was unknowingly introduced, and it was removed later on. After this, oscillations were not observed again.

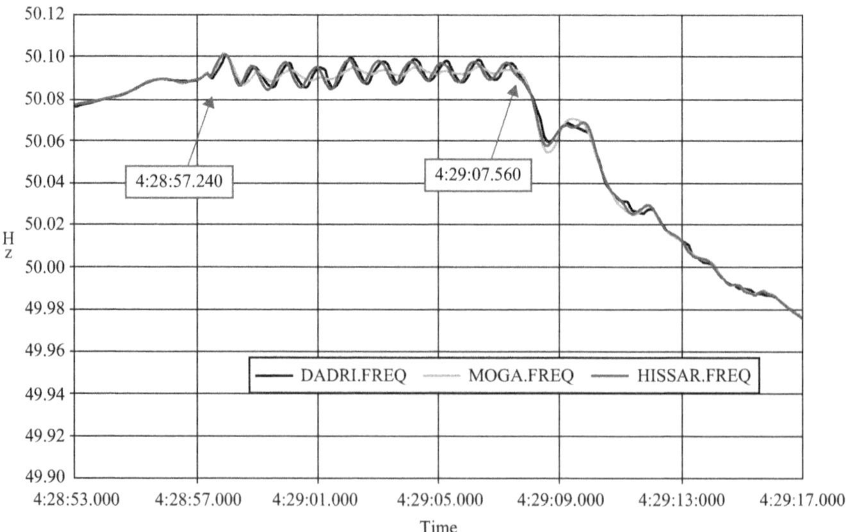

Figure 1.22 Oscillations due to need of SPS delay

1.9 Way-forward

Since the implementation of the first pilot project in India and realising its utility in managing and improving the performance of power grid, many separate pilot projects were taken up and integrated with a National level WAM system. Finally, this WAM system is being utilised as the mainstream system by the grid operators, and no more has the status of the pilot. Consequently, based on the experience gained so far in the last 5–6 years, India has ventured in a full-scale WAMS popularly known as 'Unified Real-Time Dynamic State Measurement' (URTDSM) project. Most probably, URTDSM is the single largest project in the world wherein 1,600 PMUs, and 32 PDC are being installed in an integrated hierarchical manner.

Figure 1.23 shows the architecture of the project. The project envisages to install PMUs on all 400 kV and above sub-stations, generating station with 220 kV switchyard, all HVDC systems. Furthermore, all 400-kV lines will have PMUs installed at both the ends. This will enable calculations of realistic line parameters. PDCs will be provided at all the states, regional and national load despatch centres. Also, nodal PDC will also provide if the number of PMUs is more than 40 in any one of the states. Many applications are also being developed under a research and development project specially commissioned by the POWERGRID for effectively utilising the PMU data for further improving and maintaining the performance of Indian grid. The experience gained by India will definitely be useful for power system fraternity as a whole.

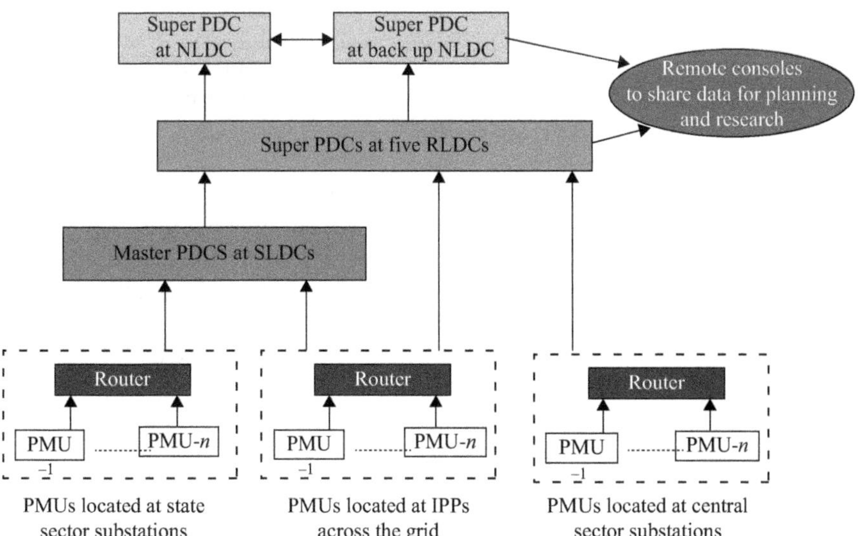

Figure 1.23 Unified real-time dynamic system measurement project scheme

Acknowledgement

The views expressed in this paper are of authors only. Power System operation Corporation does not subscribe the same in any way. The author is grateful to the management of Power System Operation Corporation Ltd for giving permission of using case studies and giving permission to publish this chapter.

References

[1] Phasor measurement unit. Available from: https://en.wikipedia.org/wiki/Phasor_measurement_unit [Accessed 15 Apr 2016].

[2] BURN – An Energy Journal. The electricity grid: A history. Available from: http://burnanenergyjournal.com/the-electricity-grid-a-history/ [Accessed 16 Apr 2016].

[3] JS. How electricity grew up? A brief history of the electrical grid .... Available from: https://power2switch.com/blog/how-electricity-grew-up-a-brief-history-of-the-electrical-grid/ [Accessed 25 Oct 2016].

[4] Agrawal, V. K., Agarwal, P. K., and Rathour, H. Experience of upscaling and integration of regional level synchrophasors pilot projects to a national level project. D2 colloquium, CIGRE, 13–15 November 2013, Bangalore, India.

[5] Hurtgen, M. and Maun, J.-C. Applications of PMU measurements in the Belgian electrical grid, Report, 9 May 2012.

[6] Stastny, L., Franek, L., and Z. Bradac. Time synchronised low-voltage measurements for smart grids, *Procedia Engineering*, vol. 100, 2015, 1389–1395, ISSN 1877-7058, http://dx.doi.org/10.1016/j.proeng.2015.01.506.

[7] Soonee, S. K., Narsimhan, S. R., Porwal, R., and Pandey, V. Application of phase angle measurement for real time security monitoring of Indian electric power system – an experience, CIGRE Paris Session, 2008, CIGRE, August 2008, Paris, France.

[8] SmartGrid.gov. Synchrophasor technologies and their deployment in the recovery act smart grid program, Report, US Department of Energy, August 2013.

[9] Giri, J. and Parashar, M. Transitioning from wide area monitoring to wide area management, *IEEE Smartgrid Newsletter*, April 2016. Available from: http://smartgrid.ieee.org/newsletters/april-2016/902-transitioning-from-wide-area-monitoring-to-wide-area-management.

[10] Shin, J., Nam, S., Lee, J., Baek, S., Choy, Y., and Kim T. A practical power system stabilizer tuning method and its verification in field test, *Journal of Electrical Engineering & Technology*, vol. 5, no. 3, 2010, 400–406.

Chapter 2

An optimal redundancy criterion index (ORC) for optimal placement of phasor measurement units (PMU) for full observability of power grid

Ranganath Vallakati[1] and Prakash Ranganathan[1]

2.1 Introduction

The dependency of human lives on electrical power is increasing exponentially. Our lives have become so much reliant on electricity that a brownout situation in the grid could mean life and death conditions. Thus, a stable power grid is important. Electricity possesses two important qualities—reliability and continuity. The goal of an electric utility is to supply power continuously and reliably. We need a reliable electricity rather than electricity without reliance. Continuity and reliability go hand in hand; one can be achieved only if the other is achieved. With increased injection of renewable energy sources or distributed generations (DGs), various problems are showing up more often than normal. The problems include deterioration in grid characteristics such as power quality, supply efficiencies, and reliability. Rerouting of excess power becomes a concern due to lack of energy transportation infrastructure. This is due to the fact that existing transmission and distribution infrastructures are designed only for unidirectional power flows rather than bidirectional power flows. This means that power flow starts from high-voltage lines to medium-voltage lines and then to low-voltages distribution lines. As penetration of DGs increases, the current unidirectional flow infrastructure does not work due to the power following into the low-voltage lines from the DGs. Thus, effective management, monitoring, and forecasting methods are needed in order to maintain the stability of the grid. A smarter grid provides management solutions for a stable operation.

A smart grid (SG) is simply a conventional power grid that is integrated with intelligent sensing and software technologies. SG contains automated substations, switch gears, sensors for intelligent monitoring and control, technologies for improvising generator and load efficiency that ultimately keeps the power system

[1]Electrical Engineering Department, University of North Dakota (UND), Grand Forks, ND 58201, USA

in a stable condition. SG framework also contains conventional sources such as coal-fired power plants and DGs such as wind farms and solar panels. There are intelligent monitoring sensors on transmission lines, buildings, and homes at DG locations. Sensors can take decisions on their own and are needed to be well connected for better monitoring and situational awareness. Data sensed in an SG are collected for various functionalities such as real-time monitoring, offline analysis, load-flow studies, generation, and load-shedding schemes.

The electric utilities currently use Supervisory Control and Data Acquisition (SCADA) for controlling and monitoring operations in a power system. There are several drawbacks with the existing SCADA system. In a SCADA system, data samples of voltage, current magnitude, and phase angle parameters are estimated than measured. The magnitudes are not time-synchronized, so it is difficult to relate the occurrence and timing of an event. The limitations of SCADA systems are overcome with the invention of Phasor Measurement Units by Dr Arun G. Phadke and Dr James S. Thorp in 1988. Synchrophasors are the modern-day measurement tools that gather key sensor parameters such as voltage, frequency, and phase angle information, which can be utilized to monitor the power grid [1,2]. With these sensors, each and every detail of the grid is observable, which would assist system operators to supply electricity reliably to consumers.

Synchrophasors provide time-synchronized measurements of the power system such as voltage and current phasors used to calculate active power, reactive power, and frequency. The data are then utilized for applications such as state estimation (SE), transient analysis, capacitor bank monitoring, load-shedding schemes, inter-area oscillations, and microgrid operations [3,4]. Typical microgrid operations include islanding, anti-islanding, and identifying fault locations on transmission lines [5–7].

2.2 Synchrophasors

To avoid blackouts, parameters such as frequency, voltage, current, and phase angle are analyzed with respect to time and location of occurrence. The key advantage of synchrophasor is its ability to time stamp the measured data with a high data rate. Synchrophasors utilize global positioning system (GPS) time clocks in order to time stamp their measurements. Synchrophasors helped independent system operators (ISOs) in realizing the wide area measurement systems (WAMSs). WAMSs are an essential upgrade for SCADA systems for they use synchrophasors to monitor power system conditions over large geographical areas with an aim to stabilize the grid. The recent model of synchrophasors has a sampling rate of around 240 samples per second. This level of high granularity is much required because the quantities such as frequency and phasor values vary at a sub-seconds time interval. ISOs utilize synchrophasor data collected from different geographical locations at the same time and take critical decisions in real time to maintain the grid.

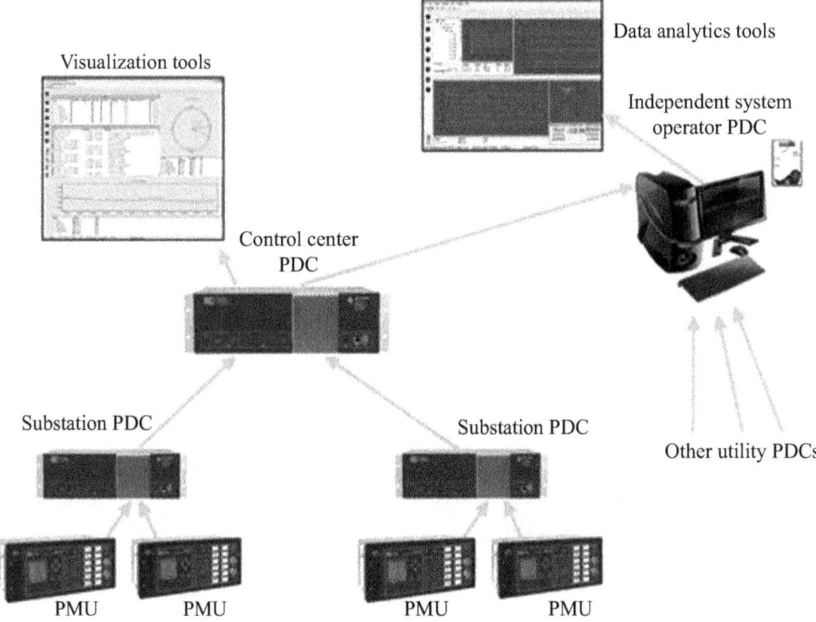

Figure 2.1 Wide area monitoring system (WAMS)

There are different components that form a WAMS network as shown in Figure 2.1. They are as follows:

1. Phasor measurement unit (PMU)
2. Phasor data concentrator (PDC)
3. GPS clock

PMU uses the most common technique for determining the phasor representation of a signal, that is, by taking the samples from the waveforms and then applying the discrete Fourier transform to compute the phasors [8]. The AC waveforms are digitized by an analog-to-digital converter for each phase. A phase-lock oscillator along with GPS source provides the needed synchronization for the samples.

Waveforms are obtained through a line of components such as transformers and filters, and these measured signals are time synchronized using the GPS clock. The GPS clocks allow synchronization of signals into a Universal Time Frame (UTC) to assist the wide area system operations. The time-synchronized phasor measurements are aggregated in PDC as per IEEE Standard C37.118 [9]. PDC acts as a centralized main-stream data server that collects and stores the streaming information from various synchrophasors across the grid as per IEEE Standard C37.244 [10]. The data stored are then extracted and used for various applications.

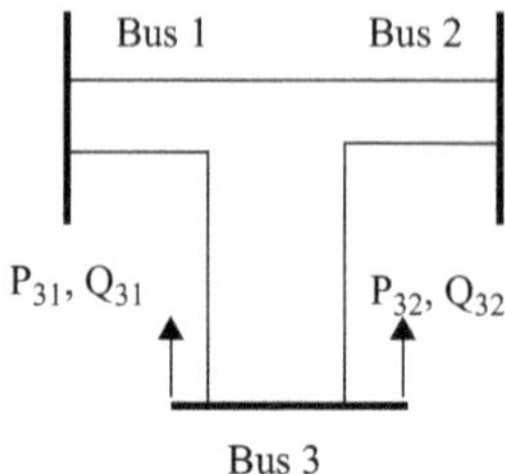

	Measured parameters at Bus 1, Bus 2, and Bus 3	Calculated
SCADA	$\lvert V_1\rvert,\lvert V_2\rvert,\lvert V_3\rvert, P_{31},Q_{31},P_{32},Q_{32}$	$\lvert\delta_{31}\rvert, \lvert\delta_{32}\rvert$
PMCU	$\lvert V_1\rvert,\lvert\ \delta_1\rvert,\lvert V_2\rvert,\lvert\delta_2\rvert,\lvert V_3\rvert,\lvert\ \lvert\delta_3\rvert$	–

Figure 2.2 SCADA vs PMU measurements [11]

2.2.1 *Advantages of synchrophasors over SCADA measurements*

SCADA provides information about voltage magnitude, active and reactive power flows, load measurements, generator outputs, the breaker status, and performances of capacitor banks and transformers as calculated estimates from SE. When power system state is changing quickly, conventional SE fails to capture certain situations making the SCADA scan inefficient.

In addition, analog measurements are proportional to the time differences between the measurements and the rate at which states of system change [11,12]. With synchronized measurements, variables such as phase angles need not be estimated as PMU captures all measurements that are synchronous to UTC. Also, the quickly changing states can be captured for accurate SE. Figure 2.2 provides a comparative analysis of SCADA and PMU measurements.

2.2.2 *Challenges with synchrophasor measurements*

1. Economics of synchrophasor units

PMUs are deployed across utilities to improve the SE of power systems. However, the major constraint for utilities is the cost of PMUs. In real-world applications, utilities need to set up a number of things before PMUs can be placed on the grid. This includes, but is not limited to, construction of communication infrastructure, procurement of PMUs, PDC for data aggregation, and synchronization clocks. Thus, it is an expensive issue for utilities to upgrade their monitoring systems. Table 2.1 shows the cost of installing a synchrophasor system [13]. The total cost is approximately US$30,000 excluding any infrastructure, operational, and labor costs. However, devices such as PDC may collect data from a number of PMUs. Therefore, a small or mid-size utility can install only a few PDCs as a typical PDC can handle data up to 40 PMUs.

However, there is a solution. By applying the fundamental Ohm's law on the transmission lines, a PMU placed on a bus allows for observation of the neighboring buses, that is, adjacent bus voltages are easily computable. Thus, the placement of PMUs in a grid became a challenge for the utilities so as to reduce the cost as well as improve the power system situation to maintain reliability and continuity in the system. This problem of reducing the cost factor as well as to increase the reliability of the grid using PMU placement is addressed in this chapter by using the method of linear programming (LP).

Table 2.1 Typical approximation of synchrophasor cost

Type of equipment	Cost (US$)
PMU with protection, automation, and controls	~15,000
PDC (up to 40 PMUs)	~8,000
Synchronization clocks	~2,000
Digital equipment (firewalls, cables, and routers)	~5,000

2. Reliability of synchrophasor data

The main objective of PMU placement problems is to minimize cost. Different optimization methods try to reduce system redundancy by making the buses measured the least number of times. During contingency situations such as failure of a PMU or a communication line, the system loses its observability and the system becomes unobservable affecting operator's ability in decision-making. Thus, optimal methods for PMU placement problem should include the concept of redundancy in its measurements to keep the system stable during contingencies. This research proposes a criterion for including such a concept of redundancy into the PMU-placement algorithm.

3. Utilization of synchrophasor data

Synchrophasors generate around 30–120 samples per second that contain information about the state of power system. Thus, ISO are bombarded with nearly 108,000 samples/h, which equates to 2,592,000 samples/day or 77,660,000 samples a month. These numbers grow exponentially, and approximately around 1.5 TB of data are accumulated in a month's span. With all such big data accumulated, ISO makes use of these data for meaningful applications. It is neither possible to monitor every data point nor unmonitored. A visualization technique had been developed using a unit-circle representation of phasor data. Unit-circle visualization type will help ISO to monitor the incoming data with ease of understanding. The data representation is stored in the database (db) prior to data mining tasks. ISO uses synchrophasor data for predictive analytics to forecast the system behavior. Any forecasting task requires knowledge of history of data. Data mining is the process of extracting useful information to understand more meaningful interpretation about the system. A density-based clustering algorithm called as Density-Based Spatial Clustering of Applications with Noise is used to cluster the synchrophasor data. The results of clustering are stored in the db after the initial stage of visualization.

2.3 Optimal placement of phasor measurement units (PMU)

PMUs provide time-synchronized voltage and current measurements when placed on a bus. These time-synchronized values are critical for continuous operation of WAMS in SGs. It is not required to place a PMU on every bus in the system to get full observability. Using Ohm's law, a PMU placed on a specific bus can be used to measure the voltage and current phasors at that point. It can also be used to calculate the voltage and current phasors on the lines that are connected to the placed bus. Also, it is not feasible for the utilities to afford a PMU for every bus in their grid due to its cost. Thus, there is a need to find a solution for optimal placement problem (OPP) for PMUs.

2.3.1 Need for OPP for synchrophasors

Case 1: In Figure 2.3(a), all four bus points are accompanied with PMUs for complete observability. The situation is shown in Table 2.2. There exists many solutions of redundant observabilities than required. Thus, it would be useless to have a PMU on every bus.

Case 2: Figure 2.3(b) shows how three PMUs on buses 1, 2, and 3 are sufficient to cover the full-observability of the system. This situation is given in Table 2.2. There is a reduced redundancy in the measurements, but the system is yet fully observable.

Case 3: Figure 2.3(c) shows a 4-bus system with only two PMUs placed on the buses 1 and 4. By analyzing case 3 column of Table 2.2, we can say that the values at the buses 2 and 3 without the PMU can be calculated using the line parameters (R, X) and measured parameters at buses 1 and 4. In this way, even with two PMUs, the system remains fully-observable.

Thus, we can conclude that placing of PMUs using known topology in a certain manner could provide complete observability of system with cost savings that include operational costs and hardware (equipment and infrastructure) costs. In addition, this research includes the zero-injection buses of the power systems to further reduce the number of PMU required for full observability. Zero-injection buses are similar to transshipment nodes in the OPP algorithm that reduce the minimum number of PMUs. Details about zero-injection buses are explained in later section of the chapter. An index, optimal redundancy criterion (ORC) is proposed to address the issue of reliability that is very important for the power systems. ORC assist in identifying the best solution for OPP, while enforcing some conditions to maintain redundancy and full-observability of critical buses.

2.4 PMU placement problem formulation using linear programming (LP)

In this research, the OPP of PMU is solved using the LP technique. In any LP method, the problem is defined by having an objective function that maximizes or minimizes a linear function subject to a set of linear constraints. The constraints include combination of equalities or inequality constraints. Our goal is to reduce the minimum number of PMUs required for a power system without losing the complete observability of the system.

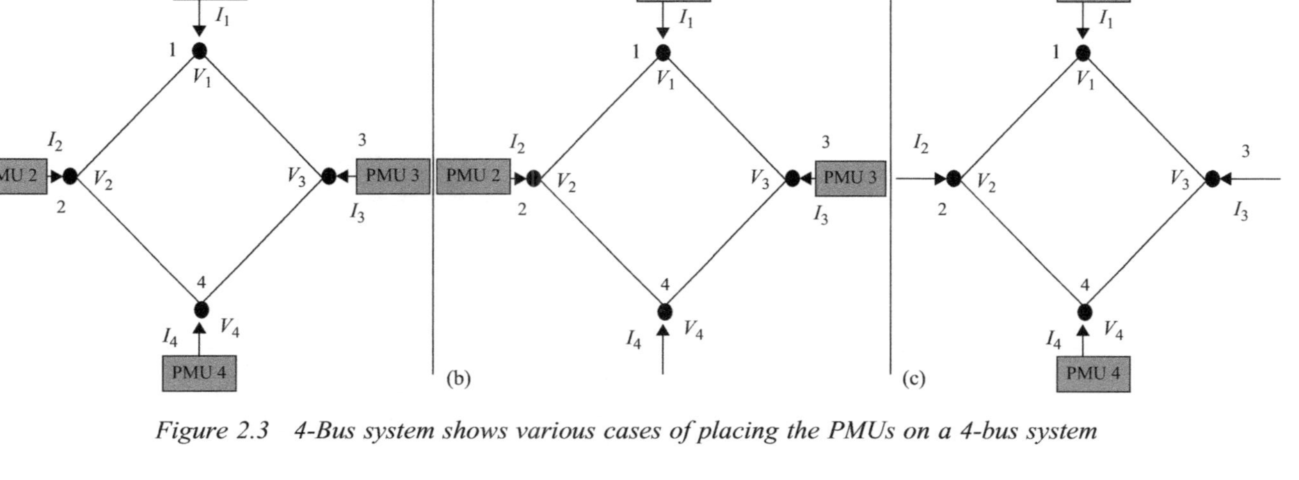

Figure 2.3 4-Bus system shows various cases of placing the PMUs on a 4-bus system

Table 2.2 Case studies of PMU placement on four bus system

		Case 1	Case 2	Case 3
Bus 1	Measured	PMU 1	PMU 1	PMU 1
	Calculated	PMU 2, PMU 3	PMU 2, PMU 3	–
Bus 2	Measured	PMU 2	PMU 2	–
	Calculated	PMU 1, PMU 3	PMU 1, PMU 3	PMU 1, PMU 4
Bus 3	Measured	PMU 3	PMU 3	–
	Calculated	PMU 2, PMU 4	PMU 2	PMU 1, PMU 4
Bus 4	Measured	PMU 4	–	PMU 4
	Calculated	PMU 2, PMU 3	PMU 2, PMU 3	–

2.4.1 Problem formulation

For a system with N buses, the OPP of synchrophasors can be formulated as shown in (2.1):

$$\begin{aligned} &\text{Minimize } \sum_{i}^{N} W_i * X_i \\ &\text{Subject to } f(X) \geq \hat{1} \end{aligned} \tag{2.1}$$

where X_i is a binary decision variable vector, whole entries are defined using the binary variables as follows:

$$X_i = \left\{ \begin{matrix} 1 & \text{if a PMU is installed at bus i} \\ 0 & \text{otherwise} \end{matrix} \right\} \tag{2.2}$$
$$\hat{1} \text{ is a vector whose entries are all ones}$$

W_i is the cost associated with the installation of PMU at that bus i. Figure 2.4 shows a single line diagram of an IEEE 14 bus system. There are five generation points at buses 1, 2, 3, 6, and 8. The generators at buses 1 and 2 generate both active and reactive power, whereas generators at 3, 6, and 8 generate reactive power. There are 11 load points connected at buses 2, 3, 4, 5, 6, 9, 10, 11, 12, 13, and 14, which consume the active and reactive power. Using the LP approach, the system is evaluated to find the PMU placement locations.

The two cases of the system that is considered to determine the OPP solution are as follows:

1. By considering regular test-bus system with any zero-injection buses
2. By considering zero-injection buses as special constraints.

2.4.2 System with no zero-injection buses

Assuming all of the 14 buses in the IEEE 14 bus system are regular ones having at least a load or a generation point. There is a requirement to measure or observe all the buses as changes can happen at each of them due to the generation and load

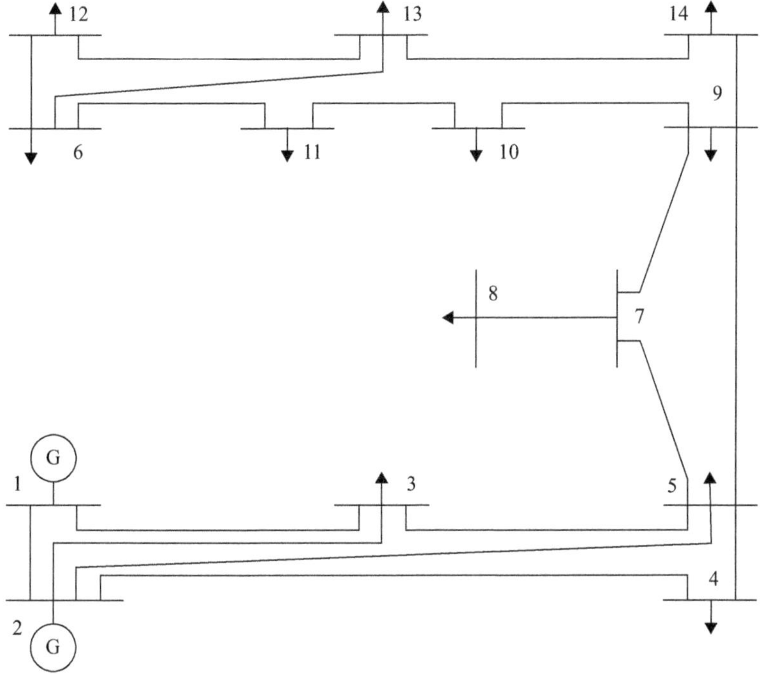

Figure 2.4 IEEE 14 bus system [16]

points. A binary connectivity matrix is formed to indicate links between the buses. The entries of $A_{i,j}$ are defined as follows in (2.3):

$$A_{i,j} = \left\{ \begin{array}{ll} 1 & \text{if } i = j \\ 1 & \text{if } i \text{ and } j \text{ are connected} \\ 0 & \text{if otherwise} \end{array} \right\} \tag{2.3}$$

Matrix $A_{i,j}$ can be directly obtained from the bus admittance matrix by transforming its entries into binary form as shown in (2.4):

$$A_{i,j} = \begin{bmatrix} 1 & 1 & 0 & 0 & 1 & 0 & 0 & 0 & 0 & 0 & 0 & 0 & 0 & 0 \\ 1 & 1 & 1 & 1 & 1 & 0 & 0 & 0 & 0 & 0 & 0 & 0 & 0 & 0 \\ 0 & 1 & 1 & 0 & 0 & 0 & 0 & 0 & 0 & 0 & 0 & 0 & 0 & 0 \\ 0 & 1 & 1 & 1 & 1 & 0 & 1 & 0 & 1 & 0 & 0 & 0 & 0 & 0 \\ 1 & 1 & 0 & 0 & 1 & 0 & 0 & 0 & 0 & 0 & 0 & 0 & 0 & 0 \\ 0 & 0 & 0 & 1 & 0 & 1 & 0 & 0 & 0 & 0 & 1 & 1 & 1 & 0 \\ 0 & 0 & 0 & 0 & 0 & 0 & 1 & 1 & 1 & 0 & 0 & 0 & 0 & 0 \\ 0 & 0 & 0 & 1 & 0 & 0 & 1 & 1 & 0 & 0 & 0 & 0 & 0 & 0 \\ 0 & 0 & 0 & 0 & 0 & 0 & 1 & 0 & 1 & 1 & 0 & 0 & 0 & 1 \\ 0 & 0 & 0 & 1 & 0 & 0 & 0 & 0 & 1 & 1 & 1 & 0 & 0 & 0 \\ 0 & 0 & 0 & 0 & 0 & 1 & 0 & 0 & 0 & 1 & 1 & 0 & 0 & 0 \\ 0 & 0 & 0 & 0 & 0 & 1 & 0 & 0 & 0 & 0 & 0 & 1 & 1 & 0 \\ 0 & 0 & 0 & 0 & 0 & 1 & 0 & 0 & 0 & 0 & 0 & 1 & 1 & 1 \\ 0 & 0 & 0 & 0 & 0 & 0 & 0 & 0 & 1 & 0 & 0 & 0 & 1 & 1 \end{bmatrix} \tag{2.4}$$

Here, a zero in the connectivity matrix derived from the admittance matrix implies that there is no connection between buses *i* and *j*. Although a value of one in the matrix implies that there is a connection between the buses. Once the binary connectivity matrix is formed, the LP constraints for all the 14 buses from the 14 connections are shown in (2.5)–(2.18):

$$\text{Bus 1: } x_1 + x_2 + x_5 \geq 1 \tag{2.5}$$

$$\text{Bus 2: } x_1 + x_2 + x_3 + x_4 + x_5 \geq 1 \tag{2.6}$$

$$\text{Bus 3: } x_2 + x_3 + x_4 \geq 1 \tag{2.7}$$

$$\text{Bus 4: } x_2 + x_3 + x_4 + x_5 + x_7 + x_9 \geq 1 \tag{2.8}$$

$$\text{Bus 5: } x_1 + x_2 + x_4 + x_5 \geq 1 \tag{2.9}$$

$$\text{Bus 6: } x_6 + x_{11} + x_{12} + x_{13} \geq 1 \tag{2.10}$$

$$\text{Bus 7: } x_4 + x_7 + x_8 + x_9 \geq 1 \tag{2.11}$$

$$\text{Bus 8: } x_7 + x_8 \geq 1 \tag{2.12}$$

$$\text{Bus 9: } x_4 + x_7 + x_9 + x_{10} + x_{14} \geq 1 \tag{2.13}$$

$$\text{Bus 10: } x_9 + x_{10} + x_{11} \geq 1 \tag{2.14}$$

$$\text{Bus 11: } x_6 + x_{10} + x_{11} \geq 1 \tag{2.15}$$

$$\text{Bus 12: } x_6 + x_{12} + x_{13} \geq 1 \tag{2.16}$$

$$\text{Bus 13: } x_6 + x_{12} + x_{13} + x_{14} \geq 1 \tag{2.17}$$

$$\text{Bus 14: } x_9 + x_{13} + x_{14} \geq 1 \tag{2.18}$$

The operator (+) in the formulation is a logical "OR" operation, whereas the inequality operator ($\geq$) ensures that at least one of the variables appearing in the sum will be nonzero. For an example, consider the constraints (2.5) and (2.6) associated with buses 1 and 2. Equation (2.5) indicates that at least one PMU must be placed on at least one of the three buses bus 1 (x_1), bus 2 (x_2), or bus 5 (x_5). Similarly, (2.6) implies that at least one PMU should be installed on any one of the buses 1, 2, 3, 4, or 5 in order to make bus 2 observable.

2.4.3 System with zero-injection measurements

Zero-injection buses are buses in which no currents are neither injected nor withdrawn into/from the system (i.e., buses which have neither load nor generation units). In the power systems, there are few bus locations where there is no requirement for measuring. The use of Kirchhoff's current law to include zero-injection buses in modeling the OPP further reduces the number of PMUs required. Each zero-injection node in a system creates an additional constraint. Thus, the number of PMUs required for complete observability of the system can be reduced. For a zero-injection bus *i*, let A_i indicate the set of adjacent buses to bus *i*. Let $B_i = A_i \cup \{i\}$. Let the number of zero injections be *Z*. As there is no change in current (i) values at a zero-injection bus, the variables related to the zero-injection buses are eliminated from all the constraining equations. The model formulation in an ILP framework, therefore, selectively allows some of the buses to be unobservable. However, there are additional constraints that need to be included for a fully observable system. They are as follows:

- The unobservable buses must be adjacent to zero-injection buses.
- The number of unobservable buses in the set (zero-injection bus and its adjacent buses) is at mostly as one (1) in the matrix.

In the IEEE 14-bus system shown in Figure 2.4, bus 7 is a zero-injection bus. Thus, the new constraining equation can be written as

$$\text{Bus 1: } x_1 + x_2 + x_5 \geq 1 \tag{2.19}$$

$$\text{Bus 2: } x_1 + x_2 + x_3 + x_4 + x_5 \geq 1 \tag{2.20}$$

$$\text{Bus 3: } x_2 + x_3 + x_4 \geq 1 \tag{2.21}$$

$$\text{Bus 4: } x_2 + x_3 + x_4 + x_5 + x_8 + x_9 \geq 1 \tag{2.22}$$

$$\text{Bus 5: } x_1 + x_2 + x_4 + x_5 \geq 1 \tag{2.23}$$

$$\text{Bus 6: } x_6 + x_{11} + x_{12} + x_{13} \geq 1 \tag{2.24}$$

$$\text{Bus 8: } x_4 + x_8 + x_9 \geq 1 \tag{2.25}$$

$$\text{Bus 9: } x_4 + x_8 + x_9 + x_{10} + x_{14} \geq 1 \tag{2.26}$$

$$\text{Bus 10: } x_9 + x_{10} + x_{11} \geq 1 \tag{2.27}$$

$$\text{Bus 11: } x_6 + x_{10} + x_{11} \geq 1 \tag{2.28}$$

$$\text{Bus 12: } x_6 + x_{12} + x_{13} \geq 1 \tag{2.29}$$

$$\text{Bus 13: } x_6 + x_{12} + x_{13} + x_{14} \geq 1 \tag{2.30}$$

$$\text{Bus 14: } x_9 + x_{13} + x_{14} \geq 1 \tag{2.31}$$

The constraints (2.19)–(2.31) are the updated (or modified) connections obtained from the bus connectivity matrix formed from the modified 14 bus system as shown in Figure 2.5. There are only 13 constraints, as the bus 7 which is a zero-injection bus is removed and thereby reduces the number of PMUs. In this way, the number of PMUs required to maintain complete observability is reduced, if a system has zero-injection nodes.

2.5 System observability redundancy index (SORI)

Identifying an optimal solution is challenging as there can be multiple solutions for an OPP. To solve this problem, an SORI is used [14]. If a bus i is observed by a PMU "n" number of times, then the SORI (γ) is given by (2.32):

$$\gamma_i = \sum n_i \tag{2.32}$$

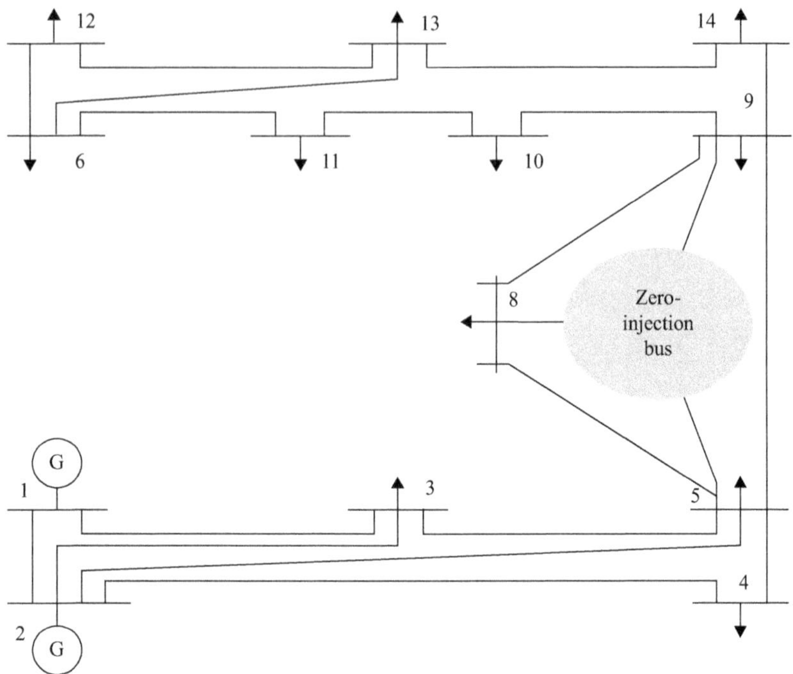

Figure 2.5 IEEE 14 bus system with zero-injection bus

The SORI will help in choosing an optimal solution that will provide a higher redundancy enabling complete observability. In general, larger redundancy values yield higher SORI values. Thus, maximizing the SORI is a good measure of success for the OPP, as this approach helps keeping the system observable during contingencies. For example, let us consider the IEEE 14-bus system in Figure 2.4. The system is made completely observable with 4 PMUs either being placed at 2, 6, 7, and 9 bus locations or at 2, 6, 8, and 9 bus locations. In both the cases, all 14 buses are fully observed. Table 2.3 provides a comparison of bus-observabilities for both the solutions. Using (2.32), the SORI is calculated by simply adding all bus observabilities in a system. In the first case (second column of Table 2.3) when the PMUs are placed at 2, 6, 7, and 9 locations, the SORI is calculated as follows:

$$\text{SORI} = 1 + 1 + 1 + 3 + 1 + 1 + 2 + 1 + 2 + 1 + 1 + 1 + 1 + 1 = 18 \tag{2.33}$$

While in the second case (third column of Table 2.3), the SORI will be

$$\text{SORI} = 1 + 1 + 1 + 2 + 1 + 1 + 2 + 1 + 1 + 1 + 1 + 1 + 1 + 1 = 16 \tag{2.34}$$

Thus, even though with the same number of PMUs, full-observability is obtained in both cases, it is recommended to place the PMUs at buses 2, 6, 7, and 9 as this

Table 2.3 Observability index for various placement solutions in IEEE 14 bus system

	PMU placement at buses: 2, 6, 7, 9	**PMU placement at buses: 2, 6, 8, 9**
Bus 1	PMU at bus 2	PMU at bus 2
Bus 2	PMU at bus 2	PMU at bus 2
Bus 3	PMU at bus 2	PMU at bus 2
Bus 4	PMU at buses 2, 7, and 9	PMU at buses 2 and 9
Bus 5	PMU at bus 2	PMU at bus 2
Bus 6	PMU at bus 6	PMU at bus 6
Bus 7	PMU at buses 7 and 9	PMU at buses 8 and 9
Bus 8	PMU at bus 7	PMU at bus 8
Bus 9	PMU at buses 7 and 9	PMU at bus 9
Bus 10	PMU at bus 9	PMU at bus 9
Bus 11	PMU at bus 6	PMU at bus 6
Bus 12	PMU at bus 6	PMU at bus 6
Bus 13	PMU at bus 6	PMU at bus 6
Bus 14	PMU at bus 9	PMU at bus 9

would yield better results and observability. In addition, a complete redundancy is attained, which is critical for contingency scenarios or events.

2.6 Optimal redundancy criterion (ORC)

Although the OPP yields solutions for full-observability, there is a likelihood of uncertainty during contingency situations which may result in partial or zero observability of the system. Contingency here refers to a situation in the power system where,

- There is a loss of PMU measurements from the observing bus. This would result in the loss of PMU measurements from all the connected buses.
- There is a loss in communication channel that is transmitting information from the observing bus.
- There is a loss of PMU measurements from the observable bus.

The above contingency cases can be caused through physical damage of equipment, infrastructure, and cyber-physical damages. This means, in certain cases, it is better to include a certain number of buses to be observed by multiple PMUs. Thus, the loss of PMU measurements from one device will not affect the observability of any other bus. The factor of percentage or observability weights of buses can be considered in different ways. Some literature studies consider high voltage buses to be more critical than other buses, whereas others consider all generating buses to be critical buses [15]. We propose a new index called ORC that is based on the number of connections to each bus. ORC is defined as the ratio of the total number of connections between buses to the total number of buses in the system. We can select the buses that have more connections or links by adapting the following criterion (2.35):

$$\mathrm{ORC} = \frac{\sum_{i=1}^{n} \sum_{j=1, i \neq j}^{n} B_{i,j}}{n} \tag{2.35}$$

where i and j represent the buses, $B_{i,j}$ represent connections between i and j, and n is the total number of buses in the system. The ORC conditions are enforced as follows:

Step 1: ORC is calculated using the formula provided in (2.35).
Step 2: Based on ORC, the critical bus constraints are selected from all the constraints.
Step 3: Increase the minimum observability requirement of the selected critical buses.

Let us consider constraints (2.5)–(2.18), in which each constraint represents a bus in the IEEE 14-bus system. Here, the total number of connections for all 14 buses is the sum of their connections:

$$B_1 + B_2 + \cdots + B_{14} = 56 \tag{2.36}$$

The number of buses in the system is 14. Thus, ORC for the 14 bus system is given by

$$\mathrm{ORC} = \frac{56}{14} = 4 \tag{2.37}$$

Therefore, from the constraints formed earlier, we consider (2.6), (2.8), and (2.13) and increase the weights, as these buses have a greater number of connections than ORC values (five, six, and five connections respectively). The weights in constraints appear on the right-hand side of the equation. Now, the updated constraints would be as shown in (2.38)–(2.40):

$$\text{Bus 2: } x_1 + x_2 + x_3 + x_4 + x_5 \geq 2 \tag{2.38}$$

$$\text{Bus 4: } x_2 + x_3 + x_4 + x_5 + x_7 + x_9 \geq 2 \tag{2.39}$$

$$\text{Bus 9: } x_4 + x_7 + x_9 + x_{10} + x_{14} \geq 2 \tag{2.40}$$

For full-observability, a "1" is placed on the right-hand side of the constraint equation of each critical bus, indicating that each bus is observed at least once. Based on system and individual utility need or requirements, one can adjust the observability weights. For our test system, we chose to make the critical buses observable twice. This will provide sufficient redundancy and help increase the reliability of the system.

2.7 OPP on standard test systems

This section discusses the results obtained for the OPP using LP approach on Standard IEEE test cases. To validate the approach, the following test cases have been considered:

1. IEEE 14 bus system
2. IEEE 30 bus system
3. IEEE 57 bus system
4. IEEE 118 bus system
5. SRIPG 208 bus system
6. IEEE 300 bus system

All the required data for IEEE test cases have been taken from the University of Washington webpage [16]. The data for the SRIPG test system have been considered from [17]. The Advanced Modeling and Programming Language (AMPL) IDE platform is used to develop the LP models. IBM ILOG CPLEX Optimization Studio solver is used to solve the formulated linear constraints obtained from the test cases.

2.7.1 Advanced Modeling and Programming Language (AMPL)

AMPL is an algebraic modeling language for mathematical programing. It was designed and implemented in 1985 and has been evolving ever since [18]. This is a sophisticated modeling tool that supports the entire optimization modeling life-cycle: development, testing, deployment, and maintenance. AMPL integrates

1. Modeling language for describing optimization data, variables, objectives, and constraints.
2. A command language for browsing models and analyzing results and a scripting language for gathering and manipulating data and for implementing iterative optimization schemes. All use the same concepts and syntax for streamlined application-building.

The main advantages using AMPL are as follows:

1. It has a general and easy to use syntax for arithmetic, logical, and conditional expressions. It uses familiar conventions for summations and other iterated operators.
2. A wide range of solvers such as CPLEX, Gurobi, Xpress (Linear and convex quadratic solvers), CONOPT, Ipopt, KNITRO, MINOS, SNOPT (nonlinear solvers), Bonmin, Couenne, KNITRO (mixed-integer solvers) are available. In addition, open-source solvers can also be hooked to AMPL.

AMPL is notable for the similarity of its arithmetic expressions to customary algebraic notation, and for the generality and power of its set and subscripting expressions. AMPL also extends algebraic notation to express common mathematical programing structures such as network flow constraints and piecewise

Table 2.4 Zero-injection buses

Bus systems	Zero-injection buses	Number of zero injections
14	{7}	1
30	{6,9,11,25,28}	5
57	{4,7,11,21,22,24,26,34,36,37,39,40,45,46,48}	15
118	{5,9,30,37,38,63,64,68,71,81}	10
300	{4,7,12,16,19,24,34,35,36,39,42,45,46,60,62,64, 69,74,78,81,85,86,87,88,100,115,116,117,128, 129,130,131,132,133,134,144,150,151,58,160, 164,165,66,168,169,174,193,194,195,210,212, 219,226,237,240,244,1201,2040,9001,9005, 9006,9007,9012,9023,9044}	65

linearities. AMPL has a flexible interface that enables several solvers to be available at once so a user can switch among solvers and select options that may improve and optimize the solution. All of the general set and arithmetic expressions of the AMPL modeling language can also be used for displaying data and results; a variety of options are available to format data for browsing, printing reports, or preparing input to other programs. Using AMPL, mathematical algebraic problems can be directly converted into a small system of inequalities and an objective function. The availability of an algebraic modeling language makes it possible to emphasize the kinds of general models that can be used to describe large-scale optimization problems.

The model (.mod) files are fed into the AMPL program that works like a compiler. The model and the input data are put into a format that is understood by the solver. The solver actually finds an optimal solution to the problem by reading the intermediate file produced by AMPL and applying an appropriate algorithm. Based on the topology information, the constraint formulations can be directly embedded into the AMPL. In this work, the methodology of formulating the algebraic topological constraints in AMPL and CPLEX linear optimization method is used to solve the problem.

2.7.2 Zero-injection buses

Table 2.4 shows the various zero-injection buses in the test cases considered. It provides various bus numbers that act as zero-injection buses in the test systems to model the zero-injection constraints for the OPP.

The LP formulations are carried for all the test systems in AMPL. First, solutions are obtained with normal test cases without considering any zero-injection buses. Second, zero-injection constraints are added as additional constraints to study the number of PMUs required. Then, an ORC condition is enforced on both the scenarios (i.e., with and without zero injection) bus constraints to determine the reliability of the model. Table 2.5 shows the number of PMUs

Table 2.5 Optimal number of PMUs without zero-injections

Test systems	ILP method	Other methods
IEEE 14 bus	4 {2,6,7,9}	4
IEEE 30 bus	10 {1,5,6,9,10,12,15,18,25,27}	10
IEEE 57 bus	17 {2,6,12,19,22,25,27,32,36, 39,41,45,46,49,51,52,55}	17
IEEE 118 bus	32 {1,5,9,12,15,17,21,23,29,30,36,40,44,46,50, 51,54,62,64,71,75,77,80,85,86,90,94,101,105, 110,115,116}	32
SRIPG 208 bus	55 {5,6,7,12,22,25,27,28,30,33,47,48,50,54,58,60, 63,67,73,75,88,89,92,95,99,104,107,110,111,115, 117,118,120,122,127,128,131,132,136,138,145, 146,152,154,157,159,163,165,181,183,188,190, 191,196,201}	58
IEEE 300 bus	87 {1,2,3,11,12,15,17,20,23,24,26,33,36,39,43,44,49, 55,57,61,62,63,70,71,72,74,77,81,84,86,97,102, 104,105,108,109,114,119,120,122,130,132,133,134, 137,139,140,143,153,154,159,160,164,166,173,178, 181,184,189,194,204,208,210,211,214,217,221,225, 229,231,233,234,237,238,240,244,245,249,9003,9004, 9005,9006,9006,9006,9006,9006,9533}	87

required for the test systems without zero-injection constraints. The column two in Table 2.5 identifies the number of PMUs required for each of the test systems. The table also lists the bus locations, on which the PMUs needed to be placed, and the column three shows the optimal number of PMUs obtained in various literatures.

By comparison of column two and three, the minimal set of PMUs required for the test cases yielded similar results that have been previously proposed in the literatures [19]. However, the SRIPG system seems to provide a better result of 55 PMUs when compared to the results provided in [20]. Table 2.6 provides the results obtained for the OPP when zero-injection buses are considered. Generally, the number of PMUs required decreases with zero-injection considerations, due to the increase in the constraints in the LP approach and this can be seen in Table 2.6.

While considering zero-injection buses for solving the OPP for PMUs, there is a significant decrease in the number of PMUs required. This can be seen in by comparing Tables 2.5 and 2.6. Using the LP formulations, the number of PMUs required for complete observability is smaller than the solutions provided in the literatures (IEEE 57 and 118 bus systems). To the best of authors' knowledge, there is no study that has solved the OPP for the IEEE 300-bus system by considering zero-injection buses. This is carried as an additional task in this research. This research has successfully implemented LP formulations in AMPL on IEEE 300 bus system to obtain the minimal set of PMUs required.

Table 2.6 Optimum number of PMUs while considering zero-injection constraints

Test systems	ILP method	Other methods
IEEE 14 bus	3 {2,6,9}	3
IEEE 30 bus	7 {2,3,10,12,19,24,27}	7
IEEE 57 bus	10 {1,13,18,25,29,32,38,41,51,54}	11
IEEE 118 bus	27 {2,8,11,12,19,22,27,28,32,34,40,45,49,53, 56,62,70,76,77,80,85,87,91,94,101,105,110}	28
SRIPG 208 bus	N/A	N/A
IEEE 300 bus	68 {1,2,3,11,15,21,23,26,33,43,44,49,55,57,61, 63,664,70,72,73,77,7071,97,102,104,105,108, 109,7166,114,119,120,122,126,137,139,140,143, 152,155,167,175,178,183,184,188,198,204,205, 214,217,223,225,229,231,233,234,238,245,249, 7017,7039,9002,9003,9004,9006,9006,9006}	N/A

2.7.3 Enforcing optimal redundancy criterion (ORC)

Another case was to find optimal number of PMUs with ORC condition enforced. Table 2.7 shows the result when ORC condition is enforced. This is done simply by modifying the right-hand side of the inequality constraint. Here, the redundancy of the critical buses is set to 2. By enforcing ORC, all critical buses are observed at least twice. This means that any failure in PMU measurements will not affect the system's observability. For more clear observation on ORC's performance, consider the results in Table 2.8. The column two from Table 2.8 shows the minimum number of PMUs required, and the column three shows the number of PMUs required with ORC enforced. The column four gives us an idea of percentage increase in number of PMUs when ORC is enforced. It is observed that with around 10% increase in number of PMUs, more redundancy in measurements is seen.

A similar table is prepared for test cases while considering zero-injection buses. Table 2.9 shows the percentage of increase in number of PMUs considering zero-injection constraints.

2.7.4 SORI

Table 2.10 shows the SORI obtained for various test systems considering both with and without zero-injection constraints. It also shows a comparison of the results obtained through proposed approach and other methods used in literatures [21]. It is seen that the LP formulated SORIs are higher than the ones in the existing literatures. This indicates that using the LP approach, the test case systems are much more redundant and the full-observability of the buses is higher, compared to the values indicated in other literatures.

Table 2.7 Optimal number of PMUs enforcing ORC

Test systems	Without zero injections with ORC	With zero injections with ORC
IEEE 14 bus	5 {1,4,8,10,13}	4 {4,5,10,13}
IEEE 30 bus	10 {1,2,6,9,10,12,15,18,25,27}	8 {2,3,10,12,13,19,23}
IEEE 57 bus	19 {1,4,6,9,10,11,15,20,24,25, 28,32,36,38,41,46,49,53,57}	13 {3,6,13,16,17,20,25, 29,32,38,41,51,54}
IEEE 118 bus	37 {3,5,9,12,15,17,19,22,23,26, 28,34,37,40,45,49,52,56,59, 61,66,68,71,75,77,80,85,86, 90,93,94,100,104,107,109, 110,114}	29 {3,8,12,15,19,21,27,29, 32,34,42,45,49,53,56,57, 62,70,75,77,80,85,86,90, 94,100,101,105,110}
SRIPG 208 bus	60 {1,6,9,12,20,24,25,27,28,30,33, 47,48,50,55,57,58,60,63,67,73, 75,83,88,89,92,95,99,102,103, 107,110,111,115,117,118,120, 122,127,128,132,136,138,142, 145,146,152,153,154,157,159, 163,165,181,184,188,190,191, 196,201}	N/A
IEEE 300 bus	96 {1,2,3,11,12,13,15,17,21,23,24, 26,323,39,40,41,42,43,44,55,57, 61,62,63,69,70,71,72,74,77,84,88, 94,102,104,105,108,109,110,117, 119,120,122,127,130,132,133,134, 137,139,145,146,148,153,154,160, 164,166,173,175,177,184,188,189, 194,196,205,206,210,211,214,217, 219,221,225,229,231,232,236,237, 238,240,245,246,249,7049,9003, 9004,9005,9006,9006,9006,9006, 9006,9006,9006}	76 {1,2,3,11,13,15,17,21,23, 26,37,41,43,44,55,57,61,63, 664,70,71,72,73,77,94,102, 104,105,108,109,110,7166, 119,120,121,125,126,137, 139,140,145,148,153,156, 163,167,173,176,180,184, 188,196,198,201,206,214, 217,221,225,229,231,233, 234,238,245,247,249,7039, 7049,9002,9003,9004,9006, 9006,9006,9006}

Table 2.8 Number of PMUs with ORC enforced without zero-injection buses

Bus system	Number of PMUs for full observability	Number of PMUs with ORC	% Of increase in PMUs with ORC
IEEE 14 bus system	4	5	25%
IEEE 30 bus system	10	10	0
IEEE 57 bus system	17	19	11.764%
IEEE 118 bus system	32	37	15.625%
SRIPG 208 bus system	55	60	7.272%
IEEE 300 bus system	87	96	10.344%

Table 2.9 Number of PMUs with ORC enforced with zero-injection buses

Bus system	Number of PMUs for full observability	Number of PMUs with ORC	% Of increase in PMUs with ORC
IEEE 14 bus	3	4	25%
IEEE 30 bus	7	8	14.29%
IEEE 57 bus	10	13	30%
IEEE 118 bus	27	29	7.41%
SRIPG 208 bus	N/A	N/A	N/A
IEEE 300 bus	68	76	11.76%

Table 2.10 System observability redundancy index

Bus system	SORI			
	Without zero injections	With zero injections	Without zero injections	With zero injections
	ILP method		Other methods	
IEEE 14 bus	18	14	19	15
IEEE 30 bus	51	46	N/A	N/A
IEEE 57 bus	72	52	72	61
IEEE 118 bus	167	182	164	152
SRIPG 208 bus	320	NA	N/A	N/A
IEEE 300 bus	427	355	N/A	N/A

Table 2.11 Costs associated with PMU placement

Test system	Without zero injections (in US $)	Without zero injections with ORC (in US $)	With zero injections (in US $)
IEEE 14 bus	96,000	96,000	74,000
IEEE 30 bus	2,28,000	2,28,000	1,62,000
IEEE 57 bus	3,82,000	4,26,000	2,28,000
IEEE 118 bus	7,12,000	8,22,000	6,02,000
SRIPG 208 bus	12,26,000	13,36,000	NA
IEEE 300 bus	19,38,000	21,36,000	15,12,000

2.7.5 PMU placement costs

One of the main objectives of the PMU placement problem is maintaining the reliability in measurements of system parameters with the least cost to utility companies. Table 2.11 lists the approximate cost of placing a PMU in the system. Based on the approximate costs, the costs associated with the PMU placement for

Table 2.12 System observability redundancy index with ORC

Test bus systems	Without ORC	With ORC
IEEE 14 bus system	18	18
IEEE 30 bus system	51	51
IEEE 57 bus system	72	85
IEEE 118 bus system	167	215
SRIPG 208 bus system	320	358
IEEE 300 bus system	427	490

test systems are estimated. Table 2.11 provides the cost information for a utility to place the PMUs in a grid. The column two shows the dollar amount required for placing PMUs without considering zero injections, whereas the third column indicates the amount required without considering zero injections and ORC. The fourth column shows the dollar amount required for systems considering zero injections. By observation, using zero-injection constraints in the LP formulations, the total cost could be reduced significantly. The ORC increases the cost of PMUs by a fraction of 9 (for larger test systems). Nevertheless, the cost increase is judiciously justified by trading reliability for full observability for critical infrastructure in the power grid.

2.7.6 Justification of ORC index

ORC increases the total cost for an electric utility. However, it also increases the redundancy in observability of the critical buses required to keep a system stable. Table 2.12 shows an increase in SORI values with ORC enforced. For example, IEEE 57 bus system requires 17 and 19 PMUs for full observability with and without enforcing ORC respectively as seen in Table 2.8. Comparing to Table 2.12, the SORI rose from 72 to 85 in the same system when ORC is enforced. This indicates that with just an increase of two PMUs in the system, full-observability is increased to around 13 buses in the 57 bus system; thus, increasing the observability for critical buses in the system. The other test cases also show positive results with an increased value in SORI. Table 2.8 shows that the number of PMUs required for full-observability with and without ORC index for an IEEE 118 bus system are 32 and 37, respectively. Now, let us consider Table 2.14 for some contingency conditions in IEEE 118 bus system.

Scenario 1: During normal operating conditions, a PMU at bus location 40 in IEEE 118 bus system will observe buses 37, 39, 40, 41, and 42. If the PMU at bus 40 fails for some reason (weather, planned, or unplanned failure) to observe the bus 37, then this failure will affect the observability of 7 other buses that are interconnected to the bus 37. This can be visually understood with the help of case 1 in Figure 2.7. This is the main reason why bus 37 is selected as a critical bus when ORC is enforced. On application of ORC, the critical bus 37 will be observed by 3 other PMUs as shown in Figure 2.7 and Table 2.13.

Table 2.13 Contingency cases in IEEE 118 bus system with and without ORC

Scenario	Critical buses	PMU failures (number of un-observed buses)	No. of unobservable buses	% Of unobservability without ORC	PMU placed buses with ORC enforced (critical buses are observed at least twice)	% Of unobservability with ORC
1	37	Case 1: PMU at bus **40**: {37,39,40,41,42}	5	4.237	Bus 37 is observed thrice: {34,37,40}	0
2	37,59	Case 1: PMU at bus **40**: {37,39,40,41,42} Case 2: PMU at bus **54**: {49,53,54,55,56,59}	11	9.32	Bus 37 is observed thrice: {34,37,40} Bus 59 is observed thrice: {56,59,61}	0
3	37,59, 105	Case 1: PMU at bus **40**: {37,39,40,41,42} Case 2: PMU at bus **54**: {49,53,54,55,56,59} Case 3: PMU at bus **105**: {104,105,106,107,108}	16	13.55	Bus 37 is observed thrice: {34,37,40} Bus 59 is observed thrice: {56,59,61} Bus 105 is observed twice: {104,108}	0
4	37,59, 92,105	Case 1: PMU at bus **40**: {37,39,40,41,42} Case 2: PMU at bus **54**: {49,53,54,55,56,59} Case 3: PMU at bus **105**: {104,105,106,107,108} Case 4: PMU at bus **94**: {92,93,94,95,96,100}	22	18.6	Bus 37 is observed thrice: {34,37,40} Bus 59 is observed thrice: {56,59,61} Bus 105 is observed twice: {104,108} Bus 92 is observed twice: {93,94}	0
5	5,37,59, 92,105	Case 1: PMU at bus **5**: {3,4,5,6,8,11} Case 2: PMU at bus **40**: {37,39,40,41,42} Case 3: PMU at bus **54**: {49,53,54,55,56,59} Case 4: PMU at bus **105**: {104,105,106,107,108} Case 5: PMU at bus **94**: {92,93,94,95,96,100}	28	23.72	Bus 5 is observed twice: {3,5} Bus 37 is observed thrice: {34,37,40} Bus 59 is observed thrice: {56,59,61} Bus 105 is observed twice: {104,108} Bus 92 is observed twice: {93,94}	0

6	5,12,37, 59,92,105	Case 1: PMU at bus **5**: {3,4,5,6,8,11} Case 2: PMU at bus **12**: {2,3,7,11,12,14,16,117} Case 3: PMU at bus **40**: {37,39,40,41,42} Case 4: PMU at bus **54**: {49,53,54,55,56,59} Case 5: PMU at bus **105**: {104,105,106,107,108} Case 6: PMU at bus **94**: {92,93,94,95,96,100}	36	30.5	Bus 5 is observed twice: {3,5} Bus 12 is observed twice: {3,12} Bus 37 is observed thrice: {34,37,40} Bus 59 is observed thrice: {56,59,61} Bus 105 is observed twice: {104,108} Bus 92 is observed twice: {93,94}	0
7	5,12,17,37, 59,92,105	Case 1: PMU at bus **5**: {3,4,5,6,8,11} Case 2: PMU at bus **12**: {2,3,7,11,12,14,16,117} Case 3: PMU at bus **17**: {15,16,17,18,30,31,113} Case 4: PMU at bus **40**: {37,39,40,41,42} Case 5: PMU at bus **54**: {49,53,54,55,56,59} Case 6: PMU at bus **105**: {104,105,106,107,108} Case 7: PMU at bus **94**: {92,93,94,95,96,100}	43	36.4	Bus 5 is observed twice: {3,5} Bus 12 is observed twice: {3,12} Bus 17 is observed twice: {15,17} Bus 37 is observed thrice: {34,37,40} Bus 59 is observed thrice: {56,59,61} Bus 105 is observed twice: {104,108} Bus 92 is observed twice: {93,94}	0

Scenario 2: Similarly, with normal conditions, PMUs at buses 40 and 57 will observe buses 37, 39, 40, 41, 42 and 49, 53, 54, 55, 56, 59, respectively. However, if any of these PMUs fails, there will be an unobservability in IEEE 118 bus system including two critical buses 37 and 59 that are interconnected to 14 other buses. This counts toward 9.32% unobservability in the total system. But when ORC is enforced, this unobservability is prevented. Both the PMUs at 37 and 59 will observe the critical buses redundantly zeroing down the unobservability from 9.32%. There are seven case studies that are conducted using the IEEE 118 bus system as shown in Figure 2.6 in which the reliability of 100% is obtained by enforcing ORC in our formulations. There are many other scenarios in which the similar contingency cases can happen, and applying ORC rule will prove useful and reliable. The contingency scenarios are listed in Table 2.13 with 1, 2, 3, 4, 5, 6, and 7 number of PMU failures. Table 2.13 provides the following information:

1. What if certain PMU(s) are lost and which buses are getting affected?
2. If there is a loss of critical bus and which buses are unobservable?
3. Percentage of unobservability due to the loss of PMU.
4. Reliability results with ORC are enforced.
5. Percentage of unobservability during the loss of PMU(s), and when ORC is enforced.

This is another example that demonstrates the effectiveness of ORC index in assisting the utilities to keep the full-observability of critical buses.

2.7.7 Time computation using LP

In addition, the LP run time is calculated by measuring the computational run time (CPU) for each of the test cases. Table 2.14 shows the time taken by the LP method using AMPL and CPLEX solvers. The computational results of LP approach with and without ORC were compared with existing literatures [14]. From Table 2.14, it is seen that it takes virtually zero-seconds to compute the optimal solutions for all test cases. The other methods take some measureable amount of time to find the optimal solutions. All the test cases ran on the following configured PC system:

- **Processor**: Intel Xeon W3503 @ 2.4 GHz
- **Installed memory**: 8 GB
- **Hard drive**: 465 GB

2.7.8 Forecasting the scalability of OPP using linear regression

In [23], the authors have made a conclusion that only one fourth to one-third of the system buses are need to be provided with PMUs in order to make a system fully observable. Based on our LP results, the estimation of the number of PMUs required for random bus systems was done by utilizing the well-known forecasting linear regression (LR) method. LR is a statistical approach for modeling the relationship between the dependent and independent variables. It is a simple, yet

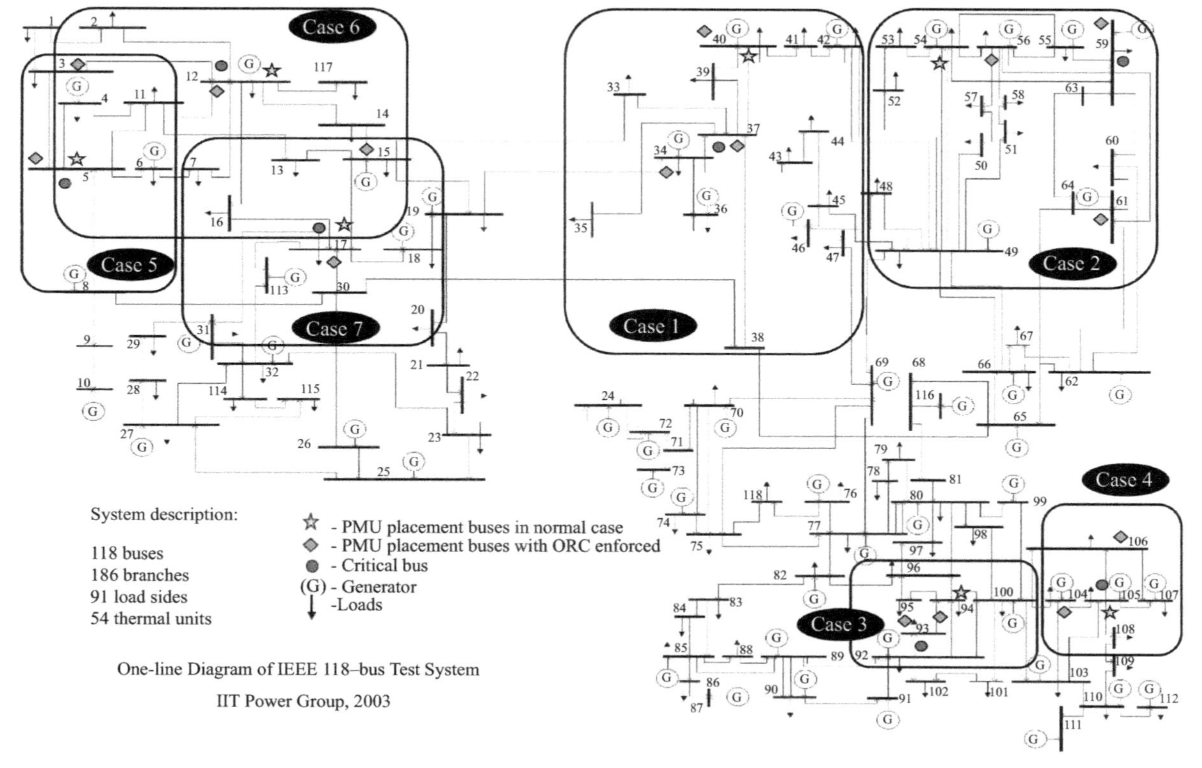

Figure 2.6 IEEE 118 bus system with contingency cases

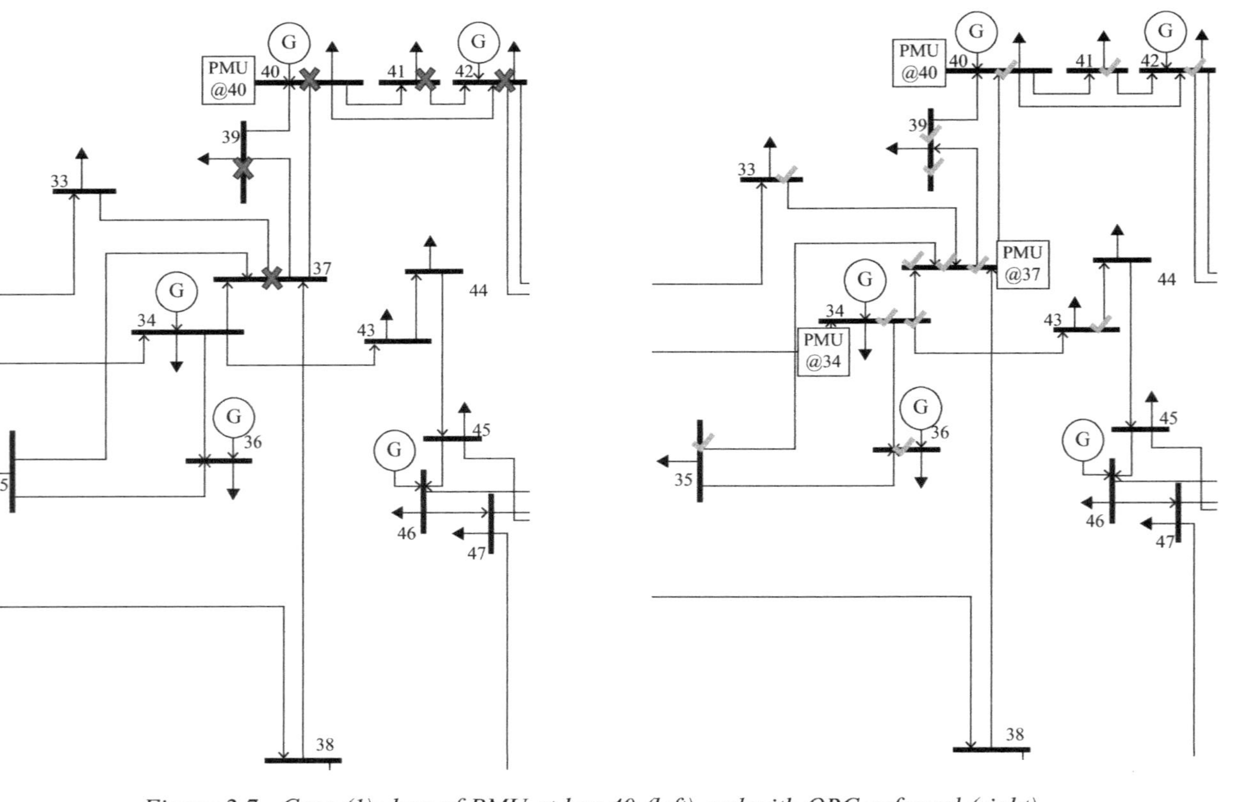

Figure 2.7 Case (1): loss of PMU at bus 40 (left) and with ORC enforced (right)

Table 2.14 Computation time for LP method

Test bus systems	ILP method (ms)	ILP method with ORC (ms)	Other methods [22] (ms)
IEEE 14 bus	0.0	0.0	660
IEEE 30 bus	15.6	15.6	830
IEEE 57 bus	15.6	31.2	870
IEEE 118 bus	15.6	46.8	1,340
SRIPG 208 bus	15.6	46.8	–
IEEE 300 bus	15.6	46.8	–

powerful approach to predict the value of a variable based on the value of another variable.

Using the technique of LR with least squares method, the number of PMUs required for different bus systems is estimated using (2.41)–(2.43):

$$Y = \alpha + \beta \times X \tag{2.41}$$

$$\alpha = \overline{Y} - \beta \times \overline{X} \tag{2.42}$$

where,

$$\beta = \frac{\sum_{i=1}^{n} (X_i - \overline{X}) \times (Y_i - \overline{Y})}{\sum_{i=1}^{n} (X_i - \overline{X})^2} \tag{2.43}$$

$\overline{X}$ and $\overline{Y}$ are the averages of X and Y variables, respectively. The values of X and Y have been considered from our results in Tables 2.5 and 2.6. X refers to the number of buses in the systems, whereas Y refers to the number of PMUs required. Table 2.15 provides the estimation of the PMUs required with respect to the number of buses in the system.

The column two shows the number of PMUs estimated through LR, whereas the column three provides the actual results obtained through ILP approach. From the column four, the percentage of full-observability in any system is around 30% which indicates that approximately 30% of the buses in a system need to be equipped with PMUs to make the system fully observable.

2.8 Conclusion

The chapter provides an ORC for placement of synchrophasors. Test cases were evaluated for IEEE systems for scalability, and the criteria provide promising results for the placement problem.

Table 2.15 Estimation of number of PMUs required

No. of buses in a system (X)	Minimum observability		Percentage of observability (no. of PMU placed buses in the system)
	Estimated no. of PMUs using linear regression (Y)	Through ILP approach	
7	2	2	28.5
14	4	4	28.5
30	9	10	30
57	16	17	28.07
100*	28	*	28
118	33	32	27.11
200*	56	*	28
208	58	55	26.44
300	84	87	29
500*	140	*	28
1,000*	280	*	28
1,500*	421	*	28.06
2,000*	561	*	28.05
3,000*	842	*	28.06

*No bus systems are practically available so there are the no solutions.

Acknowledgments

The authors would like to acknowledge the grant supports from Division of Research & Economic Development at the University of North Dakota (UND-21418-4010-02294), and NSF award number 1537565. Dr Ranganathan would also like to acknowledge the support from his graduate students Ranganath Vallakati and Arun Sukumaran Nair for their contributions.

References

[1] "How Smart Grid Works." [Online]. 2008. Available: http://www.nature.com/news/2008/080730/images/454570a-6.jpg. [Accessed: 01-Jan-2015].

[2] A. G. Phadke, "Synchronized phasor measurements in power systems," *IEEE Comput. Appl. Power*, vol. 6, no. 2, pp. 10–15, Apr. 1993.

[3] B. Mohammadi, A. Rabiei, and S. Mobayen, "Online inter-area oscillation monitoring in power systems using PMU data and Prony analysis," vol. 6, no. 22, pp. 5267–5272, 2011.

[4] K. S. Shim, S. T. Kim, J. H. Lee, E. J. Choi, and J. H. Choi, "Detection of low-frequency oscillation using synchrophasor in wide-area rolling blackouts," *Int. J. Electr. Power Energy Syst.*, vol. 63, pp. 1015–1022, Dec. 2014.

[5] K. P. Lien, C. W. Liu, C. S. Yu, and J. A. Jiang, "Transmission network fault location observability with minimal PMU placement," *IEEE Trans. Power Deliv.*, vol. 21, no. 3, pp. 1128–1136, 2006.

[6] A. Reddy, "PMU based Real-Time Short Term Voltage Stability Monitoring – Analysis and Implementation on a Real-Time Test Bed," North American Power Symposium, Pullman Washington, 2007.

[7] L. Huang, Y. Sun, J. Xu, W. Gao, J. Zhang, and Z. Wu, "Optimal PMU placement considering controlled islanding of power system," *IEEE Trans. Power Syst.*, vol. 29, no. 2, pp. 742–755, Mar. 2014.

[8] J. D. La Ree, S. Member, V. Centeno, J. S. Thorp, L. Fellow, and A. G. Phadke, "Synchronized phasor measurement applications in power systems," vol. 1, no. 1, pp. 20–27, 2010.

[9] K. Martin and G. Brunello, "An overview of the IEEE Standard C37.118.2 – Synchrophasor Data Transfer for Power Systems," in *2014 IEEE PES General Meeting / Conference & Exposition*, pp. 1–1, 2014.

[10] "IEEE Guide for Phasor Data Concentrator Requirements for Power System Protection, Control, and Monitoring," 2013.

[11] E. O. Schweitzer and D. E. Whitehead, "Real-time power system control using synchrophasors," in *2008 61st Annual Conference for Protective Relay Engineers*, pp. 78–88, 2008.

[12] M. V Mynam, A. Harikrishna, and V. Singh, "Synchrophasors Redefining SCADA Systems," in *13th Annual Western Power Delivery Automation Conference*, 2011.

[13] "SynchroPhasor Products." [Online]. 2016. Available: https://www.selinc.com/synchrophasors/products/.

[14] S. Dambhare, D. Dua, R. Gajbhiye, and S. Soman, "Optimal Zero Injection considerations in PMU placement: An ILP approach," *16th PSCC, Glas. July* 14, pp. 14–19, 2008.

[15] A. Pal, G. Sanchez-Ayala, V. Centeno, and J. Thorp, "A PMU placement scheme ensuring real-time monitoring of critical buses of the network," Ph.D. Dissertation, Virginia Tech, vol. 29, no. 2, pp. 510–517, 2014.

[16] "IEEE 14 bus system." [Online]. 2010. Available: http://www.ee.washington.edu/research/pstca/pf14/pg_tca14bus.htm. [Accessed: 08-Aug-2014].

[17] P. Gopakumar, G. S. Chandra, M. J. B. Reddy, and D. K. Mohanta, "Optimal redundant placement of PMUs in Indian power grid–northern, eastern and north-eastern regions," *Front. Energy*, vol. 7, no. 3, pp. 413–428, May 2013.

[18] R. Fourer, D. M. Gay, M. Hill, B. W. Kernighan, and T. B. Laboratories, "AMPL : a mathematical programming language," *Manage. Sci.*, vol. 36, pp. 519–554, 1990.

[19] M. J. Alvarez, F. S. Sellschopp, and E. Vazquez, "A PMUs placement methodology based on inverse of connectivity and critical measurements," *Int. J. Electr. Power Energy Syst.*, vol. 68, pp. 336–344, Jun. 2015.

[20] P. Gopakumar, G. Surya Chandra, and M. J. B. Reddy, "Optimal placement of phasor measurement units for Tamil Nadu state of Indian power grid," *2012 11th Int. Conf. Environ. Electr. Eng.*, pp. 80–83, May 2012.

[21] D. Dua, S. Dambhare, R. K. Gajbhiye, and S. A. Soman, "Optimal multi-stage scheduling of PMU placement: an ILP approach," *IEEE Trans. Power Delivery*, vol. 23, no. 4, pp. 1812–1820, 2008.

[22] B. K. Saha Roy, A. K. Sinha, and A. K. Pradhan, "An optimal PMU placement technique for power system observability," *Int. J. Electr. Power Energy Syst.*, vol. 42, no. 1, pp. 71–77, Nov. 2012.

[23] T. L. Baldwin, L. Mili, M. B. Boisen, and R. Adapa, "Power system observability with minimal phasor measurement placement," *IEEE Trans. Power Syst.*, vol. 8, no. 2, pp. 707–715, May 1993.

Chapter 3
Wide-area measurement-based power network protection

Pratim Kundu[1], *Saumendra Sarangi*[1] *and Ashok Kumar Pradhan*[1]

3.1 Introduction

Industries today depend heavily on electric power. Reliability of electric service is the major concern to electric power industry. Fault in a power system component disrupts the power flow and if not cleared quickly can lead to component damage and other associated problem in the system. Protective relaying is concerned with the design and operation of equipment which detects abnormal power system condition and initiates action as quickly as possible to isolate the faulted component [1,2]. A wide-area disturbance is the power outage over large geographical region that affects the power system reliability, causing interruption of the electric supply to thousands of users. Analysis of major disturbances has revealed maloperation of protection schemes as the most important reason [3–5].

Wide-area measurement system (WAMS) is a promising technology in smart grid initiative. To be able to monitor, operate, and control power systems in wide geographical area, WAMS combines the functions of metering devices (i.e., new and traditional) with the abilities of communication systems [6]. Phasor measurement unit (PMU) otherwise known as synchrophasor is a monitoring device that provides time tagged phasors, frequency, and rate of change of frequency. Data from PMUs with dedicated communication system are recorded at a central location known as phasor data concentrator (PDC). Such a measurement system, WAMS, provides a dynamic view of the system and helps deriving accurate protection decision. With use of PMUs, improved monitoring, protection, and control of power network are being achieved [7–11]. Over the last decade, applications of PMU data in power system have gained importance [12].

Discriminating a fault condition from other stressed power system events such as power swing, load encroachment, and voltage instability condition, especially

[1]Department of Electrical Engineering, Indian Institute of Technology Kharagpur, Kharagpur 721302, India

3.1.2.1 Communication delay causes

Although more and more control systems are being implemented in a distributed fashion with networked communication, the unavoidable time delays in such systems impact the achievable performance. Delays due to the use of PMUs and the communication link involved occur primarily for the following reasons:

1. Transducer delays: Voltage transducers and current transducers are used to measure the rms voltages and currents, respectively, at the instant of sampling. Window size of the discrete Fourier transform (DFT): Number of samples required to compute the phasors using DFT.
2. Processing time: The processing time required in converting the transducer data into phasor information with the help of DFT.
3. Data size of the PMU output: Data size of the PMU message is the size of the information bits contained in the data frame, header frame, and the configuration frame.
4. Multiplexing and transitions: Transitions between the communication link and the data-processing equipment lead to delays that are caused at the instances when data are retrieved or emitted by the communication link.
5. Communication link involved: The type of communication link and the physical distance involved in transmitting the PMU output to the central processing unit can add to the delay.
6. Data concentrators: Data concentrators are primarily data collecting centers located at the central processing unit and are responsible for collecting all the PMU data that are transmitted over the communication link.

The current research work focuses on the issues of maloperation of distance relays at various system conditions like power swing and protection component failure. It proposes usage of WAMS data to support protection schemes during such conditions. The latency associated with communication scheme restricts the application of WAMS data for network protection only to backup schemes (slower protection). In this work, maximum latency associated with WAMS data is considered to be 200 ms. Methods developed enhances the backup relay decision during various conditions such as fault, voltage instability, and power swing. Also the method for identification of failure of protection components is obtained using WAMS data. To test the techniques, simulations were carried out for different systems with electromagnetic transient design and control (EMTDC)/power system computer aided design (PSCAD). The sampling rate considered in the study is 1 kHz and phasors are estimated using DFT technique to obtain synchronized data from the system.

3.2 Improved network protection during stressed conditions

With continuously increasing power demand and evolution of power industry restructuring, power systems now-a-days operate more often closer to their stability limits. During steady state, a balance is maintained between generation and load. However, unexpected events such as fault and loss of large load disrupt the balance and give rise to electromechanical oscillation known as power swing [19]. Severe power swing at times leads to loss of synchronism between a generator and the rest of the utility system. This situation is referred to as out-of-step condition. All cases of power swing are not so severe; most of them result in impedance seen by distance relay entering into its operating zone and cause undesired line tripping. System disintegration can be averted if the relays are blocked during a stable event and trip signal issued for fault situations and during threat of instability.

3.2.1 Protection support during power swing

The objective of this work is to support relay decision by classifying power swing according to its severity using WAMS data [20]. This classification provides blocking signals for avoiding maloperations and detecting out-of-step condition in the system that can help initiating control actions. For this purpose, four indices specially derived from synchronized data are used to assess the dynamic system condition. During a disturbance, all the variables namely angle, voltage, and frequency in a system undergo changes [21]. Information obtained from such variable can be resourceful to evaluate the effect of a disturbance. Mentioned algorithm assesses the system in the post-disturbance period by integrating indices using a fuzzy logic–based scheme. Each area in the system is classified to a state in accordance with its response to the disturbance. This classification aids in adjusting relay operation during power swing. The output provides signal to prevent maloperation of relay and to command specific relay to segregate an area from the system during out-of-step condition. An eastern India system is used to validate the methods. Results show the accuracy of the method in supporting relay decision during power swing.

3.2.1.1 Wide-area severity indices

Rotor angle–based prediction of loss of synchronism between two areas can be derived when the angle difference between them exceeds a predefined threshold [19]. When the technique is applied for assessment of a disturbance and predicting its behavior, this criterion has limitation as it offers little pre-emption time. There are other criteria derived from various power system quantities to address stability of the system [22,23]. Assessment for all conditions of power system based on a single criterion is not possible. To address this problem, a combination of different indices is desired. In this work, four indices derived from angle, frequency, voltage, and damping information are used to estimate the system condition for use in protection decision. Synchronized data obtained from the generating stations in an area are used for calculation of the following indices.

Angular deviation referred to center of inertia (COI)

The COI-based rotor angle disturbance provides a robust index to observe the state of synchronism between the areas during a disturbance. Under normal condition, a generator in a power system operates at a fixed angular difference with respect to a reference [24]. During a power swing, the angular differences among the generators start oscillating with respect to the reference. The COI reference for the system is defined as

$$\theta^{\mathrm{COI}} = \frac{\sum_{i=1}^{n} H_i \theta_i}{\sum_{i=1}^{n} H_i} \tag{3.1}$$

where n is the number of areas, H_i is the equivalent moment of inertia of ith area, θ_i is the equivalent angle of ith area.

COI_index provides a good indication of the level of stress over a part of the system induced by the contingency. The index is calculated for all areas and a high value of the index shows that the impact of the disturbance is severe in that area.

Frequency-based power imbalance index

For a single-machine system, the overall generator-load dynamic relationship between the incremental mismatch power ($\Delta p = \Delta p_m - \Delta p_l$), and the frequency deviation (Δf) can be expressed by a simplified frequency response model [21]:

$$\Delta p = 2H \frac{d(\Delta f(t))}{dt} + D(\Delta f(t)) \tag{3.2}$$

where Δf is the frequency deviation, Δp_m is the mechanical power change, Δp_l is the load change, H is the inertia constant, and D is the damping coefficient of load in the system. The disturbance power Δp for a case can be approximated as

$$\Delta p = 2H \frac{d(\Delta f(t))/dt}{f_n} \tag{3.3}$$

where f_n is the rated frequency, 50 Hz in this case.

Bus voltage deviation

Voltage stability criterion derived from rms values of bus voltages is used to assess whether the system is able to achieve post-disturbance equilibrium as regard to voltage [23]. The steps to obtain the index are as follows:

Step 1. Calculate moving average of positive sequence voltage at any instant k using values of previous N (50 in our case) samples obtained from WAMS:

$$V_{\mathrm{avg}_k} = \begin{cases} \dfrac{\sum_{m=1}^{k} V_m}{k} & k \le N \\[2ex] \dfrac{\sum_{m=k-N-1}^{k} V_m}{N} & k > N \end{cases} \tag{3.4}$$

Step 2. Calculate diversity (d_k) between the moving average value V_{avg_k} and the voltage value at kth instant (v_k):

$$d_k = \frac{\left(V_{\text{avg}_k} - v_k\right) \times 100}{v_k} \tag{3.5}$$

Step 3. Calculate the voltage-based index vol_k:

$$\text{vol}_k = \begin{cases} \dfrac{\sum_{m=1}^{k}(d_m + d_{m-1})}{2k} & k \le N \\ \dfrac{\sum_{m=k-N-1}^{k}(d_m + d_{m-1})}{2N} & k > N \end{cases} \tag{3.6}$$

where Δt is the subinterval in the above trapezoidal integral rule and in this study $\Delta t = 1$ s.

Damping estimation using Hilbert Huang transform (HHT)

System damping is responsible for mitigating any oscillating condition of a system. An estimation of the damping coefficients should provide insight into the development of a particular disturbance and help in identifying whether the condition is critical or not. The HHT is a powerful signal-processing tool to analyze nonlinear and nonstationary signals. The signals obtained from WAMS data consist of various frequency components and are not real zero signals in which the difference between local maxima and local minima is at most 1. These signals need to be decomposed into components which can be processed using HHT. Empirical mode decomposition (EMD) provides a way to decompose a signal into various mono-frequency components known as integral mode functions (IMFs) [25,26]. Damping of individual IMFs is estimated using HHT. The average damping of all the components is the total damping value of the signal. The expression for Hilbert transform applied to a real-valued signal $w(t)$ is

$$w_H(t) = \frac{1}{\pi} P \int_{-\infty}^{\infty} \frac{\omega(\tau)}{t - \tau} ds \tag{3.7}$$

where P is the Cauchy's principal value of the integral. Hilbert transform defines the analytical form of the signal as

$$W(t) = w(t) + jw_H(t) \tag{3.8}$$

Equation (3.8) can be rewritten as

$$W(t) = x(t)e^{\theta(t) + j\phi(t)} \tag{3.9}$$

where $\theta(t)$ represents the decay function: $\theta(t) = -\int_0^t (\alpha t)dt$, with α being the damping constant.

The expression for the instantaneous modal parameters as a function of the first and the second derivatives of the signal envelope and the instantaneous frequency [26] is

$$\alpha(t) = -\left(\frac{\dot{A}}{A} + \frac{\dot{\omega}}{\omega}\right) \tag{3.10}$$

The modified signal in this case has been decomposed into a monofrequency signal by EMD, we have $\dot{\omega}(t) = 0$ and hence (3.10) simplifies into:

$$\alpha(t) = -\left(\frac{\dot{A}}{A}\right) \tag{3.11}$$

Application of HHT to the signal provides a number of modes, each having a distinct frequency and its respective damping. Damping constants of selected modes are considered, whose frequency is greater than 1, where t is the time frame of the observation window. In the simulation, an observation window of 5 s is used as the input to the EMD function.

3.2.1.2 Simulated test system

The eastern India power system considered in the study is shown in Figure 3.2. There are six areas in the system. The main interconnecting lines between various areas have been named Lines 1–10. Synchronized data obtained from each area are used for calculating the severity indices. In this work, division of system into areas is based on the location of generating stations. However, for a large system, such a division is accomplished using coherency-based studies. The method for area segregation based on coherency is proposed in [27,28].

3.2.1.3 The method

Block diagram for the method mentioned is shown in Figure 3.3 which aims at supporting conventional relay decision using WAMS data.

The severity classifier block takes decision regarding the blocking of zones 2 and 3 relays in an area and the out-of-step classifier assesses whether an area is prone to losing its stability. This assessment is carried out mainly for assisting the operation of distance relays connected to tie lines between various areas in the system.

3.2.1.4 Case study for eastern India test system: out-of-step condition

For this situation, a disturbance in the system is created with a three-phase fault in line 2 connecting bus 1103 (area HWH) and 1104 (area STH) of Figure 3.2 and subsequent operation of breakers on both sides of the line. Following the line clearance, there is a growing oscillation in the system. The results obtained from simulation are for a period of 4 s. During normal operation, the active power flow through line 1 is 140.59 MW from HWH to STH, but during swing period, the line power oscillates between 360 MW and 76 MW which suggests that area STH is under high power imbalance state during the period. This situation leads to overload of lines and may cause relays to maloperate further degrading the condition of

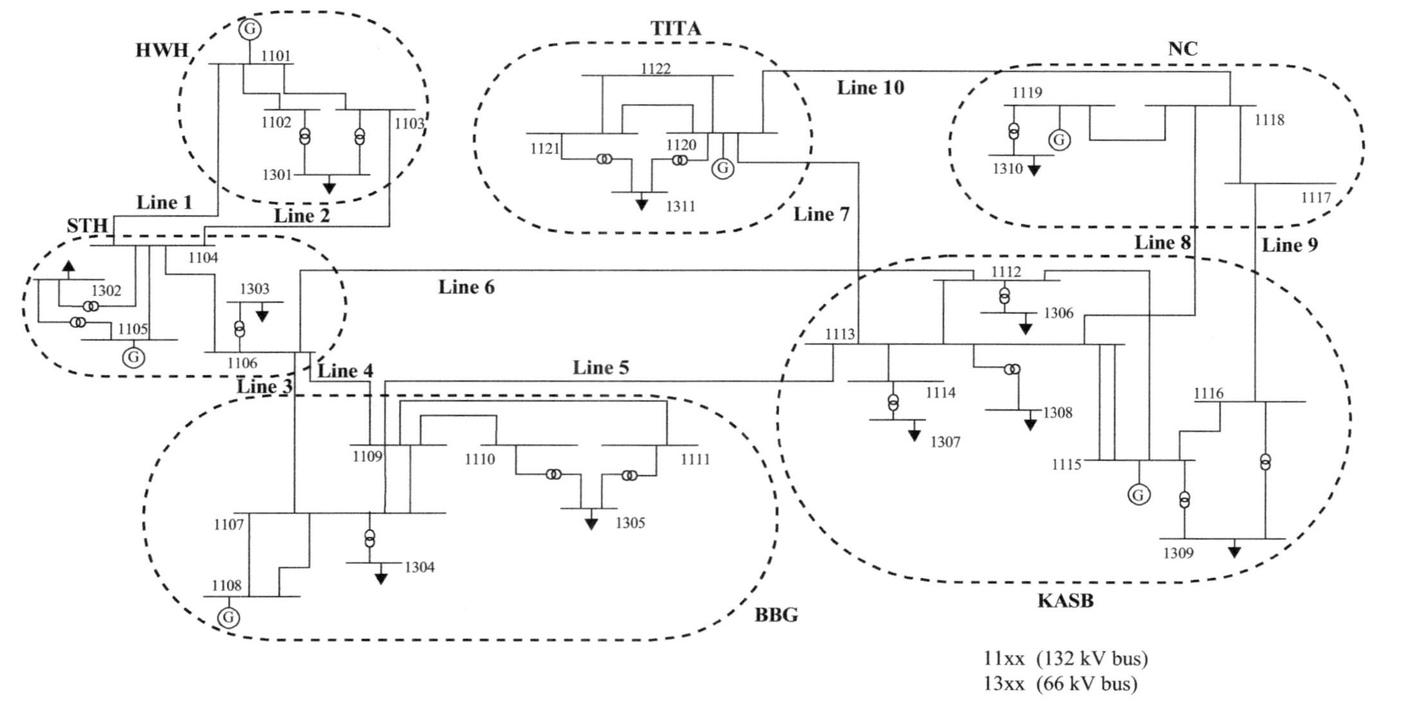

Figure 3.2 Block diagram of eastern India system

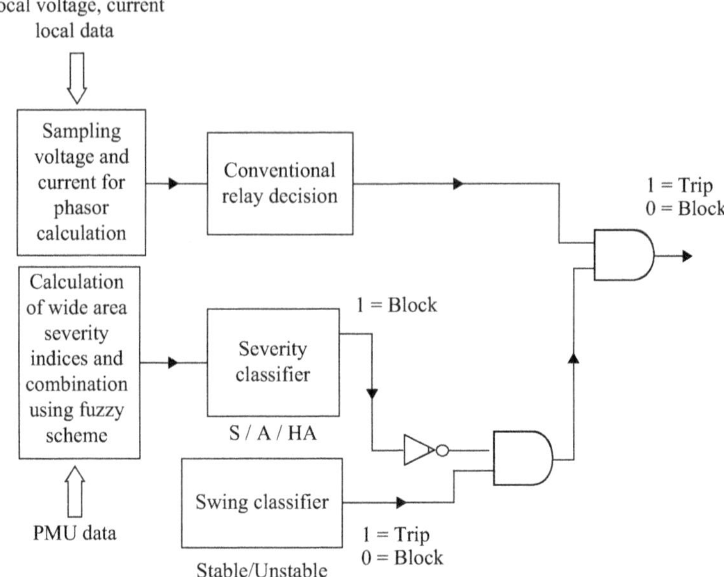

Figure 3.3 Block diagram of the protection support scheme

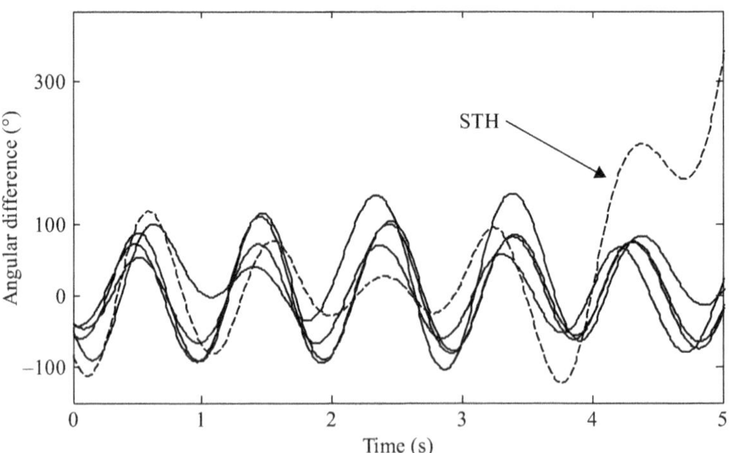

Figure 3.4 Angular plot for all areas

the system. The response of the algorithm for area STH during the disturbance is shown in Figure 3.6. The figure depicts a gradual increase in the swing severity in the system and the algorithm output moves from S to U. The damping in the area STH is unable to contain the situation which ultimately leads to out-of-step condition with respect to the rest of the system.

The angular plot for all the areas are shown in Figure 3.4 shows that area STH (dotted line) goes out-of-step after 4 s from the clearance of the faulty line.

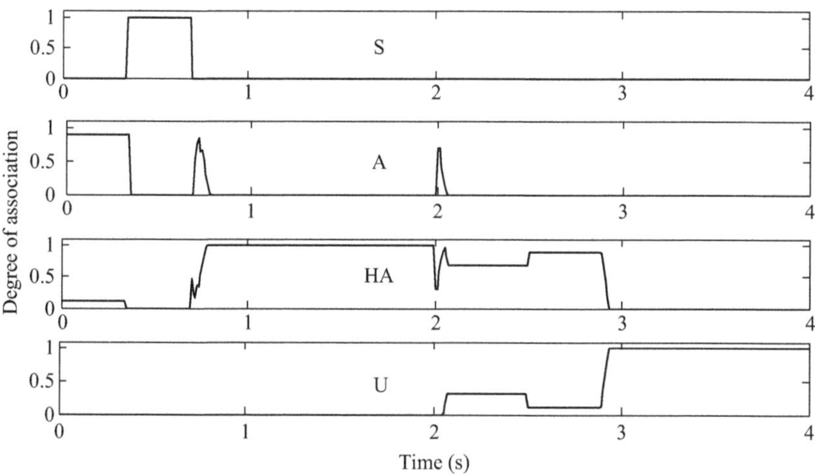

Figure 3.5 FLS output for area STH during the disturbance

To avoid this, the area should be isolated beforehand. Initially the output is in the safe region but as the disturbance aggravates, the FLS output unit in Figure 3.5 shows alarming region and finally enters the unstable region at about 3 s showing that the area has a high chance of collapse. The initiation of controlled relay operation strategy at this situation is desirable to isolate the most affected area, STH in this case, and maintain balance in the rest of the system. This prevents aggravation of disturbance and a possible blackout. The output of the algorithm predicts an out-of-step condition providing a window of 1 s to isolate the area before the loss of synchronism occurs (Figure 3.5).

3.2.2 Improved zone 3 operation

Incorrect zone 3 operation occurs in a power system as at times relay cannot distinguish a fault from other events like power swing, load encroachment, and voltage instability. The impedance seen by a relay enters into the operating zone even during nonfault situation and may cause it to maloperate. One of the reasons for this is that the relay uses local current and voltage data to take decision. Similarly, using local information relays find difficulty while discriminating a fault, especially three-phase faults which are balanced in nature, during such stressed events.

Operating regions for various relays in an interconnected power system overlap and during a fault, the number of relays will see it in their different zones. For all other stressed conditions, such as power swing, load encroachment, and voltage instability, the impedance trajectories for relays do not enter their operating regions simultaneously. The technique utilizes this fact and uses synchronized voltage and current data from strategic locations in the system to support zone 3 operation [29].

3.2.2.1 Synchrophasor placement

A bus with synchrophasor measurement in the system is known as P_bus, and the other buses are known as N_bus. The measurement points in the system are

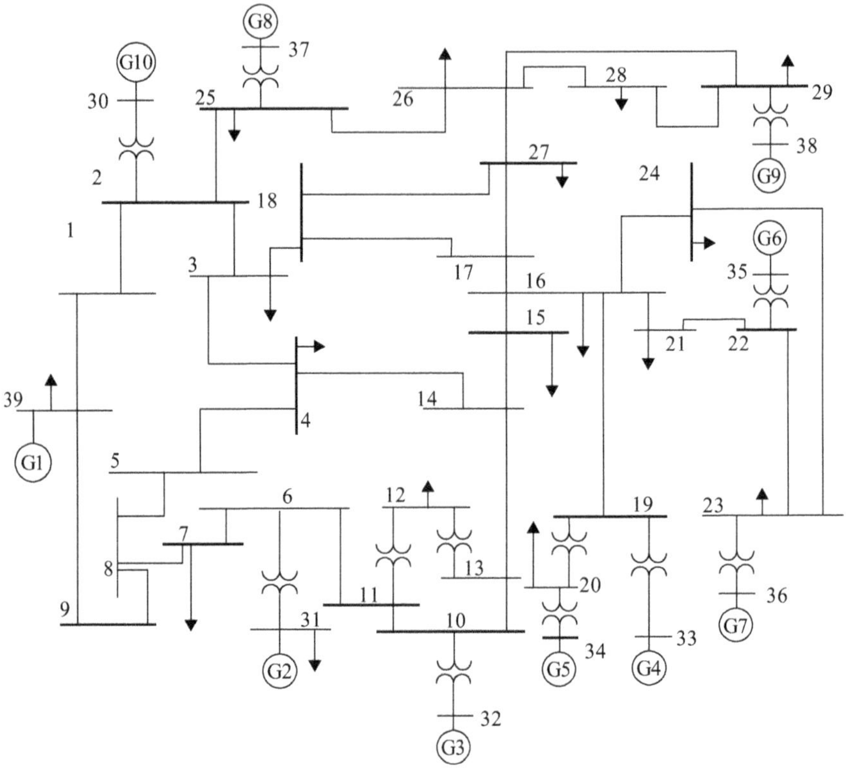

Figure 3.6 IEEE 39 bus system

Table 3.1 Measurement buses

Synchrophasor buses	2, 4, 7, 9, 10, 11, 15, 18, 19, 22, 24, 25, 27, 29, 34

selected in such a way that the lines in the system are either connected to a P_bus or should be seen in zone 3 of relays connected to a minimum of two P_buses in the system. Incidence of a fault in any line results in impedance trajectory entering into relay characteristics of two or more P_buses. During any change in the system condition, voltage and current signals provided by the synchrophasors are used to determine the affected region and to examine whether it is a fault situation or not. Table 3.1 lists the synchrophasor buses for the 39 bus New England system, Figure 3.6.

3.2.2.2 The method

Any disturbance in the system leads to change in-line current values that are proportional to the severity of the event. A line current–based detection technique is suitably applied for detection here. During normal condition, the phasor sum of line currents obtained from synchrophasors in a group is constant. The change in phasor sum is referred to as the disturbance detection index I_{dis}:

$$I_{sum}(k) = \sum_{n} I_p(k) \tag{3.12}$$

where $I_p(k)$ is the positive sequence current phasor measurement at kth instant, n is the number of buses.

The disturbance detector index for each group is calculated as

$$I_{dis}(k) = I_{sum}(k) - I_{sum}(k-1) \tag{3.13}$$

The disturbance index is calculated continuously using synchrophasor measurements. For normal condition, the index value is 0. Any change in the system affects the I_{dis} value.

Following the detection of a disturbance, the impedance based fault detector (f_d) is used to identify a fault in the network.

The index Z_d is calculated for each selected measurement bus using relation (3.14). Z_d has a value less than 1 if the apparent impedance calculated using synchrophasor measurement is less than the corresponding relay setting. During a fault, the index value for all the measurement buses will be less than 1:

$$Z_{d_j} = \frac{Z_{app}}{Z_{set_j}} \tag{3.14}$$

where

$$j = \begin{cases} 1, & \text{if} \quad Z_{set_1} \geq Z_{app} \\ 2, & \text{if} \quad Z_{set_1} < Z_{app} \leq Z_{set_2} \\ 3, & \text{if} \quad Z_{set_2} < Z_{app} \end{cases}$$

and Z_{set_1}, Z_{set_2}, and Z_{set_3} correspond to settings of zones 1, 2, and 3, respectively.

The fault detector index f_d is obtained as follows:

$$f_d = \sum Z_{d_j} \times W_j \tag{3.15}$$

$$W = \begin{cases} -10 & \text{if} \quad j = 1 \\ -5 & \text{if} \quad j = 2 \\ -2 & \text{if} \quad j = 3 \;\&\; Z_{d_3} \leq 1 \\ +1 & \text{if} \quad j = 3 \;\&\; Z_{d_3} > 1 \end{cases}$$

A weight W is associated with each measurement that depends on the apparent impedance. A negative weight is associated with the index Z_d if the apparent impedance is less than the relay setting. The absolute value of the weight is highest if the apparent impedance calculated from synchrophasor measurements is less than zone 1, followed by zone 2 and then zone 3 characteristics of relay connected to those lines. The weight is positive when the calculated apparent impedance is greater than the relay zone 3 characteristics. During a fault condition, the value of f_d becomes negative as all the relays in the selected group see the apparent impedance less than

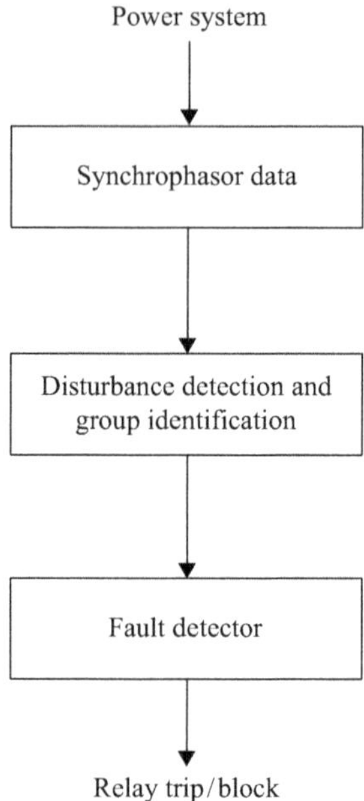

Figure 3.7 Block diagram of the zone 3 support algorithm

their relay settings. In the case of other events the index value is positive. The block diagram in Figure 3.7 shows the main steps involved in the algorithm. Line current data are used to detect any abnormal situation in the system and following that the impedance-based fault detector data are used to ensure whether it is a fault situation or not. Line currents and voltages are obtained from synchrophasor data from strategic locations in the system. The decision block then issues signal for blocking or unblocking of zone 3 relay operation in the system.

3.2.2.3 Case study

During a power swing condition current and voltage in the system oscillate. As a result, the impedance seen by the relay enters into different operating regions resulting in malfunction. To avoid such an event, the relays need to be blocked. A blocking scheme should be able to verify whether it is a fault situation or not. In this case, clearance of line connecting bus 26 and 29 at 1.2 s, following a three-phase fault at 1 s in the line, has led to oscillation in the system. The initiation and clearance of the fault results in index I_{dis} to have sudden change in its value. The two spikes in Figure 3.8 show the change in the index. A change in index I_{dis} using synchrophasor measurements at bus 25 for line 25-26, at bus 27 for line 27-26 and at bus 29 for line 29-28 indicates a disturbance. The fault detector index (f_d) is calculated using relation (3.15). The impedance trajectories during the power swing condition are shown in Figure 3.9, and the fault detector index value is plotted in Figure 3.10.

The impedance plots are obtained using synchrophasor measurements, whereas the operating characteristics are for relays corresponding to the measurement buses.

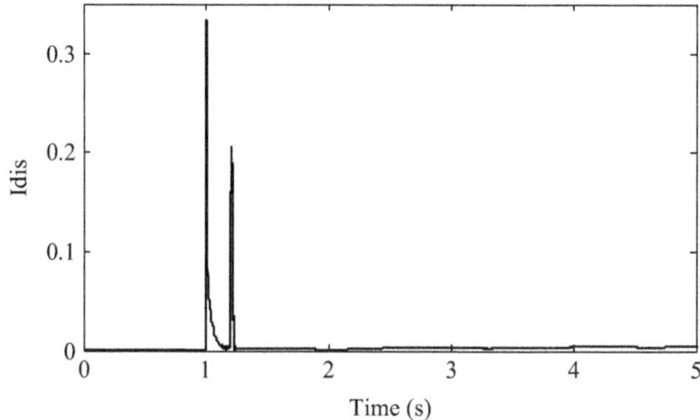

Figure 3.8 Result for disturbance detector during power swing

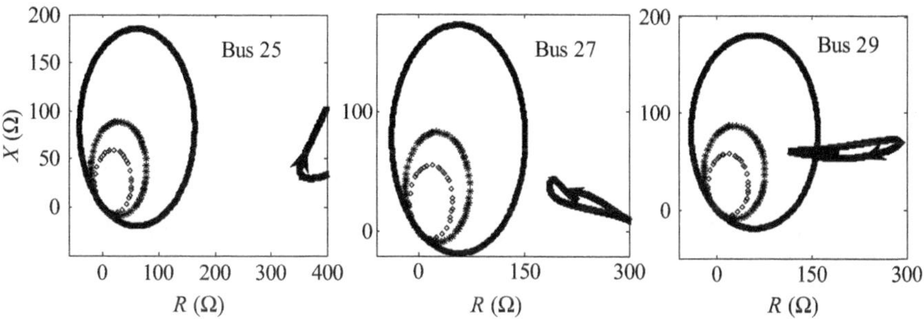

Figure 3.9 Impedance trajectory for power swing case

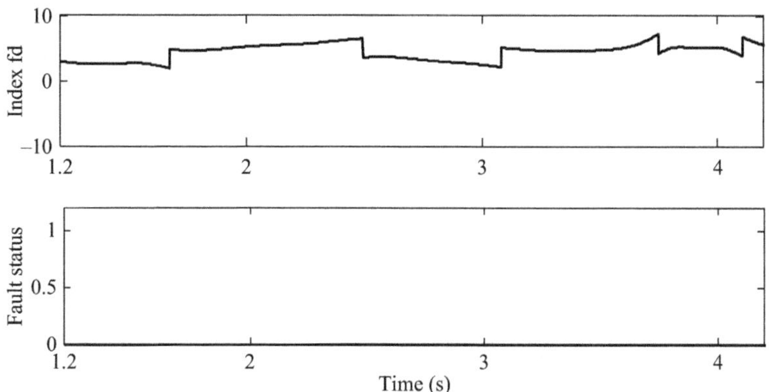

Figure 3.10 Fault detector result during power swing case

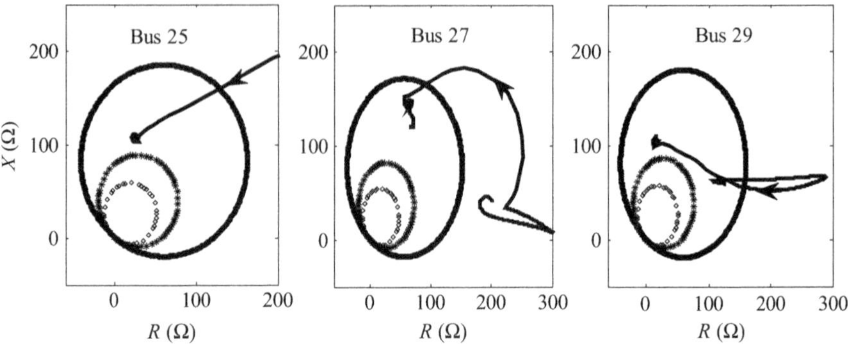

Figure 3.11 Impedance trajectory for three-phase fault during power swing

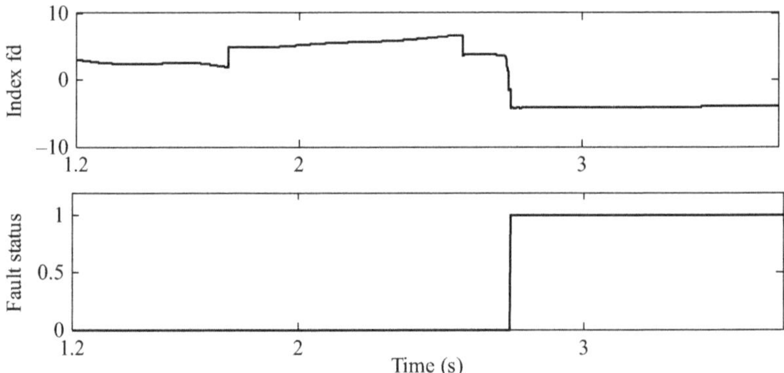

Figure 3.12 Fault detector result for three-phase fault during power swing

The impedance trajectory for bus 29 enters into its zone 3 region due to the power swing condition.

This may result unwanted zone 3 operation if not blocked. Z_d calculated using relation (3.14) for measurement at bus 29 for line 29-28 is less than 1 and for the other two measurements in the group it is greater than 1. Fault detector index f_d calculated using Z_d of measurements from buses 25, 27, and 29 is positive which confirms that there is no fault.

A fault in line 26-28 results in impedance trajectories calculated by synchrophasor measurements entering the zone 3 regions for corresponding relays at bus 25 for line 25-26, at bus 27 for line 27-26, and at bus 29 for line 29-28, Figure 3.11. This results in Z_d of each measurement calculated using (3.14) to be less than 1. For power swing case, Z_d calculated for measurement bus 29 is less than 1, but in this case Z_d for all measurements are less than 1. The weight associated with each measurement for calculating f_d thus becomes negative, Figure 3.12. The fault detector index f_d was positive during power swing.

3.3 Online identification of protection element failure

Fault in a power system is cleared by different protection schemes comprising different relays, CBs and accessories. Collective action of all these elements in a protection scheme is necessary to isolate a fault. Failure of an element in such a protection arrangement may lead to unwanted line trip risking further outages in the network. This work proposes a WAMS data–based technique for online identification of protection component failure of transmission line. Voltage current phasors along with relay decision and CB status signals are used to identify the faulted line in the system followed by identification of the protection component failure, if any, in the selected section.

3.3.1 Faulted line identification

Initiation of fault at any point in a network results in huge current flow. This causes severe drop of voltage at buses close to the fault point. Also due to the inductive nature of fault, phase angle between voltage and current phasors at these buses deviate. Hence, change of voltage and phase angle at buses in a network is used to identify fault point in a system. For this purpose, two indices are applied, which are obtained using synchronized voltage and current phasors as defined below:

1. v_abs (n): absolute value of positive sequence voltage deviation during fault for bus-*n*.
2. vi_phase (nm): deviation of phase angle between the positive sequence voltage of a bus and current in a line connected to the bus during fault where *n* and *m* correspond to buses connected by the line.

The bus with maximum deviation in positive sequence voltage phasor is selected. Next, the change in phase angle between the voltage and current phasors for the lines connected to the selected bus is observed. The line with highest deviation in the phase angle value is identified as the faulted line.

3.3.2 Component failure indices

Correct operation of various components such as relay, CB, and accessories is imperative for error-free functioning of a protection scheme. Failures of any such component can affect the power system stability. A simple description of a protection scheme is shown in Figure 3.13.

It shows the basic elements involved in detection and removal of faults. Various signals on status of monitored devices such as CB, protective relay decision signal can be obtained in real time through WAMS [30].

The algorithm proposes four indices that are based on the relay decisions and CB status signals obtained for a system. They are defined as

1. Relay operation signal (rel_*i*): $i = 1, 2, 3$ for zones 1, 2, and 3, respectively. A value of 1 for operation and 0 for nonoperation is assigned. If zone 1 has operated then rel_1 = 1 and rel_2 and rel_3 are 0.

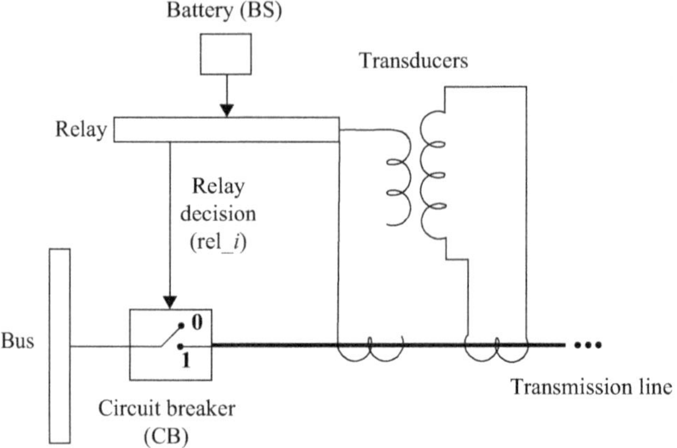

Figure 3.13 Elements of the protection arrangement

2. CB status signal: 1 if breaker is open and 0 if closed.
3. Battery status (BS): index value is 0 if battery has defect and 1 if functioning properly. A defective battery indicates that the relay operation is not active [31].
4. Timer signal (tim_sig_*i*): value depends on rel_*i*. For zone 1, its value is tim_sig_1 = 1 always and tim_sig_2, tim_sig_3 = 0. For zone 2 operation, the index value is 0 till 15 cycles and for zone 3 operation till 50 cycles after the inception of fault and becomes 1 thereafter. The time delay values are selected based on standard backup protection operation time.
5. Relay status (RS): 0 for relays connected to the faulted line. For backup relays, it is 2 for the first 15 cycles (300 ms) after fault initiation and 0 thereafter. For reverse operating relays, it has a value of 2.

3.3.3 Signal communication

Protection operation is accomplished at high speed; therefore, an online assessment and correction of these schemes require high-speed data communication. Today's wide-area communication topologies (such as Synchronous Optical Network) are capable of delivering messages from one area of a power system to multiple nodes in few milliseconds. IEEE synchrophasor standard [32] uses binary format for data package transfer through communication channels. In such system, a configuration frame is a machine-readable binary data set containing information and processing parameters for a synchrophasor data stream. The data fields in a configuration frame are used for communicating information. In this work, along with synchronized phasor data, protection element operation signals are obtained from WAMS. For this purpose, a 4-bit binary signal on the status of the elements is used. This

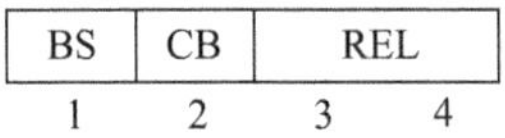

BS	CB	REL	
1	2	3	4

Figure 3.14 4-bit signal information obtained from WAMS data for the method

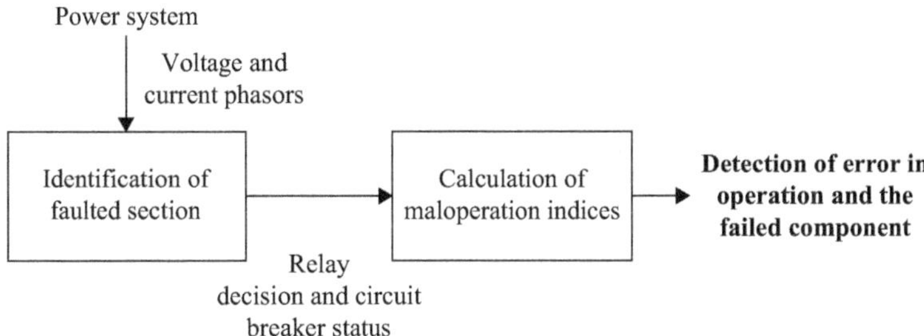

Figure 3.15 Flow diagram of the method

4-bit data can be transmitted through one of the data fields of a configuration frame. The binary data structure for the scheme is shown in Figure 3.14. The first two bits are for BS and CB signals, respectively, bits 3 and 4 are information on relay operation signals.

3.3.4 The method

Identification of protection maloperation for distance relay components during a fault condition in a power system is being proposed in [33]. The steps for the method are shown in Figure 3.15. In the first step the algorithm identifies the line affected by fault situation in the system. For this, synchronized positive sequence voltage and current data are used. Depending on the fault location, system topology and zones of operation, relays in the system are assigned weights (rel_w) [31,34]. In the second part of the algorithm, indices described are computed using data obtained from WAMS to identify the failed component, if any, for a protection operation.

After fault-affected line is detected, the relays connected to the faulted line and its adjacent lines are weighted according to their zone of operation for the concerned fault situation. Relays connected to the faulted line are assigned each a weight of 1. For adjacent lines, relays located at far end from faulted line are each weighted 0.5, and relays at near end are assigned weights of −3 each. The algorithm uses WAMS data as shown in Figure 3.17 to identify if there is any protection component maloperation. This information is obtained through WAMS from relays that are weighted after identifying the line affected by fault. The process has two levels of computation.

For identification of maloperating protection elements, two *protection component failure* indices are used as defined below:

1. Malfunction index (mal_fn): It determine whether a failure of any protection scheme operation has taken place and the class of maloperation, that is, whether a zone 1 or zone 2 or zone 3 maloperation has occurred.
2. Failure index (fail_in): It pinpoints the component for which the protection scheme has maloperated.

The details for the computation process for the identification are provided below:

Stage 1: To check any protection maloperation: This level detects whether maloperation of any protection scheme in the network has taken place or not. For this the mal_fn index is computed using the following relation (3.16):

$$\text{mal\_fn} = \sum_{i=1}^{m} \text{rel}_{w_i} \times w_i \tag{3.16}$$

where m is the buses connected to the faulted line and its adjacent lines and rel_{w_i} is the weight associated to each relays.

Zone 1 relays are weighted with a value of 1. Thus if CBs corresponding to relays connected to faulted line operate then mal_fn value becomes 2 which corresponds to a correct protection operation.

Stage 2: To identify failed component: After the identification of class of maloperation and computation of the failure index, the next step is to pinpoint the protection component failure. Using the indices defined in Section 3.3.2, failure index (fail_in) is computed as

$$\text{fail\_in} = \text{BS} + \text{rel\_}i + \text{CB} \times 0.5 + \text{tim\_sig\_}i \times 1.5 + \text{RS} \tag{3.17}$$

where $i = 1$, 2, and 3 for zones 1, 2, and 3, respectively.

Table 3.2 lists the range of values of the two indices (mal_fn and fail_in) and the remarks therefrom. Some of the simulated cases are presented in the following section for a test system.

3.3.5 Case study: relay underreaching

A single line to ground fault is simulated in the line connecting bus 26 and 28 at 0.1 s. The relay setting at bus 28 is such that it is underreaching for the particular situation as a result of which fault could not be cleared at bus 28 for system in Figure 3.6. CB opens at bus 26 successfully but fault current continues from bus 28 because of relay malfunction. Synchronized measurements are available from all nearby buses close to the fault. Deviations in synchronized voltage-phasor measurements of selected buses are shown in Figure 3.16. It is observed that buses 26 and 28 have the maximum voltage deviation. The plots for deviation of phase angle between voltage and current for different lines connected to buses 26 and 28 are shown in Figure 3.17, respectively. It is observed that for measurements at both

Table 3.2 Indices and corresponding inference

Malfunction index	**0 < mal_fn ≤ 1 (zone 1)**			**1 < mal_fn < 2 or mal_fn > 2 (zone 2/zone 3)**					**Mal_fn ≤ 0**
Failure index	fail_in ≤ 0	fail_in = 1	fail_in = 3.5	fail_in = 1	fail_in = 1.5	fail_in = 3.5	fail_in = 4.5 or fail_in = 2.5	fail_in = 6	fail_in = 6
Inference	Transducer error	Relay under reach	Circuit breaker error	Relay under reach	Transducer Error	Circuit breaker error	Timer failure	Relay over-reach	Unwanted reverse operation
	fail_in = 4: **correct operation**								

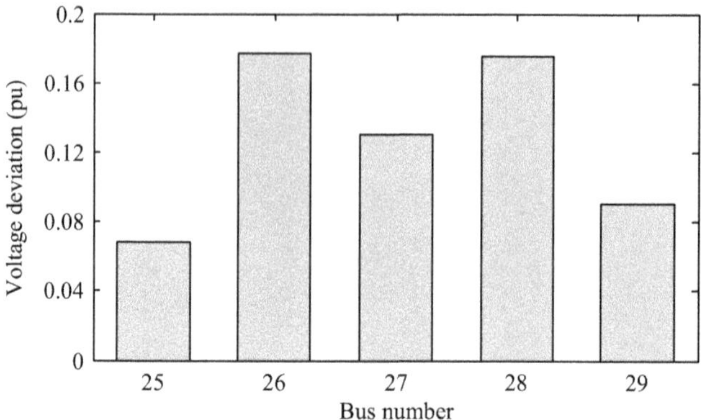

Figure 3.16 Bus with significant voltage deviation for fault in line 26-28

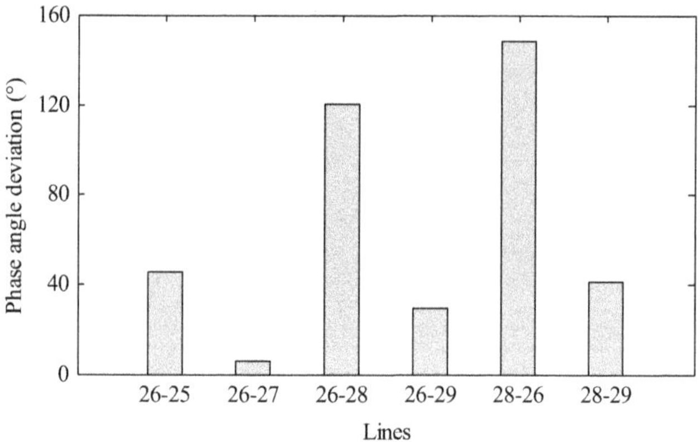

Figure 3.17 Phase angle deviation for lines connected to bus 26 for fault in line 26-28

buses 26 and 28, phase angle deviation for line 26-28 is maximum. From this, the technique concludes that the fault is in line 26-28, which is correct.

Depending on the fault location, relays in the system are assigned weights. Relays connected to the faulted line are weighted 1. Relays that provide backup for this fault condition are weighted 0.5. The relays for which the fault is in the reverse direction are weighted −3. For the case relay weights are as follows:

rel_*w* (26-28(26)) = 1, rel_*w* (28-26(28)) = 1, rel_*w* (25-26(25)) = 0.5,
rel_*w* (27-26(27)) = 0.5, rel_*w* (29-26(29)) = 0.5, rel_*w* (29-28(29)) = 0.5,
rel_*w* (26-25(26)) = −3, rel_*w* (26-27(26)) = −3, rel_*w* (26-29(26)) = −3,
rel_*w* (28-29(28)) = −3.

After faulted line identification and assigning relay weights, the bitwise information for protection schemes of weighted relays are obtained. The information obtained

Table 3.3 Information received for fault in line 26-28

Bus	Line	Information		
		BS	CB	Rel *i*
26	26-28	1	1	01
28	28-26	1	0	10
25	25-26	1	0	11
27	27-26	1	0	10
29	29-26	1	0	11
29	29-28	1	0	11

for protection schemes for fault in line 26-28 are shown in Table 3.3 (information for reverse operating relays are not shown).

The next step is to assess the protection operation and identify failure, if any, in the protection scheme:

1. Depending on the faulted line identification, relays at bus 26 and 28 for line 26-28 have been given a weight of 1 each. As CB status at bus 26 is 1 and for all other breakers in the system, CB is 0. Calculation of malfunction index using (3.16) gives mal_fn = rel_w (26-28(26)) × CB (26-28(26)) = 1 × 1 = 1.

   Value of mal_fn confirms a zone 1 maloperation, that is, one of the zone 1 protection components has failed.
2. Calculating the failure index for protection operation at bus 26 using (3.2) we get fail_in (26-28(26)) = BS + rel 1 + CB × 0.5 + tim sig 1 × 1.5 + RS = 1 + 1 + 1 × 0.5 + 1 × 1.5 + 0 = 4

   The value of BS, CB, and Rel_*i* is 1, obtained from Table 3.3. Timer sig value is also 1, and RS is 0 as the relay is connected to the faulted line. A value of 4 for fail_in shows a correct operation at bus 26 for fault in line 26-28.
3. Calculating the failure index for protection operation at bus 28 using (3.17), we get fail_in (26-28(28)) = BS + rel 1 + CB × 0.5 + tim sig 1 × 1.5 + RS = 1 + 0 + 0 × 0.5 + 0 × 1.5 + 0 = 1

   A value of 4 for failure index shows that there is no maloperation at bus 27 for fault in line 26-28.

The protection failure indices for the system clearly conclude that the relay at bus 28 has underreached. As a result of which no tripping operation took place in bus 28 for a fault in the line connecting bus 26 and 28. Protection scheme at bus 26 for the line operates properly. The algorithm is able to identify the protection component failure correctly.

With fast communication availability for WAMS data, the detection can be done in about eight cycles (0.16 s) after fault inception. The backup relay for this case is at bus 29 connected to line 29-28. This relay sees the fault in its zone 3 region. So the timer is set to operate after 1 s. After detection of the discrepancy in the relay operation at bus 28, the relay at bus 29 can be redirected to clear the fault thus speeding up the backup process.

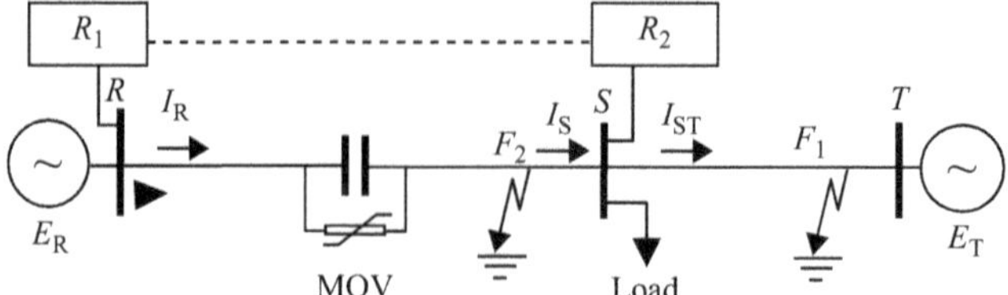

Figure 3.18 A two-source system with series compensation at mid of the line

3.4 Adaptive protection

Relays in power systems are set according to the certain assumption about the power systems. As the system conditions change, the relay settings may not be appropriate at that instant that causes its maloperation. The changing system condition requires adaptive setting for reliable operation. With the advent of the digital relays, protection functions can be changed to make the settings more appropriate. New technology like PMU provides a way to gather information on system condition and paves a way for wide-area synchronized data–based protection mechanism. WAMS data–based adaptive relay settings of two systems are proposed in this section.

3.4.1 Series compensated line

Series compensation is utilized to improve the power flow of a line. However, its use imposes protection challenges like voltage and current inversion, overreaching and coordination problem in between distance relays. Due to the presence of series compensation apparent impedance seen by the relay is modulated and when the compensation is present in the fault loop, the relay overreaches. At times, series compensation may be bypassed due to maintenance purpose or by the operation of its overvoltage protective devices. For accurate operation of the relay information regarding the compensation level in the transmission line is required.

Consider the 60% series compensation at the middle of the line RS as shown in Figure 3.18. If the relay at the bus R for line RS is set without considering the compensation level in the line then there are chances that relay will operate for fault at F_1, where F_1 comes in the zone 2 of the line ST. Similarly if the relay is set considering the compensation level, and the capacitor is bypassed before fault for maintenance purpose and by its overvoltage protective device during fault the relay at R for line RS will not detect a fault at F_2 that comes in zone 2 of the line RS. This is tested with a line-to-ground fault with $R_f = 1\ \Omega$ at F_2 and the impedance trajectory with capacitor bypassed condition during fault is plotted in Figure 3.19. It is observed from figure that the apparent impedance settles outside the boundary set considering series compensation and inside the boundary with compensation is not considered. This reveals that relay setting needs a change if the compensation level varies during fault. Such maloperation can be prevented if the compensation

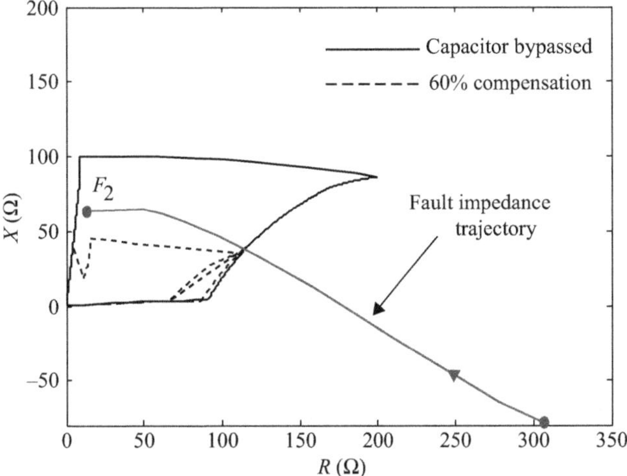

Figure 3.19 Comparison of zone 2 trip boundaries with 60% compensation at middle and capacitor bypassed

level present in the line can be communicated to the relay at R. Synchronized data available at both end of the line can be applied to compute the compensation level.

3.4.1.1 The adaptive method

A 400 kV, 50 Hz series compensated power system as shown in Figure 3.18 is considered. For this case, measurement units (distance relays with global positioning system (GPS) facility) [35] are considered to be available at the two buses R and S. The buses are connected with a communication system that may be available for any other purpose in the system. Adaptive zones 2 and 3 setting of relay at bus R in the system is to be accomplished using power system condition and the level of series compensation. The positive sequence phasor data at a rate of 50 fps from the ends are processed at the relay end to obtain the compensation level. The impedance between two buses R and S can be calculated by

$$Z_{1\mathrm{RS}} = \frac{(\overline{V}_{\mathrm{R}})2 \; - \; (\overline{V}_{\mathrm{S}})2}{\overline{V}_{\mathrm{S}}\overline{I}_{\mathrm{R}} \; + \; \overline{V}_{\mathrm{R}}\overline{I}_{\mathrm{S}}} \tag{3.18}$$

where $\overline{V}_{\mathrm{R}}, \overline{I}_{\mathrm{R}}$ and $\overline{V}_{\mathrm{S}}, \overline{I}_{\mathrm{S}}$ are positive sequence phasors at bus R and S for RS section, respectively. The series compensation X_{C} present in the line is obtained from the imaginary parts of the calculated impedance between bus R and S ($Z_{1\mathrm{RS}}$) and positive sequence impedance of the line (Z_{1L}):

$$X_{\mathrm{C}} = \mathrm{Im}(Z_{1L}) - \mathrm{Im}(Z_{1\mathrm{RS}}) \tag{3.19}$$

Using bus-S synchronized data, source $\overline{E}_{\mathrm{T}}$ can be written as

$$\overline{E}_{\mathrm{T}} = A_{\mathrm{T}}\overline{V}_{\mathrm{S}} - B_{\mathrm{T}}\overline{I}_{\mathrm{ST}} \tag{3.20}$$

where A_T and B_T are the ABCD parameters for the section between bus S and T-bus source. Similarly $\overline{E}_R$ is obtained from the measurements at bus R:

$$\overline{E}_R = \overline{V}_R - Z_{SR1}\overline{I}_R \tag{3.21}$$

where Z_{SR1} is the source impedance at bus R.

Once compensation level is known using (3.19), zones 2/zone 3 setting are obtained at bus R. Zone 2, being the primary protection for the line beyond zone 1, cannot detect faults with a reduced reach when compensation is bypassed. Similarly zone 3 which provides backup for the next line will also face problems. Such problems will be severe when CB fails in the next line, and fault is not detected by backup relay for capacitor bypassed condition. As the delay in data transfer is less than the time setting of zones 2 and 3, this method is feasible for zones 2 and 3.

The conduction of MOV is adjudged from the capacitor current level and corresponding equivalent impedance of the combination is computed [34]. The equivalent impedance offered by MOV and capacitor combination is calculated by iterative way which converges in maximum seven iterations, and the extra computation time is not an issue for setting zone 2/zone 3. On the other hand for a MOV-capacitor combination bypassed condition, mutual coupling is not present, and the adaptive setting is obtained accordingly. The flow diagram of the algorithm for setting the relay characteristic is shown in Figure 3.20. In the first three steps,

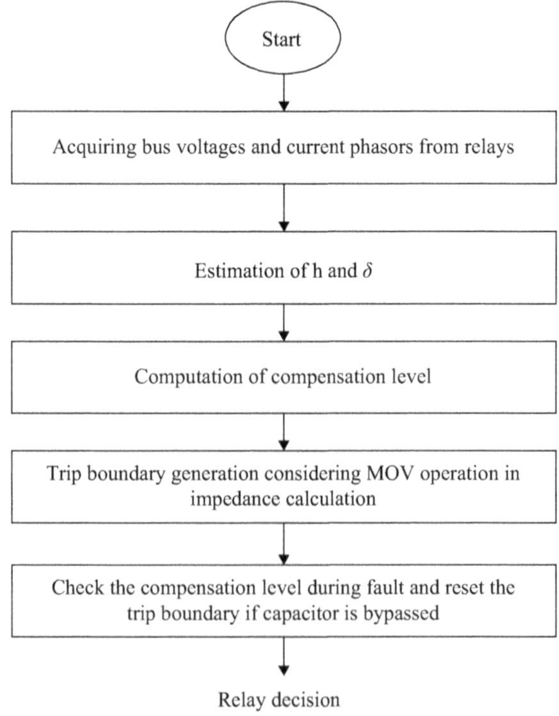

Figure 3.20 Flow diagram for the adaptive technique

the compensation level is calculated, and in the fourth step trip boundary with MOV operation is generated. Different ways can be applied to implement the scheme. One approach, being a protection decision, adaptive distance relay should take the synchronized data as input and do the setting calculation required in the relay. Otherwise, the system condition and level of compensation can be calculated outside the relay and the setting can be computed which should be transferred to the distance relay for proper decision.

3.4.1.2 Result

Trip boundary with series compensation bypassed at the middle for a different system condition: Sometimes the capacitor may be bypassed due to the unbalance in capacitor banks or for maintenance purpose. When the capacitor bank protective devices operate, it cannot be reinserted to the system immediately. In each case, it will lead to an uncompensated line condition. A different example is simulated without series compensation in the line of the system in Figure 3.18. Synchronized phasor data collected from the two ends are provided in Table 3.4.

The calculated compensation level $Z_C = -j0.61\ \Omega$ (a low value) shows that the capacitor is bypassed. Using this information, the trip boundary is generated and plotted in Figure 3.21. The generated boundary is compared with the seen impedance obtained from the simulation using EMTDC/PSCAD for line-to-ground faults at different locations varying the fault resistance 0–200 Ω. The results also show the strength of the method in providing trip boundary correctly.

3.4.2 Three terminal line

Transmission lines are tapped in between two terminals to extract economical and technical benefits. It creates protection challenges for the line as the fault current flowing from the tapping affects the voltage and current at other two terminals. For this reason, three terminal lines are protected by communication-assisted protection schemes like DUTT, permissive overreach transfer trip, directional comparison blocking. Among them, DUTT scheme for protection of three-terminal line uses zone 1 distance element at each end which finds underreach problem due to the infeed from T-point. The protection issue of three-terminal line with unequal branch lengths is more critical as they have limited overlapping regions between the zone 1 elements. These overlappings of zone 1 elements decrease with increase

Table 3.4 Synchronized data for capacitor bypassed with $h = 0.99$, $\delta = -5°$

Relay	**Bus voltage (kV)**	**Line current (kA)**		**Calculated**		
				Z_C (Ω)	h (°)	δ (°)
Bus-R	210.11 − j90.9	I_R	0.045 + j0.179	−j0.61	0.987	−5.06
Bus-S	217.60 − j84.68	I_S	−0.126 + j0.011			
		I_{ST}	−0.332 + j0.068			

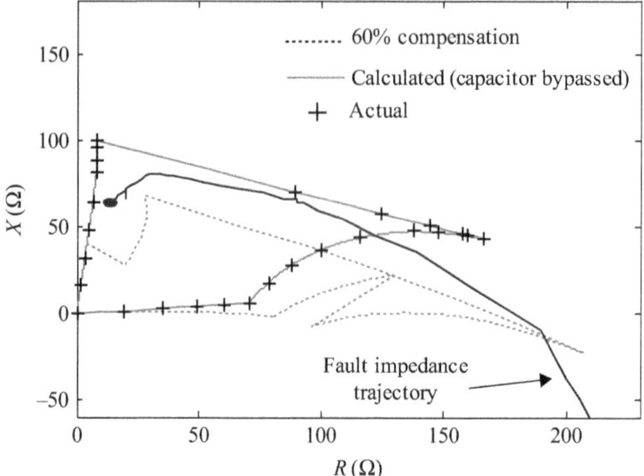

Figure 3.21 Trip boundary with capacitor bypassed at middle with $h = 0.99$, $\delta = -5°$

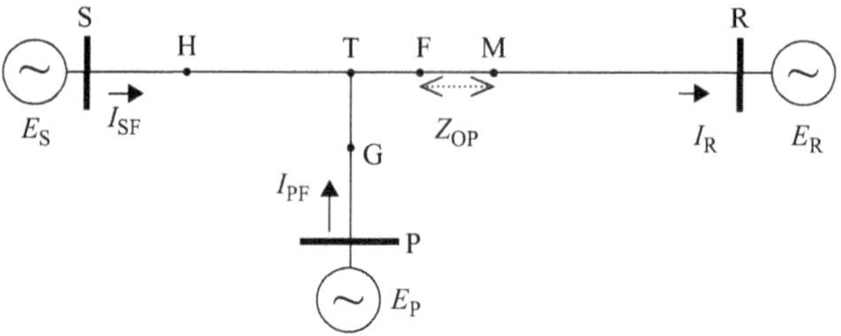

Figure 3.22 Three-terminal line

in infeed current from T-point. For high infeed currents, the overlappings with distance relay settings may vanish, which will lead to an unprotected line segment with DUTT scheme.

3.4.2.1 The adaptive method

A three-terminal line with unequal branch lengths as shown in Figure 3.22 is considered, where the line lengths are in the increasing order of TP, TS, and TR. The DUTT scheme protecting the lines uses zone 1 trip signals to be shared by the three distance relays at P, R, and S buses. Zone 1 overlappings for relay settings ensure that at least one of the three relays will be able to detect any internal fault in the lines. As an example, zone 1 of R bus relay is set with 80% of Z_{RP} which also covers up to point F. P bus relay is set with 80% of Z_{SP} that covers up to M. In a similar way, the settings at S bus relay are 80% of Z_{SP} that covers beyond F.

The overlapping impedances for S and P with R zone 1 relay setting can be formulated as

$$\begin{aligned} Z_{OP} &= kZ_{SP} - (Z_{PT} + Z_{TF}) \\ Z_{OS} &= kZ_{SP} - (Z_{ST} + Z_{TF}) \end{aligned} \tag{3.22}$$

where Z_{OP} and Z_{OS} represent for overlappings from relay of S and P buses, respectively, where k is 0.8 (80% setting). The overlapping impedances between relays (S and R) and (P and R) depend on infeed from the tapped line. The apparent impedance seen at S-bus relay for a three-phase fault at F in Figure 3.1 can be expressed as

$$Z_{app} = Z_{ST} + \left(1 + \frac{I_{PF}}{I_{SF}}\right) Z_{TF} \tag{3.23}$$

where I_{SF} and I_{PF} are relay currents at S and P buses during three phase fault, respectively. The expressions for I_{SF} and I_{PF} are available in [36]. Thus, the error to the apparent impedance caused by infeed current will be

$$Z_{err} = \frac{I_{PF}}{I_{SF}} Z_{TF} \tag{3.24}$$

Z_{err} relates to underreach amount of the distance relay and reflects the decrease in overlapping between S and R bus zone 1 elements. The underreach effect caused by the infeed current varies in accordance with system condition including source impedance. Under certain infeed conditions, the overlapping vanishes, which will result in DUTT maloperation.

For proper DUTT decision, the overlappings between the three relays have to be ensured. For this time, synchronized prefault data area availed from the three buses to fix the settings [36]. As three-phase fault current is more than current for any other types of faults for fault at the same location, the error caused by infeed current is maximum for former fault. Thus, if the overlapping is ensured for three-phase fault, it would satisfy for all other faults. R-bus zone 1 setting protects up to F and is not affected by infeed from P-bus. If a fault in the TF section is not detected by any of S or P-bus relays (due to underreach), DUTT scheme will fail for this case. Therefore, decrease in overlapping is checked for three-phase fault at F (reach of R-bus relay) to ensure proper relay decision. With the infeed effect, the amounts of overlappings present between relay settings of S and R (C_{OS}) and P and R (C_{OP}) can be formulated as

$$C_{OS} = |Z_{OS}| - \left| \frac{|\bar{I}_{PF}|}{|\bar{I}_{SF}|} Z_{TF} \right| \tag{3.25}$$

$$C_{OP} = |Z_{OP}| - \left| \frac{|\bar{I}_{SF}|}{|\bar{I}_{PF}|} Z_{TF} \right| \tag{3.26}$$

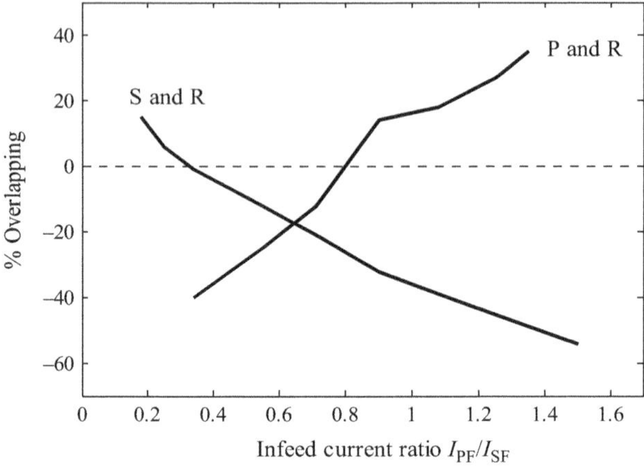

Figure 3.23 Variation in percentage of overlappings

where Z_{OS} and Z_{OP} are as defined in [36]. The percentage of overlappings between S and P buses with R bus are defined as

$$K_S = \frac{C_{OS}}{|Z_{OS}|} \times 100 \tag{3.27}$$

$$K_P = \frac{C_{OP}}{|Z_{OP}|} \times 100 \tag{3.28}$$

As the infeed current increases, the overlapping decreases and in some operating conditions, the overlappings of three relays vanish.

A 400 kV, 50 Hz three terminal system as of Figure 3.22 is simulated for different operating conditions to observe the variation in percentage of overlapping. The magnitude and angle of voltage at R-bus is maintained constant for all cases. The voltage magnitude and angle at S and P buses are varied in each case. The variation in overlapping between zone 1 relays at (S and R) and (P and R) buses with infeed current ratio is plotted in Figure 3.23. The dotted line in the two plots separates overlapping and nonoverlapping zones. The percentage of overlapping reduces with the increase in infeed current and for a range of infeed current ratio, portion of three-terminal line remains unprotected by DUTT scheme.

The performance of DUTT scheme is tested for a three-phase fault at 85% of the line RP section from R-bus (which is outside the zone 1 setting at R bus). The prefault loading condition of the three-terminal line is fixed *at* $h_1 = 0.98$, $\delta_1 = -15°$, $h_2 = 1.0$, and $\delta_2 = -5°$. The apparent impedances seen at S, R, and P relays are plotted in Figure 3.24. It is observed that the trajectory of apparent impedance settles outside the set zone 1 characteristics of these S, P, and R relays. Thus, the fault will not be detected by any of the zone 1 element of the DUTT protection scheme, and it fails to identify the internal fault. This happens due to the infeed. For proper operation of DUTT scheme, the zone 1 setting at S and P buses should be

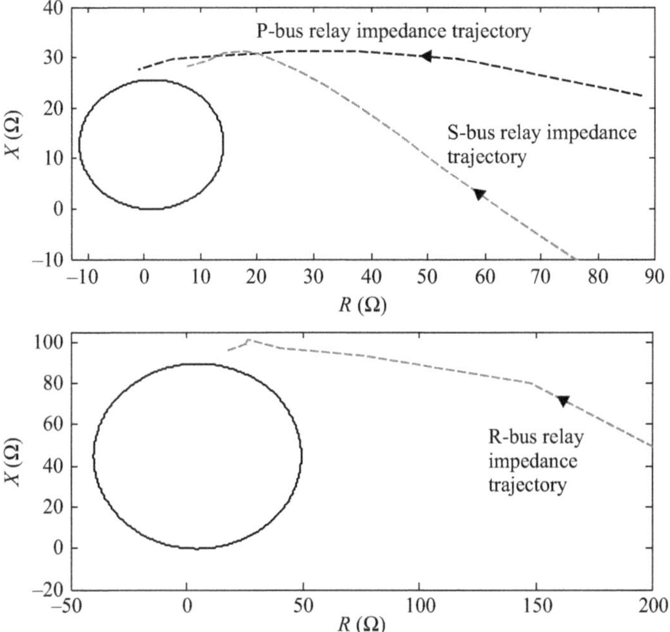

Figure 3.24 Apparent impedance trajectory for three-phase fault

monitored continuously. For this, the overlappings are estimated and then relay settings are changed when required to ensure proper decision by DUTT scheme.

The system of Figure 3.22 is considered, in which distance relays (GPS time synchronized) are available at the three buses and are linked with a communication system. For each relay, zone 1 is set 80% of the shortest line section in the forward direction. Zone 1 of R-bus relay remains within the T-point and is not affected by the infeed current. The relays at buses S and P face underreach problem and need to be adapted in accordance with the infeed condition. For the 50 Hz system, the positive sequence phasor data available at a rate of 50 fps from the three buses are processed at each relay end to estimate the operating condition:

$$\overline{E}_S = \overline{V}_S + Z_{1ss}\overline{I}_S \tag{3.29}$$

$$\overline{E}_R = \overline{V}_R - Z_{1sr}\overline{I}_R \tag{3.30}$$

$$\overline{E}_P = \overline{V}_P + Z_{1sp}\overline{I}_P \tag{3.31}$$

where Z_{1ss}, Z_{1sr}, and Z_{1sp} are the source impedance at S, R, and P buses. The relation between voltage sources are defined by

$$\frac{\overline{E}_R}{\overline{E}_S} = h_1 e^{-j\delta_1}, \quad \frac{\overline{E}_P}{\overline{E}_S} = h_2 e^{-j\delta_2} \tag{3.32}$$

where h_1 and h_2 are voltage magnitude ratios, and δ_1 and δ_2 are phase angle differences of source voltages.

Using this information, percentage of overlapping for S-bus relay is calculated by (3.27). Zero or negative value of K_S implies that the overlapping is lost. In such a case, the overlapping between S and R bus zone 1 characteristics can be retained by increasing the zone 1 setting at S-bus by ΔZ_S as below:

$$\Delta Z_S = \frac{|\bar{I}_{PF}|}{|\bar{I}_{SF}|} Z_{TF} \tag{3.33}$$

80% setting of S-bus relay protects up to G (say) in TP section as shown in Figure 3.22. The increase in S-bus relay setting may overreach TP section which is not desirable. To ensure that the setting of S-bus relay does not go beyond P, the adaptive setting has to be modified. This requirement can be assured if the reach of adaptive setting at S-bus relay is restricted up to G. This is only possible when the increased setting ΔZ_S is less than the error in seen impedance of S-bus relay caused by the infeed from R-bus for a three-phase fault at G. The error in seen impedance can be expressed as

$$Z_{Ser} = \frac{|\bar{I}_{RG}|}{|\bar{I}_{SG}|} Z_{TG} \tag{3.34}$$

where $\bar{I}_{RG}$ and $\bar{I}_{SG}$ are fault currents from R and S buses (refer [36]). When $|\Delta Z_S| > |Z_{Ser}|$, adaptive setting at S bus may reach beyond TP section. In order to maintain overlapping without overreach, setting at S-bus is decreased and R-bus setting is increased simultaneously by the same amount. The adaptive setting reach for TP section is checked by

$$C_S = |\Delta Z_S| - |Z_{Ser}| \tag{3.35}$$

In case, C_S is positive, the setting for S-bus relay should be updated with

$$Z_{S\_New} = Z_{S\_SET} + Z_{Ser} \tag{3.36}$$

Simultaneously, R-bus relay setting is changed to

$$Z_{R\_New} = Z_{R\_SET} + (\Delta Z_S - Z_{Ser}) \tag{3.37}$$

where Z_{S_SET} and Z_{R_SET} correspond to the fixed 80% settings for S and R relays, respectively. If $|\Delta Z_S| \leq |Z_{Ser}|$, S-bus setting is increased by ΔZ_S as in (3.33).

Similarly, the percentage of overlapping between P-bus zone 1 relay with R-bus zone 1 is calculated as in (3.28). Zero or negative overlapping implies that adaptive setting is required. The overlapping between P and R buses is maintained by increasing the setting by

$$\Delta Z_P = \frac{|\bar{I}_{SF}|}{|\bar{I}_{PF}|} Z_{TF} \tag{3.38}$$

Adaptive P-bus relay setting may overreach TS section (say beyond S). The 80% setting of P-bus covers up to H on TS section. As mentioned earlier for S-bus relay reach, P-bus setting should be limited up to H. For this, the error in P-bus apparent impedance caused by infeed from R-bus can be obtained for a three phase fault at H as

$$Z_{\mathrm{Per}} = \frac{|\bar{I}_{\mathrm{RH}}|}{|\bar{I}_{\mathrm{PH}}|} Z_{\mathrm{TH}} \tag{3.39}$$

$\bar{I}_{\mathrm{RH}}$ and $\bar{I}_{\mathrm{PH}}$ are the three-phase fault currents. The reach of adaptive P-bus setting, for TP section is checked by

$$C_{\mathrm{P}} = |\Delta Z_{\mathrm{P}}| - |Z_{\mathrm{Per}}| \tag{3.40}$$

In case, C_{P} is positive, the setting for P-bus relay should be

$$Z_{\mathrm{P\_New}} = Z_{\mathrm{P\_SET}} + Z_{\mathrm{Per}} \tag{3.41}$$

Simultaneously, R-bus relay setting is modified to:

$$Z_{\mathrm{R\_New}} = Z_{\mathrm{R\_SET}} + (\Delta Z_{\mathrm{P}} - Z_{\mathrm{Per}}) \tag{3.42}$$

where $Z_{\mathrm{P\_SET}}$ corresponds to 80% setting for P-bus relay. For $|\Delta Z_{\mathrm{P}}| \leq |Z_{\mathrm{Per}}|$, the P-bus setting is increased by ΔZ_{P}. It is observed that both the adaptive settings never overreach simultaneously.

The flow diagram for DUTT adaptive setting scheme is provided in Figure 3.25. For a system condition, percentage of overlapping is calculated using (3.27) and (3.28).

For any negative or zero value of percentage overlapping, increments in impedance required for S and P-bus relays are calculated. Reach of adaptive setting is checked and adaptive settings are modified in the case of any violation in reach. This ensures proper DUTT protection decision. The scheme uses other end prefault data, and the settings are updated in each cycle if required. During a fault it need only 1 or 0 signal from the other end. So the communication burden is not significant as in the differential scheme. In the case of communication failure, it switches to the conventional fixed setting mode.

3.4.2.2 Results

The method for adaptive setting of three-terminal line, is tested for a 400 kV, 50 Hz 3-bus system as shown in Figure 3.22. Few important results are demonstrated in this section. Adaptive setting is checked for three phase and unsymmetrical faults at different system conditions. One-cycle DFT is used to compute the phasors from sampled data. The system is simulated using EMTDC/PSCAD. Synchronized voltage and current data are used to estimate the overlappings, and the adaptive trip boundary is adapted for required line.

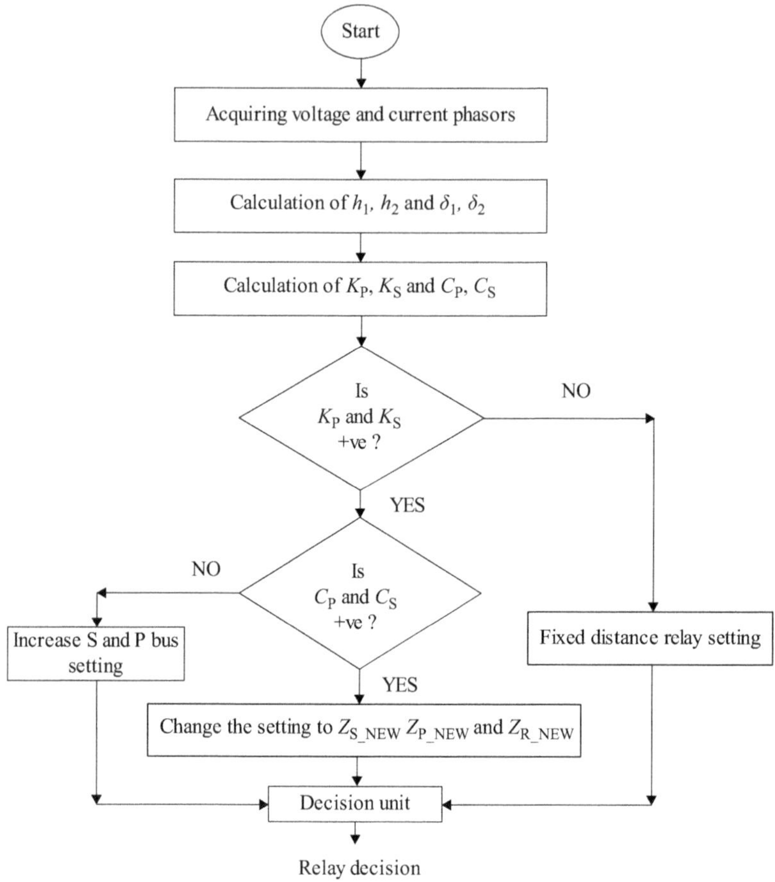

Figure 3.25 Flow diagram for the adaptive technique

Test for variation in system condition
Power system condition changes with time. To test the adaptive DUTT scheme, the system is initially operating at a condition with $h_1 = 1.01$, $\delta_1 = 0°$, $h_2 = 1.01$, and $\delta_2 = 5°$. *The condition is changed to* $h_1 = 0.98$, $\delta_1 = -15°$, $h_2 = 1.0$, and $\delta_2 = -5°$. The prefault synchronized data obtained from three buses are used to calculate the voltage magnitude ratio and phase angle differences. Using this in relations (3.27) and (3.28), the percentage overlappings for S and P bus zone 1 relays are calculated (refer Table 3.5). Negative K_S and K_P infer that overlapping of the three zone 1 relays is lost. For proper operation of DUTT scheme, adaptive settings for S and P buses are obtained by increasing the settings. Both fixed and adaptive settings are plotted in Figure 3.26 for S and P buses.

The accuracy of the adaptive setting is checked for a three-phase fault created at 85% of line RP from the R-bus with negligible fault resistance during the second operating condition. The fault impedances seen by relay at S-bus and P-bus are

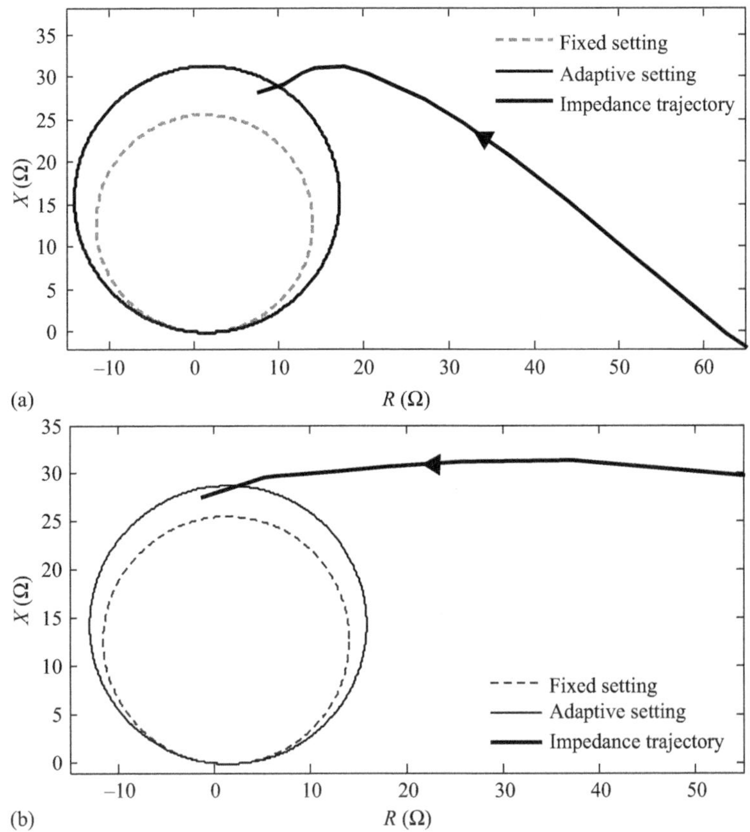

Figure 3.26 *(a) S-bus zone 1 adaptive setting for DUTT scheme. (b) P-bus zone 1 adaptive setting for DUTT scheme*

Table 3.5 *Synchronized data and different indices ($h_1 = 0.98$, $\delta_1 = -15°$, $h_2 = 1.0$, and $\delta_2 = -5°$)*

Bus	Voltage (kV)	Line current (kA)	System condition				Indices			
			h_1	h_2	δ_1 (°)	δ_2 (°)	K_S	K_P	C_S	C_P
S	236.87 ∠−1.66°	0.69 ∠−10.4°	0.98	1.0	−14.13	−4.7	−67.5	−54.7	−1.8	−1.3
P	238.81 ∠−2.89°	0.21 ∠166.1°								
R	232.69 ∠−15.3°	0.46 ∠−21.7°								

plotted in Figure 3.26(a) and (b), respectively. It is observed from the figures that the fault impedances settle outside the fixed zone 1 characteristic and inside the adaptive settings of the two relays. Thus, fault is identified correctly; it shows the accuracy of the adaptive method. Increased setting at S and P buses may overreach for TP and TS sections. From Table 3.5, it is observed that C_S and C_P are negative, which confirms that the new S and P buses settings do not have an overreach problem, and therefore adaptive setting for R bus relay is not required.

3.5 Conclusion

Maloperation of power system protection schemes can result in escalation of a small disturbance to a severe situation in a power system. Events such as power swing, load encroachment, failure of components involved in the protection schemes leads to incorrect operation. This causes unwanted tripping or delay in clearing of faults from the system. This work attempts to use wide-area measurement data to support power system protection decisions in order to avoid maloperations. A combined feature-based decision for relay operation during power swing is developed to derive correct protection decision. Method for supporting relay decisions during stressed conditions such as power swing, voltage instability, and load encroachment by discriminating fault from these events is presented. Indices obtained from WAMS data in a power system are used for identifying malfunction of any element in a protection scheme. Further such informative data improve series compensated and three terminal line protections by applying adaptive settings for distance relays. The WAMS-based techniques enhance protection schemes by making them more reliable and speed up protection for power networks.

Nomenclature

WAMS	wide-area measurement system
PMU	phasor measurement unit
PDC	phasor data concentrator
GPS	global positioning system
DFT	discrete Fourier transform
SPS	special protection scheme
RAS	remedial action scheme
POTT	permissive overreaching transfer trip
DUTT	direct underreaching transfer trip
DCB	directional comparison blocking
PSB	power swing blocking
OST	out-of-step trip
SCV	swing center voltage
COI	center of inertia
HHT	Hilbert Huang transform
EMD	empirical mode decomposition
IMF	integral mode function

FLS	fuzzy logic system
θ	phase angle
t	time
Z	impedance
H	moment of Inertia

References

[1] A. G. Phadke and S. H. Horowitz, *Relaying, Power System*. Taunton, UK: Research Studies Press, 2008.

[2] J. L. Blackburn, *Protective Relaying Principles and Applications*. New York: Marcel Dekker, 1987.

[3] A. G. Phadke and J. S. Thorp, "Expose hidden failures to prevent cascading outages [in power systems]," *IEEE Comput. Appl. Power*, vol. 9, no. 3 (Jul.), pp. 20–23, 1996.

[4] S. Tamronglak, S. H. Horowitz, A. G. Phadke, and J. S. Thorp, "Anatomy of power system blackouts: preventive relaying strategies," *IEEE Trans. Power Deliv.*, vol. 11, no. 2 (Apr.), pp. 708–715, 1996.

[5] G. Andersson, P. Donalek, R. Farmer, *et al.*, "Causes of the 2003 major grid blackouts in North America and Europe, and recommended means to improve system dynamic performance," *IEEE Trans. Power Syst.*, vol. 20, no. 4 (Nov.), pp. 1922–1928, 2005.

[6] D. Junce and C. Zexiang, "Mixed measurements state estimation based on wide-area measurement system and analysis," in *2005 IEEE/PES Transmission & Distribution Conference & Exposition: Asia and Pacific*, 2005, pp. 1–5.

[7] M. Gol and A. Abur, "A fast decoupled state estimator for systems measured by PMUs," *IEEE Trans. Power Syst.*, vol. 30, no. 5 (Sep.), pp. 2766–2771, 2015.

[8] L. Hu, Z. Wang, I. Rahman, and X. Liu, "A constrained optimization approach to dynamic state estimation for power systems including PMU and missing measurements," *IEEE Trans. Control Syst. Technol.*, vol. PP, no. 99, pp. 1–1, 2015.

[9] A. Rendon Salgado, C. R. Fuerte Esquivel, and J. G. Calderon Guizar, "SCADA and PMU measurements for improving power system state estimation," *IEEE Lat. Am. Trans.*, vol. 13, no. 7, pp. 2245–2251, 2015.

[10] M. M. Eissa, M. E. Masoud, and M. M. M. Elanwar, "A novel back up wide area protection technique for power transmission grids using phasor measurement unit," *IEEE Trans. Power Deliv.*, vol. 25, no. 1, pp. 270–278, 2010.

[11] M. G. Adamiak, A. P. Apostolov, M. M. Begovic, *et al.*, "Wide area protection—technology and infrastructures," *IEEE Trans. Power Deliv.*, vol. 21, no. 2 (Apr.), pp. 601–609, 2006.

[12] J. De La Ree, V. Centeno, J. S. Thorp, and A. G. Phadke, "Synchronized phasor measurement applications in power systems," *IEEE Trans. Smart Grid*, vol. 1, no. 1, pp. 20–27, 2010.
[13] H. J. Altuve, J. B. Mooney, and G. E. Alexander, "Advances in series-compensated line protection," in *2009 62nd Annual Conference for Protective Relay Engineers*, 2009, pp. 263–275.
[14] D. Erwin, M. Anderson, R. Pineda, D. A. Tziouvaras, and R. Turner, "PG&E 500 kV series-compensated transmission line relay replacement: design requirements and RTDS® testing," in *2011 64th Annual Conference for Protective Relay Engineers*, 2011, pp. 191–202.
[15] IPSR Committee, *Protection aspects of multi terminal lines*. IEEE Publ. 79 TH0056-2-PWR, 1979.
[16] NERC, "A technical document prepared by the systems protection and control task force of the NERC planning committee—the complexity of protecting three terminal transmission line," 2006.
[17] R. Hasan, R. Bobba, and H. Khurana, "Analyzing NASPInet data flows," in *2009 IEEE/PES Power Systems Conference and Exposition*, 2009, pp. 1–6.
[18] A. G. Phadke and J. S. Thorp, "Communication needs for wide area measurement applications," in *2010 5th International Conference on Critical Infrastructure (CRIS)*, 2010, pp. 1–7.
[19] P. Kundur, *Power System Stability and Control*. New York: McGraw Hill, 1994.
[20] P. Kundu and A. K. Pradhan, "Wide area measurement based protection support during power swing," *Int. J. Electr. Power Energy Syst.*, vol. 63, pp. 546–554, 2014.
[21] A. R. Messina, *Inter-area Oscillations in Power Systems, A Non-linear and Non-stationary Perspective*. New York, NY: Springer, 2009.
[22] V. Vittal, "Self-healing in power systems: an approach using islanding and rate of frequency decline-based load shedding," *IEEE Trans. Power Syst.*, vol. 18, no. 1 (Feb.), pp. 174–181, 2003.
[23] D. J. Kim, "System and method for calculating voltage stability risk-index in power system using time series data," 2007.
[24] I. Kamwa, S. R. Samantaray, and G. Joos, "Development of rule-based classifiers for rapid stability assessment of wide-area post-disturbance records," *IEEE Trans. Power Syst.*, vol. 24, no. 1 (Feb.), pp. 258–270, 2009.
[25] N. E. Huang, Z. Shen, S. R. Long, *et al.*, "The empirical mode decomposition and the Hilbert spectrum for nonlinear and non-stationary time series analysis," *Proc. R. Soc. A Math. Phys. Eng. Sci.*, vol. 454, no. 1971 (Mar.), pp. 903–995, 1998.
[26] Z. Wu and N. E. Huang, "Ensemble empirical mode decomposition: a noise-assisted data analysis method," *Advances in Adaptive Data Analysis*, vol. 1, no. 1, pp. 1–41, 2009.
[27] I. Kamwa, A. K. Pradhan, G. Joos, and S. R. Samantaray, "Fuzzy partitioning of a real power system for dynamic vulnerability assessment," *IEEE Trans. Power Syst.*, vol. 24, no. 3 (Aug.), pp. 1356–1365, 2009.

[28] I. Kamwa, A. K. Pradhan, and Gé. Joos, "Automatic segmentation of large power systems into fuzzy coherent areas for dynamic vulnerability assessment," *IEEE Trans. Power Syst.*, vol. 22, no. 4 (Nov.), pp. 1974–1985, 2007.

[29] P. Kundu and A. K. Pradhan, "Synchrophasor-assisted zone 3 operation," *IEEE Trans. Power Deliv.*, vol. 29, no. 2 (Apr.), pp. 660–667, 2014.

[30] K. Behrendt and K. Fodero, "The perfect time: an examination of time-synchronization techniques," 2006.

[31] S. H. Horowitz and A. G. Phadke, "Third zone revisited," *IEEE Trans. Power Deliv.*, vol. 21, no. 1, pp. 23–29, 2006.

[32] "IEEE Std C37.118.1-2011 (Revision of IEEE Std C37.118-2005)," *IEEE Std C37.118.1-2011 (Revision of IEEE Std C37.118-2005)*. pp. 1–61, 2011.

[33] A. K. Pradhan and P. Kundu, "Online identification of protection element failure using wide area measurements," *IET Gener. Transm. Distrib.*, vol. 9, no. 2 (Jan.), pp. 115–123, 2015.

[34] D. L. Goldsworthy, "A linearized model for MOV-protected series capacitors," *IEEE Trans. Power Syst.*, vol. 2, no. 4, pp. 953–957, 1987.

[35] A. K. Pradhan and S. Sarangi, "Synchronised data-based adaptive backup protection for series compensated line," *IET Gener. Transm. Distrib.*, vol. 8, no. 12, pp. 1979–1986, 2014.

[36] S. Sarangi and A. K. Pradhan, "Adaptive direct underreaching transfer trip protection scheme for the three-terminal line," *IEEE Trans. Power Deliv.*, vol. 30, no. 6 (Dec.), pp. 2383–2391, 2015.

Chapter 4

Synchrophasor-assisted visualization and protection of power systems

Sukumar Brahma[1]

Traditional visualization of power system is performed through supervisory control and data acquisition (SCADA) systems and state estimator algorithm. Though decades of development and implementation efforts coupled with modern computing resources have resulted in a robust and dependable visualization tool, the measurements fed to the SCADA system are scalar values and are not synchronized at a precise timestamp. Typically, a new set of measurements is available every 2–4 s, which is sufficient for steady-state analysis. With recent advances in sensing, it is possible to get measurements of better quality at a faster rate. Recent generation of phasor measurement units measure and transmit synchronized phasor values of voltages and currents at 60 fps. The purpose of this chapter is not to show how to assimilate these measurements with the current SCADA system, but to explore the possibility of creating another visualization layer independent of SCADA that can provide more insight in to real-time dynamic events taking place in power systems. A concept of supervisory protection is also explored to increase security of protection schemes by detecting relay misoperations. Results are presented from published research papers to facilitate critical evaluation of the feasibility of these concepts.

4.1 Traditional visualization with SCADA

A block diagram showing supervisory control and data acquisition (SCADA) processes is shown in Figure 4.1. SCADA acquires data related to system state, network topology, and control settings from remote terminal units at substations. These consist of breaker statuses, transformer taps and phase shifts, scalar values of voltages and currents, real and reactive power, and settings of control devices. Using these data where measurements outnumber the system states, and stochastically modeling errors in measurements, a state estimator algorithm solves an overdefined set of equations to create the best estimate of the state of the power

[1]Klipsch School of Electrical and Computer Engineering and Electric Utility Management Program (EUMP), New Mexico State University, Las Cruces, NM 88003, USA

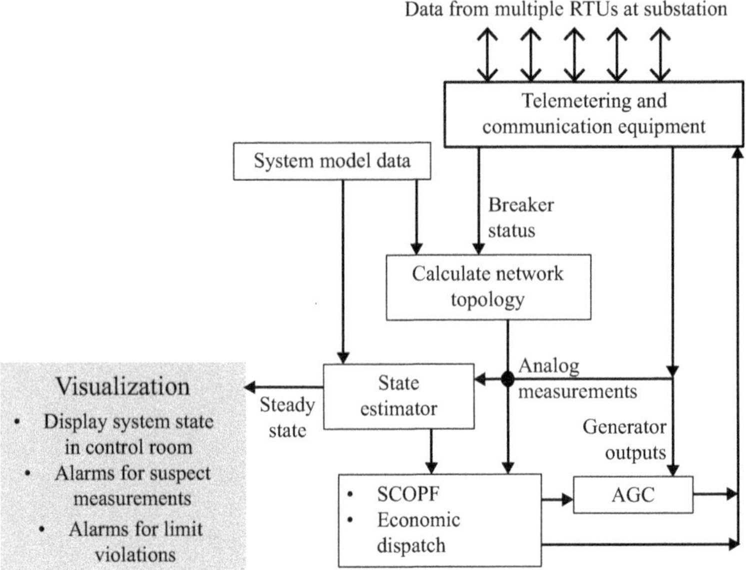

Figure 4.1 SCADA processes

system—phasor values of voltages at all buses and power flow through all lines are estimated and displayed on large screens in control centers for visualization. Probable bad data are also identified. Alarms for bad data and overloads are generated. The estimated state is then used for further analysis to run a security constrained optimal power flow (SCOPF) and economic dispatch to finally generate new settings for automatic generation control. With recent computing resources, a SCOPF can take up to 15 min. If SCADA system is lost, power system operations would be seriously, and probably fatally affected, as was the case during the August 13, 2003 blackout in the northeastern United States.

4.2 State monitoring with synchrophasors

Synchrophasors are phasor measurements that are synchronized using an accurate time reference. Phasor measurement units (PMUs) measure voltages and currents in power systems as synchrophasors using global positioning system to provide time synchronization. Number of PMUs has grown at a rapid rate in North America, increasing from 200 research-grade PMUs in 2009 to 1,700 production-grade PMUs in 2014. Typically, several PMUs transmit their measurements to a central server called phasor data concentrator (PDC) through communication channels. Compared to SCADA systems reporting loosely synchronized *scalar* values of grid state once every 2–4 s, PMU data, though sparser, capture the physics of the grid with much higher precision and speed. These measurements therefore open a window of opportunity for finer visualization of power systems by detecting and identifying *dynamic* events in real time. Figure 4.2 shows a voltage instability event in Figure 4.2(a) and an angle instability event in Figure 4.2(b). If the system state were monitored with PMUs, providing 60 measurements every second, these dynamic events could be vividly visualized. It takes approximately 5 min for precipitation of the final phase of voltage instability and 3 min for angle instability, which would have been observed through 300 and 180 PMU measurements, respectively. Such high-fidelity visualization and direct measurement of angles can provide fast and accurate insight into dynamic events and can facilitate timely response from operators to avert system-wide instability.

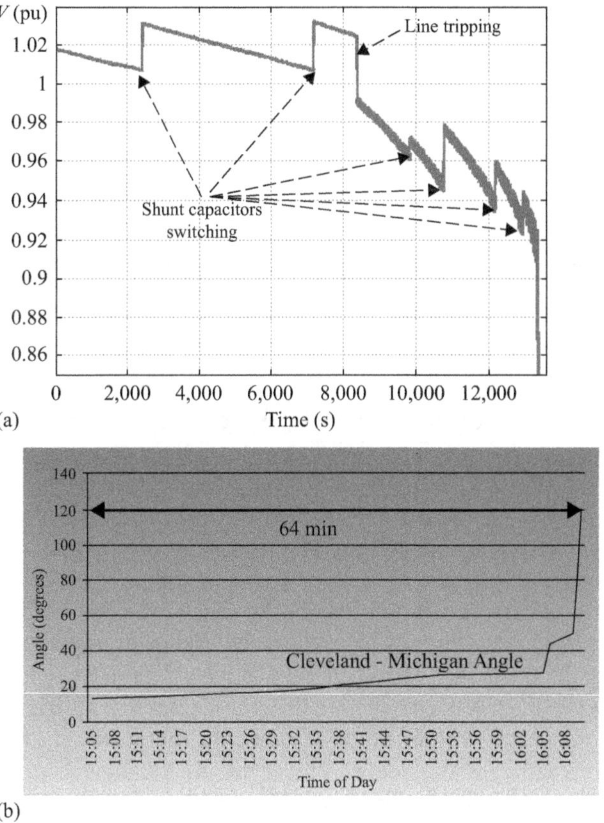

Figure 4.2 Dynamic events showing (a) voltage instability (reprinted with permission from IEEE Power System Relaying Committee) and (b) angle instability (reprinted with permission from Schweitzer Engineering Laboratories, Inc.)

4.3 Disturbance monitoring and identification using synchrophasors

Current generation of PMUs measure three-phase voltages and currents at 60 or 120 frames per second (fps); these rates are likely to increase in near future. PMUs also measure frequency and rate of change of frequency, which are important parameters for detecting a disturbance. Such data rates are unprecedented for electric utilities. As of now, these data are simply archived and only used for postmortem analysis. PMUs do have the capability of detecting a disturbance and creating separate files storing disturbance data to facilitate such analysis. North American Reliability Council (NERC), the regulating entity in North America, has published document PRC 002-2 that mandates disturbance monitoring and reporting requirements. This document mandates either of the following to detect a disturbance: (i) frequency $<$59.55 Hz or $>$61 Hz, (ii) rate of change of frequency $df/dt > 0.124$ Hz/s, (iii) undervoltage trigger set no lower than 85% of the normal operating voltage for a duration of 5 s. When a disturbance is detected using these criteria, a disturbance file is created that includes at least two cycles of predisturbance data as mandated by NERC. NERC mandates at least 30 cycles worth of data including the predisturbance data to be stored in the disturbance file. Utilities have been known to adopt more stringent criteria for detection and longer disturbance files than the mandated norms. For example, disturbance files created by a set of PMUs for a North American utility are 3-min long. It has 55 s (3,300 cycles at 60 Hz) of predisturbance data and 125 s of data (7,500 cycles) after disturbance was detected. The trigger criteria selected by this utility were voltage change of $\pm 10\%$ change in the nominal voltage, frequency change of ± 0.05 Hz, and $df/dt = \pm 5\ \text{Hz}/s$.

Figure 4.3 shows field recordings of PMU data for different types of disturbances obtained from four PMUs owned by this utility. Voltage and frequency values are plotted in per unit based on 345 kV and 60 Hz, respectively. These are from older generation of PMUs and show positive sequence voltages and frequency measured at 30 fps. The plots show truncated data files covering 0.5 s before and 1.5 s after a disturbance is detected. It can be seen that most of the dynamics are captured within half a second or 30 cycles (for 60-Hz system) after the disturbance, but voltages for line trip and capacitor switching events do take longer to settle down. Fault and generation loss events in the figure are confirmed through corresponding entries in the utility log, the rest of the events are educated guesses made by comparing the disturbance patterns with simulated data [1].

It can be seen that each type of disturbance leaves a certain signature on voltage and frequency waveforms. The other important observation is that data recorded by all four PMUs show similar trends. This is because PMUs are capturing the same physics during the dynamic event. The PMU electrically closest[1] to a disturbance would show the "strongest" signature or largest deviations. Measurement noise is also present on the waveforms. A recent study of measurement

[1]A metric for electrical closeness could be the bus impedance matrix element corresponding to the bus with PMU and the fault point.

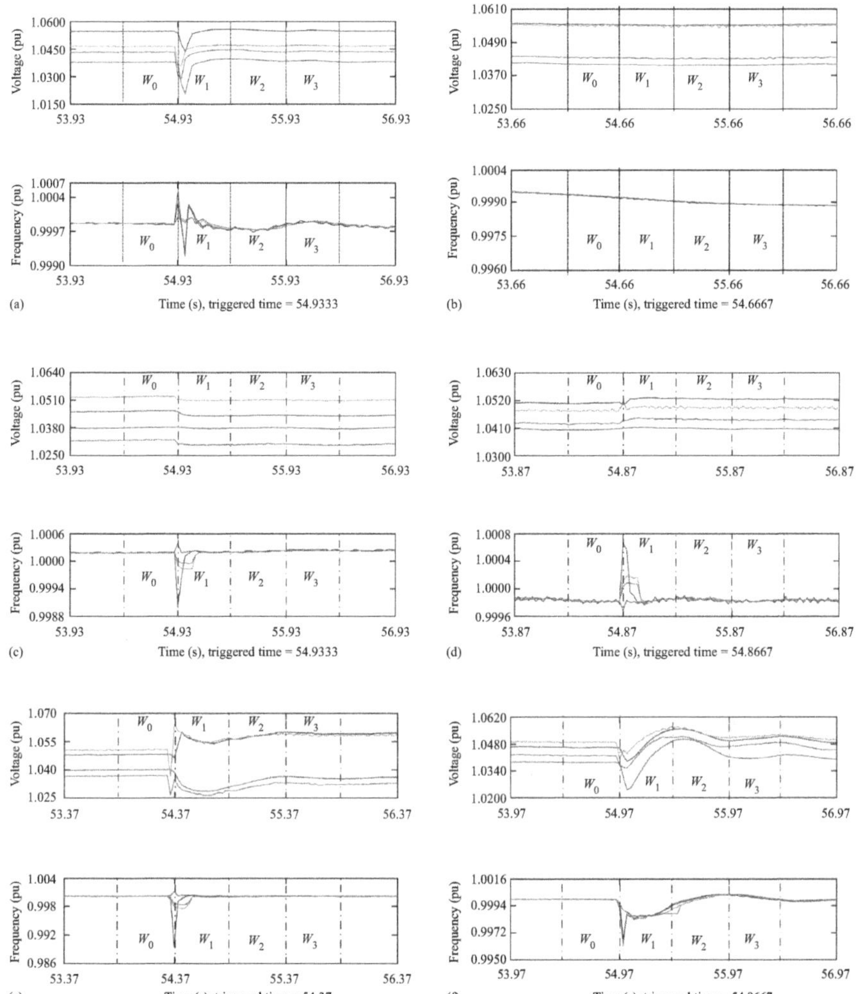

Figure 4.3 Plots of field-recorded PMU data for (a) fault, (b) generation loss, (c) large load switched on, (d) large loaded switch off, (e) line trip, and (f) series capacitor switched on

noise in field PMU data from transmission, primary distribution and secondary distribution networks show that the noise follows a Gaussian distribution with 0 mean and can be approximated with a signal-to-noise ratio of 45 db [2].

Based on the observations made so far, it is reasonable to ask the question—"Instead of simply storing the disturbance files for postmortem use, can we classify them in real time to identify the associated disturbance, thus creating a visualization tool for dynamic disturbances?" The answer to this question is explored in this section.

This essentially is a classification problem in which classes are the types of disturbance. For any classification problem, the first stage is feature extraction in which data are typically transformed to a domain in which representative features are extracted while reducing the volume/dimensionality of original time-series data. Classifiers would use these features to identify the type of the associated disturbance. The science and engineering of feature extraction and classification forms an area in itself. Exhaustive discussion of this area is not within the scope of this chapter. In this section, methods that have been tried on disturbance data recorded by PMUs will be briefly discussed. As the visualization is required in real time, both speed and accuracy of classification are important.

4.3.1 Feature extraction

Though many feature extraction methods exist in literature, there is no qualitative approach to determine which method will work the best for a given dataset. This section reviews some classical feature extraction methods and a relatively new method that has been found promising for PMU data. Data from different PMUs are highly correlated due to the same underlying physics. This fact can be exploited to first identify the PMU dataset that is most affected by a disturbance and then simply use that dataset for feature extraction. One such method to identify the most affected dataset is described in [3]. All of the feature extraction methods summarized here assume such dataset to be available.

4.3.1.1 Discrete Fourier transform

Discrete Fourier transform (DFT) converts data in time domain to features in frequency domain using complex exponentials as the basis.

$$X(k) = \frac{1}{N}\sum_{n=0}^{N-1} x(n)e^{-j2\pi kn/N} \tag{4.1}$$

where $X(k)$ are the DFT coefficients.

DFT coefficients form the representative features for the original time series. DFT works well with stationary signals, but is not designed to capture temporal variation of a time series.

4.3.1.2 Discrete wavelet transform

The discrete wavelet transform (DWT) decomposes a time series into multiple levels of resolution using basis functions derived from a mother wavelet. DWT coefficients form the representative features for the original time series. DWT, unlike DFT, offers a choice of large number of basis functions called mother wavelets. However, its performance is highly dependent on the choice of the mother wavelet, which requires trial and error approach. The right choice typically results in a sparse representation, meaning the least number of representative DWT coefficients. Due to data visualization at different resolution levels, temporal variations of a time series are captured by this method.

4.3.1.3 Principal component analysis

This technique represents the coordinates of data in $\Re^D$ dimension as D principal axes, differentiated by their variance. Principal component analysis (PCA) represents the coordinates of data in $\Re^D$ dimension as linear combination of D new orthogonal basis vectors or principal components (PCs). PCs capture the variance of the original correlated data in decreasing order, allowing the user to discard the PCs having low variance without losing critical information, thus allowing dimensionality reduction. PCA has been shown to be capable of capturing hidden dynamics underlying a complex dataset through the PCs.

4.3.1.4 Shapelets

DFT, DWT, and PCA employ domain conversion to create features. Shapelets have been recently introduced by Ye and Keogh [4] as simply "snippets" or subsequences of the time-series data. The essence of this method is to discover the most representative subset of a time series as its feature. Shapelets provide easy visualization and interpretation as they are not transformed to another domain. However, they are computationally expensive. The method to extract shapelets essentially requires scanning the entire time series for a subsequence of certain length that is most representative of the class associated with the time series. This would mean for a time series of length n, the number of subsequences to be tested are $(n) \times (n+1)/2$—a daunting number for a large time series. After their initial introduction, shapelets discovery has been vigorously researched [5–8]. Efficient algorithms for shapelet discovery are proposed [6,9] because the exponential nature of shapelet candidates prohibits the use of a brute-force algorithm. These algorithms may not extract the same shapelets, but they can still achieve good classification accuracies.

In order to reduce the computation burden of discovering shapelets, domain knowledge can be used. In case of PMU data, as seen in Figure 4.3, critical changes occur for a short time after the disturbance. So the search can be restricted to a subset of the waveform close to the triggering point of disturbance. Such shapelets are termed *Dshapelet* (DS in short) due to the influence of the application domain.

4.3.2 Classification

Popular classification methods work on data with known class labels. Classification of PMU-generated disturbance files technically does not fall into this category. Author has access to field data from a North American utility. The data are from four PMUs in 345-kV network reported at 30 fps reporting positive sequence quantities. Although a total of 1,013 disturbance events were recorded by these PMUs over a 4-year period, the corresponding utility logs are available only for 84 events—23 generation loss, 58 faults, and three line-trips. Thus, there is a large section of data potentially belonging to unknown types of disturbance events, in other words, the number of classes and their labels (disturbance types) are unknown. Before using traditional classifiers over the extracted features from PMU data, number of classes within the dataset needs to be established. This type of problem lends itself to unsupervised clustering, in which partition-based, hierarchical, and density-based approaches exist. A summary of these approaches can be found in [1]. This chapter explains the reasoning behind choosing an agglomerative hierarchical clustering approach that can identify natural clusters and outliers without parametric input from users and is computationally more efficient compared to divisive clustering approaches. The study reported in the paper revealed that the clusters or disturbance types hidden in the unlabeled disturbance files were (1) faults, (2) generation loss, (3) series capacitor switched on, (4) series capacitor switched off, (5) load switched on, (6) load switched off, and (7) synchronous motor switched on. Once the class labels are known, traditional classifiers can be used. Two classifiers—*k*-nearest neighbor (kNN) and support vector machines (SVMs) that performed best in our study [10] are summarized in this section.

4.3.2.1 kNN classifier

kNN algorithm is an instance-based lazy classification method. Different from most classification algorithms, which consist of a training stage to train a model by utilizing the training data and a testing (or classification) stage to classify testing data using the learned model, kNN does not have a training stage. Instead, it treats all the training data as a model. In the testing stage, this method finds the k most similar nearest neighbors of a testing instance and predicts the testing instance's class label to be the one that is shared by the majority of its kNNs. To use kNN algorithm, a proximity measure to calculate the similarity/distance between data points (e.g., Euclidian distance) is required. k is typically a small positive integer. For the special case of $k = 1$, the test data are assigned to the class of the training data that is closest to the test data. The advantage of this classification method is its easy interpretation and its flexibility of producing arbitrarily shaped decision boundaries. The disadvantage of this method is that its testing stage is expensive when there is large amount of training data.

4.3.2.2 SVM classifier

SVM performs classification based on a few support vectors derived from the training dataset. These support vectors correspond to the position vectors of the training data located near the classification boundary (a hyperplane) within a specified margin—the larger the margin, the better the performance of the classifier. The class to which a test data point belongs is determined by projecting its position vector over the stored support vectors. This basic SVM principle is applicable for classifying linearly separable data. For data with nonlinear decision boundary, an appropriate nonlinear mapping is used to transform the data to a linearly separable subspace. This mapping is achieved by introducing kernel functions for SVM. Appropriate choice of kernel function and its associated parameters can accomplish highly nonlinear classification tasks by storing only a few support vectors instead of storing the entire training data. This generalization of the training data results in faster testing response, but this approach has high memory and computational requirement during the training stage, in which support vectors are created. The SVM is applicable for a binary classification problem. The multiclass problems are decomposed into several binary classification problems using one-versus-all or one-versus-one approach to accommodate the use of SVM. Other alternatives also exist. In recent years, SVMs have gained much interest in the computer science community due to their complex computations and other application communities due to their accurate prediction. The disadvantage of SVM is that it is hard to interpret the misclassified results, especially when the decision boundary is non-linear. The nonintuitive kernel functions make it difficult to trace a misclassified result back to the data and analyze directly the reason(s) for misclassification.

4.3.3 Case study: disturbance identification using PMU-generated disturbance files

In this case study, we assume that the disturbance types (class labels) are known through unsupervised clustering described in Section 4.3.2. As there is no universally fitting set of features and classifiers, suitability of features and classifier for every classification problem needs some trials. In this study, results from using the features and classifiers described in Sections 4.3.1 and 4.3.2 on real as well as simulated PMU data are described. The purpose is to find a combination of feature extraction methods and classifier that is best suited to disturbance data generated by PMUs. The full study is reported in [11].

4.3.3.1 Data

For credibility of any classifier, use of field data and availability of ground truths are two important metrics. As mentioned in Section 4.3.2, author has access to field data for a total of 1,013 disturbance events recorded by four PMUs over a 4-year period, but the utility logs are only available for 23 generation loss events and 58 faults—we neglect the line-trip event for this study, because line-trips are intentional, and hence always known. This poses two problems: (1) the data are too few to test a classifier and (2) ground truths are available for only two events. A slightly nonideal solution had to be adopted to circumvent this difficulty. General Electric's PSLF^{TM} dynamic simulation tool has the utility's system simulated in great detail. This software was used to simulate data for more types and number of disturbances as shown in Table 4.1. Two-second long event files were generated, including 0.5 s of predisturbance data. The reporting rate was kept the same as the field data—30 fps.

Table 4.1 Events simulated to generate disturbance files

Class	Events
Fault	1,260
Generation loss	558
Load switching off	348
Load switching on	354
Reactive power switched off	497
Reactive power switched on	442
Synchronous motor switched off	36
Total events	3,495

Thus, each event file, for both field-recorded (truncated as shown in Figure 4.3) and simulated events, had 60 data points. Field data were preprocessed for bad/missing data. This process is described in detail in [12].

4.3.3.2 Selection of features and classifiers

For feature extraction, all coefficients were selected for DFT and DWT. Least asymmetric wavelet was chosen as the mother wavelet for DWT; however, other mother wavelets like Daubechies and Coiflet also gave comparable results. For the PCA-based features, the largest 20 PCs out of a possible 60 were selected based on trials.

The original disturbance files generated by PMUs are 3-min long, having 5,400 data points for each variable. For an exhaustive search for shapelets, it would take $5,400 \times 5,401/2 = 14.5827 \times 10^6$ subsequences to process. However, it is clear from Figure 4.3 that the critical information lies around the disturbance trigger. Due to this reason, a subset of the disturbance file was formed and used as mentioned in subsection 4.3.3.1 to extract DS. This substantially reduced the search space. Further, it was observed that the critical information resided around the largest deviations or peaks. Therefore, the first extreme point was selected as the anchor point for extracting DS. The algorithm first finds the global extreme points—global maximum and global minimum—from the chosen subset and chooses the first extreme point as the anchor point. Next, the algorithm extracts one subsequence that is centered at the anchor extreme point and has length $2L + 1$. If the anchor point is at index n_1, then the extracted subsequence (or DS) is $V_{n_1-L}, \ldots, V_{n_1-1}, V_{n_1}, V_{n_1+1}, \ldots, V_{n_1+L}$. For instance, if the anchor extreme point is 20, which can be the global maximum point, for length $L = 4$, the DS that we extract is the subsequence from samples 16–24. $L = 4$ was chosen in this study after experimenting with different values. Again, domain knowledge was used to determine the maximum value of L to consider by visual inspection of event waveforms.

Because the variation patterns in disturbance waveforms are different for different types of disturbances, a new feature—slope sequence of DS—denoted as (SSD) was also created as described below to capture the nature of change in DS data using slope.

For a data point $x(n)$ within DS, its K-step slope value is calculated as

$$\lambda(n) = \frac{x(n+K) - x(n-K)}{2K}. \tag{4.2}$$

For $K = 1$, given a DS with N sampled points, SSD is the sequence $\lambda(2), \lambda(3), \ldots, \lambda(N-1)$. To find SSD in this study, the DS were generated with $L = 5$, and then k was chosen as 1. The reason for choosing a larger L is that SSD reduces the length of the feature vector DS.

kNN and SVM were used as classifiers. For the kNN classifier, $k = 1$ gave good results, and a radial basis function with Gaussian kernel provided the best results with SVM. Classification was performed using features from (1) just voltage data, (2) just frequency data, and (3) both voltage and frequency data. For the third case, the features derived in the first two cases were concatenated.

4.3.3.3 Results

Due to the constraint that ground truths and field data are available only for two classes, that is, fault and generation loss, the robustness of the features were tested through different choices of training and testing data as follows:

1. use simulated data corresponding to generation loss and fault for training a two-class classifier, and use field data for testing (results in Table 4.2),
2. use simulated data corresponding to all types of disturbances for training a seven-class classifier, and use field data (two classes) for testing (results in Table 4.3),
3. use simulated data corresponding to generation loss and fault for training a two-class classifier *and* testing, using 10-fold cross-validation (results in Table 4.4),
4. use simulated data corresponding to all types of disturbances for training a seven-class classifier *and* testing, using 10-fold cross-validation (results in Table 4.5).

Table 4.2 Training (1,818 simulated events with faults and generation loss) and testing (81 actual events)

	Voltage		Voltage and frequency	
	1NN	**SVM**	**1NN**	**SVM**
Raw data	77.7	93.8	64.2	71.6
DFT	88.8	72.8	80.2	71.6
DWT	80.2	81.4	62.9	71.6
PCA	71.6	71.6	71.6	71.6
DS	87.6	**100**	69.1	71.6
SSD	95.1	97.5	88.9	**100**

Table 4.3 Training (3,495 simulated events from seven disturbance types), testing (81 actual events with FLT and GL)

	Voltage		Voltage and frequency	
	1NN	**SVM**	**1NN**	**SVM**
Raw data	49.4	81.5	61.7	71.6
DFT	55.6	71.6	72.8	71.6
DWT	50.6	75.3	60.5	71.6
PCA	71.6	71.6	71.6	71.6
DS	56.9	74.1	61.7	71.6
SSD	77.8	82.7	72.8	**100**

Table 4.4 Training (90% of 1,818 simulated events with FLT and GL), testing (10% of 1,818 simulated events with FLT and GL); 10-fold cross-validation

	Voltage		**Voltage and frequency**	
	1NN	**SVM**	**1NN**	**SVM**
Raw data	99.1	92.7	98.9	89.5
DFT	98.5	87.3	98.3	83.9
DWT	99.1	90.4	98.9	88.1
PCA	69.8	69.3	69.3	69.3
DS	**99.4**	97.9	**99.7**	96.0
SSD	**99.9**	99.3	**99.6**	**99.3**

Table 4.5 Training (90% of all 3,495 simulated events with seven disturbance types), testing (10% of all 3,495 simulated events with seven classes); 10-fold cross-validation

	Voltage		**Voltage and frequency**	
	1NN	**SVM**	**1NN**	**SVM**
Raw data	80.6	75.2	86.9	77.8
DFT	74.8	66.5	81.8	64.3
DWT	80.5	74.1	86.9	74.9
PCA	36.7	36.1	36.1	36.1
DS	78.2	69.9	86.3	79.6
SSD	79.8	71.7	**92.5**	88.5

Tables show performance with raw data to establish a baseline. As the last two cases involve only simulation data, 10-fold cross-validation was used. In this approach, dataset is partitioned into ten folds, in which each fold consists of equal (as far as possible) number of disturbance files for every disturbance type. Configuration x/y means that x% of the disturbance files are used as training data, and the remaining $100 - x\% = y\%$ are used as testing data. For each training/testing configuration, *ten* different tests are performed to guarantee that each fold is used at least once as training data and used at least once as testing data. In this study, $90/10$ configuration is used. Best results are boldfaced in tables.

Results in the tables show that concatenated features give better results. A combination of SSD and SVM performs with near perfect accuracy to classify faults and generation loss (Tables 4.2–4.4). For the experiment described through Table 4.5, in which all seven classes are being classified, the accuracy drops, although still respectable. An interesting observation is that raw data and feature extraction methods that choose subsets of raw data and their derivatives (DS, SSD) perform better than the classical methods performing change of domain. This can be explained by observing the waveforms in Figure 4.3 and noting that the changes in voltage and frequency waveforms are *visually different* for faults and generation loss. In other words, the characteristic features are already present in raw data. Finally, it is observed that simulation data used for training have no noise, but the field data do contain noise; however, SSD is robust against this noise, unlike some traditional methods that fare poorly when used with field data. For the results described in Table 4.5 in which accuracy is lower, it was observed that most misclassifications were between load/reactive power switching and generation loss, which is due to similar physics underpinning these events, resulting in similar disturbance patterns. Increase in PMU deployment can probably alleviate this problem, because the PMU nearer to the currently misclassified event may produce stronger signatures, and hence clearer features.

4.4 Use of synchrophasors for power system protection

4.4.1 Real-time constraints of power system protection

Power system protection is designed to protect the components of power system from damage and save the system from instabilities. The precipitating reason for potential damage or instability is typically a fault. Faults generate currents several times higher than rated values and must be cleared (by opening appropriate circuit breakers) within a few cycles to save both equipment and stability of power systems. As this function is crucial, it is bolstered by main and backup functions. Main protection clears the fault in typically 100 ms or less; if it fails to operate, backup protection operates and clears the fault typically in about half a second, although at the cost of selectivity—meaning some healthy part of the system will also be unnecessarily isolated. Designed and refined through more than half a century of experience, insight, and technological advances, protection function is the fastest operating component of a power system that has kept the system exceptionally safe, in light of the fact that power grid is the largest non-linear dynamic machine in the world, a large part of which is exposed to extreme weather conditions.

4.4.2 Suitability of PMU measurements for protection

In order to evaluate the possibility of using synchrophasor measurements for power system protection, it is important to assess the accuracy and real-time availability of these measurements with respect to the needs for protective relaying—dependability, security, speed, and selectivity. With current generation of PMUs, there are one or two synchrophasor measurements available every cycle; protective relays provide typically up to 64 phasor estimates per cycle. According to the guidelines in IEEE C37.118.2-2011-a2014 standard, the M Type PMUs consider a seven-cycle sliding window for phasor calculation, whereas P type PMUs consider a two-cycle sliding window. Relays consider a one-cycle window. The phasor estimates of a voltage waveform using these three window lengths are shown in Figure 4.4. The voltage waveform was generated using time-domain simulation of a transmission line, and

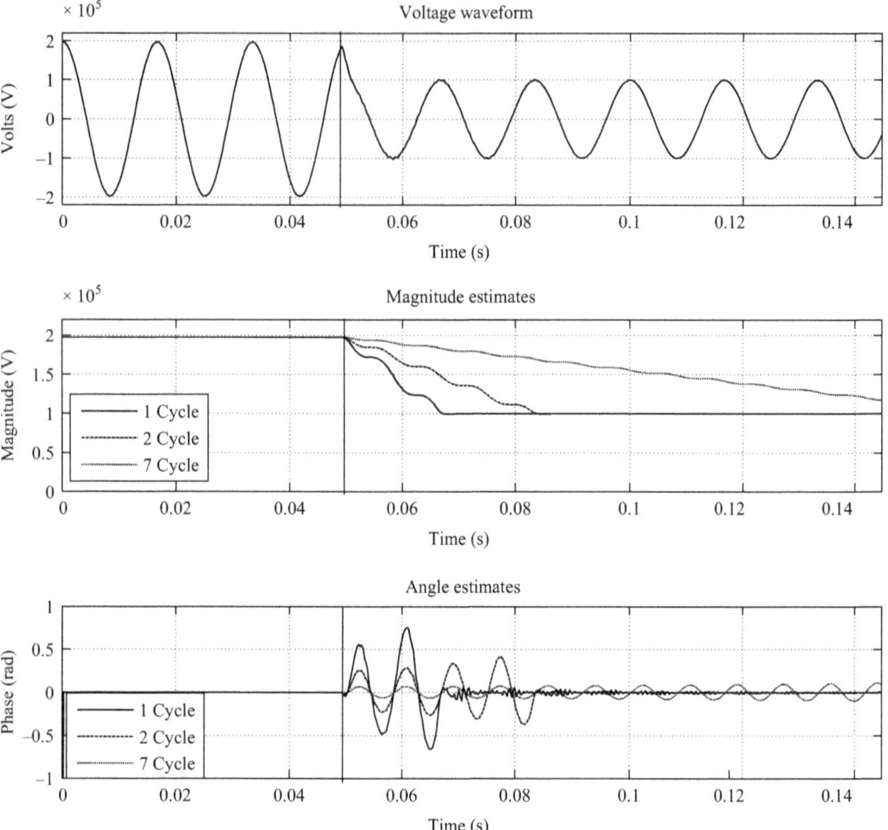

Figure 4.4 Phasor measurement using DFT with different window lengths

fault was applied at 0.05 s into the simulation. Waveform was processed by a low pass filter to avoid aliasing. Clearly, relays provide a stable measurement of the fault voltage the fastest—in one cycle after the inception of fault.

It should also be known that most PMUs are used for dynamic analysis, and state estimation are fed through metering current transformers (CTs) for accurate current measurement during normal operating conditions. These CTs will saturate heavily under faults and cannot be used for any protection applications. If PMU measurements are to be used for protective relaying, PMUs must be fed from protection CTs, which means that the metering applications will suffer from larger errors. Moreover, standards do not specify any accuracy limits for waveforms affected by faults. This means PMU estimates for faults, especially for currents with dc offset may not be accurate enough. It is therefore safer to wait for a few cycles until fault transients damp out when using PMU measurements for protective applications.

Considering these aspects of synchrophasor measurements, it is clear that their use for protection would primarily impact speed of protection schemes. Most protection schemes in power systems use local measurements, in which synchronization with a global time stamp is not required. As capabilities of a numerical relay exceed those of a PMU for protection applications, it does not make sense to use PMUs for local protection schemes. In cases in which such synchronization is desired, it is currently not advisable to use PMU measurements for main protection schemes that typically require the fault cleared in three to five cycles. The capabilities of synchrophasors would therefore be better suited to backup or wide-area-based protection schemes and for offline functions like fault location. If, however, PMU manufacturers start providing relay-grade phasor estimates at 120 or 240 fps (for a 60-Hz system) from P type PMUs, it may become feasible to use PMUs for important primary protection applications like differential protection, in which synchronized phasors can be used in relay logic. However, these applications will have to be designed differently, and a direct communication path for PMU data to be processed by the protection-logic may be necessary instead of routing them through a PDC.

4.4.3 Potential improvements in existing protection applications

Currently, there are few protection schemes in practice that rely solely on synchrophasors. Due to the direct availability of angles, there are some documented installations of improved out-of-step protection schemes implemented using synchrophasors. These schemes essentially use direct measurement across critical tie-lines to detect power swings. If a swing is detected, a prediction tool is used to predict the trajectory of the swing to determine if the swing is likely to be stable or not. A scheme in Florida, USA, and another in Tokyo, Japan, are described in [13]. The scheme in Florida uses a single machine infinite bus model and equal area criteria to predict stability of swing after observing it for up to 250 ms, whereas the scheme in Tokyo uses a time-series prediction tool to predict the swing trajectory 200 ms in future.

A secure backup distance protection for transmission lines can be cited as an example of the potential use of PMUs for backup protection. Load encroachment is a problem that often crops up with the backup protection feature of distance relays. Most blackout logs indicate this problem as a contributing factor to the progression of instability. In this phenomenon, backup protection provided by a distance relay through Zone 3 confuses extreme overload with fault and trips unnecessarily, taking out crucial power corridors at a time when they are most needed. If PMU measurements in the area are used to take over this backup function, this problem can be averted. From multiple data streams arriving at a PDC, a more informed decision can be made if there is a fault or not, and backup protection can be provided much faster than the traditional 500 ms time, even with communication latency.

One of the more ambitious applications would be to improve the performance of a system integrity protection scheme (SIPS). Such schemes acquire measurements from across a wide area of power systems. Currently, settings of such schemes are based on precalculated offline studies for foreseeable contingencies. Responses to such contingencies are also predetermined and implemented in design of SIPS. However, there can always be an unforeseen contingency that can destabilize the system. In view of the advances in computing, it may become possible to perform stability analysis in real time using synchronized information of the system state at key locations and generate responses to mitigate a particular contingency as it develops. For example, interarea oscillations, which represent swinging of one coherent group of generators in one area swinging against another group in an adjoining area, can potentially destabilize the system if they are not damped. Usually, such contingencies are envisaged at the planning stage by simulating many cases of line or generator tripping. Remedial actions to damp such oscillations are predetermined and built into SIPS logic. However, inaccuracy of dynamic system model and inability of exhaustively predicting all dangerous contingency is a drawback. Synchrophasors arriving at a PDC can be treated by a modal analysis tool (Prony Analysis for example) and damping ratios can be calculated and used for prediction of instability and subsequent implementation of remedial actions.

However, if PMU measurements are to be used for wide area protection schemes (or other schemes in which communication is required), delays introduced by communication channels also need to be considered. Such delays are not capped by standards. The following broad guidelines can be considered based on IEEE C37.118.2-2011 standard:

- Data takes 3–5 ms to travel 500 mi; this is predictable for a given network.
- Buffering, demultiplex/multiplex points, forwarding, and routing can add tens to hundreds of milliseconds; this may not be predictable.
- Error detection/correction can add hundreds of milliseconds.
- Realistically achievable delay would range between 50 and 130 ms.

4.4.4 *Supervisory protection—potential use of synchrophasors to supervise local protection*

There is one area in protection that has eluded protection engineers until now—detection of relay misoperations. If a relay fails to operate for a fault in its primary zone, there is always a backup relay to cover that fault, although at the cost of selectivity and speed. However, if a relay misoperates, meaning it operates without any fault, it cannot be detected in real time. Misoperations can be caused by faulty relay components, also known as hidden faults, incorrect settings, or misinterpretation of system conditions, for example, load encroachment

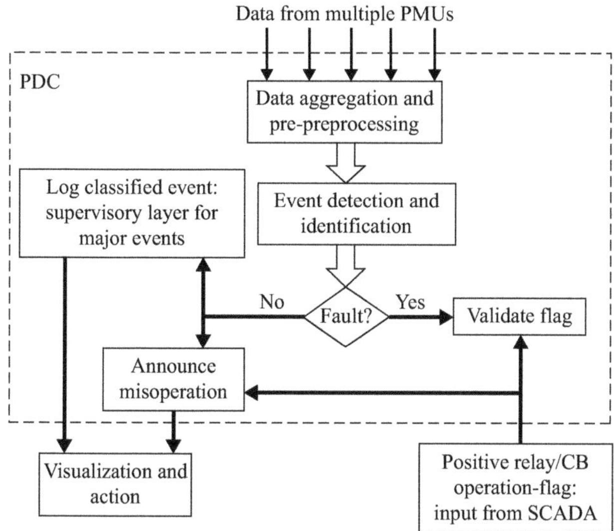

Figure 4.5 PMU-assisted visualization of dynamic events and supervisory protection

in distance relays. In the past, misoperations have been known to trigger blackouts anywhere between a few seconds to a few minutes to even hours [10]. This section explores the possibility of the concept of *supervisory protection* using synchrophasors.

As seen in Section 4.3, it is possible to identify dynamic events using PMU measurements. In the case study described in Section 4.3.3, all fault cases—field-recorded or simulated—were identified correctly using a data segment of 1.5 s after inception of fault. It has been shown that fault identification can be performed much faster in less than 15 ms using dedicated methods that do not require feature extraction or classification [3]. The block diagram of the proposed supervisory protection is shown in Figure 4.5. The scheme takes as input the circuit breaker status from SCADA. If a circuit breaker opens, and fault is *not* sensed by the supervisory layer, it is flagged as misoperation. In this case, either the breaker could be closed back and the associated relay temporarily disabled, or it could be left to the system operator as a warning in real time. In case a fault is detected, the relay action is corroborated.

4.4.5 Case study: synchrophasor-assisted out-of-step protection for generators

Generators usually have an out-of-step (OOS) detecting relay (device 78). OOS conditions occur when a generator (or a coherent group of generators) loses

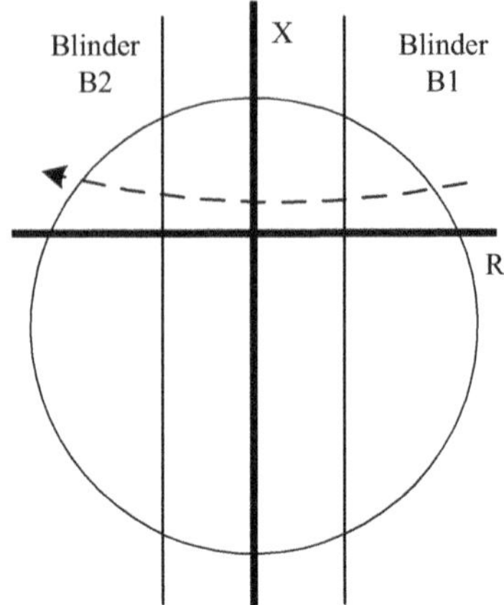

Figure 4.6 Single blinder scheme for OOS detection

synchronism with the rest of the system. This operation needs to be secure, and the most secure scheme used in practice is the single-blinder scheme with trip-on-the-way-out, as shown in Figure 4.6. It is essentially an offset Mho element that maps the angle of the generator rotor with respect to an equivalent source on to an impedance plane. If the impedance enters blinder $B1$ and leaves blinder $B2$, OOS is declared. Sometimes, the boundary of the Mho circle is also considered instead of blinder $B2$ to declare OOS. This mapping however is made with simplifying assumptions—generator controls are neglected and the rest of the system is presented as a Thevenin equivalent. If the rotor angle could be *directly* known even in a dynamic state, the OOS detection can be improved. In this case study, a scheme that uses PMU measurements at the generator bus to estimate the rotor angle is presented.

The scheme is shown in Figure 4.7. It is assumed that the voltage at the generator bus is measured with a PMU, and the measurements are available to a dynamic state estimator (DSE) tool. As the voltage magnitude V and angle θ are *both* available, the state estimator estimates the internal parameters of the generator—in this case the rotor angle δ. The angular difference $\tilde{\delta} = \delta - \theta$ is a direct indicator of swing dynamics. There are no simplifying assumptions needed in this model, and generator controls are fully taken into account. When $\tilde{\delta}$ crosses a certain threshold, OOS can be declared.

In this study, the threshold is taken as 120°, which is consistent with general consensus. Particle filter (PF) is used as a DSE. Details of this filter can be found in [14]. Compared with other widely used DSE algorithms (extended Kalman filter and unscented Kalman filter), the PF is not restricted by model assumption (e.g., probability distribution of measurement noise is Gaussian) and yields superior results on nonlinear/non-Gaussian systems at the expense of

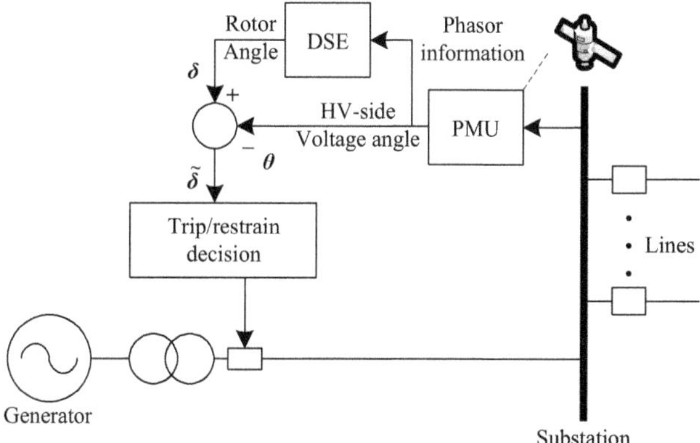

Figure 4.7 Block diagram showing direct estimation of rotor angle to detect OOS

increased computational effort. PF is also robust against measurement noise. The proposed scheme is tested on the New England 10-generator system [15] shown in Figure 4.8 simulated on SIMULINK platform. For all test cases, the measurement reporting rate of the PMU is set to a conservative value of 30 fps.

A three-phase fault was created on line *25–26* at $t = 6$ s. The faulted line is removed by opening the circuit breaker at both ends of the line. The curves of angular difference ($\tilde{\delta}$) for different clearing times are shown in Figure 4.9. The clearing times are unusually large simply for the purpose of creating swing conditions. Clearing time of 14 cycles results in an unstable swing experienced by *generator #8*, and the worst stable swing is created by choosing to trip the breaker one cycle earlier.

A 3% white Gaussian noise is added to PMU measurements. The generators are modeled accounting for subtransient dynamics (sixth-order model) [16]. The prime mover dynamics (steam-turbine governor) and the excitation system (IEEE DC1A) models are considered for each generator except for *generator #10*, which has constant excitation input. Number of particles were selected to be 80. Figures 4.10 and 4.11 show the performance of the conventional single-blinder trip-on-the-way-out scheme and the DSE-based scheme described in this section. It can be observed that DSE is generating usable estimates of the rotor angle, but unlike the conventional relays, its formulation does not make any simplifying assumptions about system topology or controls. This study is described in more details in [17]. It is also possible to introduce swing analysis and prediction processes tailored for real-time implementation. This means all the calculations for estimation and prediction should be completed during the time interval between consecutive phasor measurements. Students are encouraged to reproduce the study with different DSE algorithms.

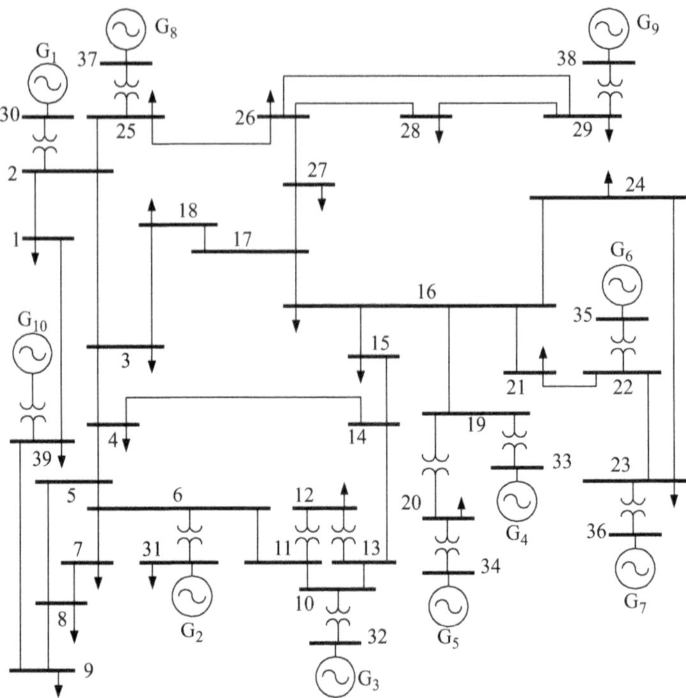

Figure 4.8 New England 10-generator 39-bus system

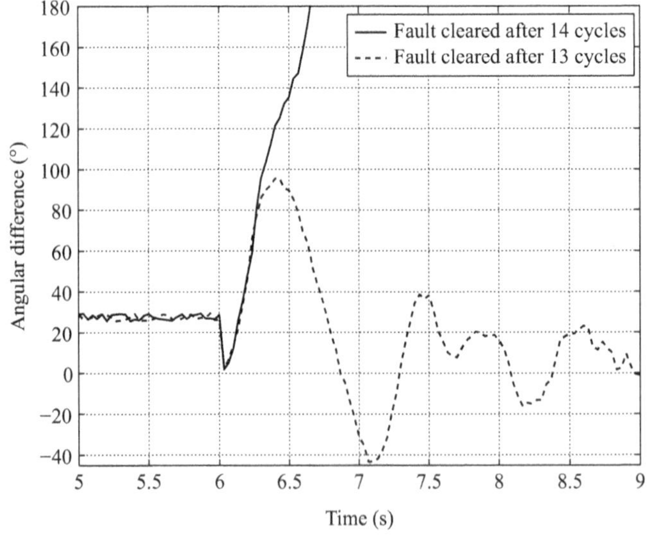

Figure 4.9 Angular difference curves for different clearing time

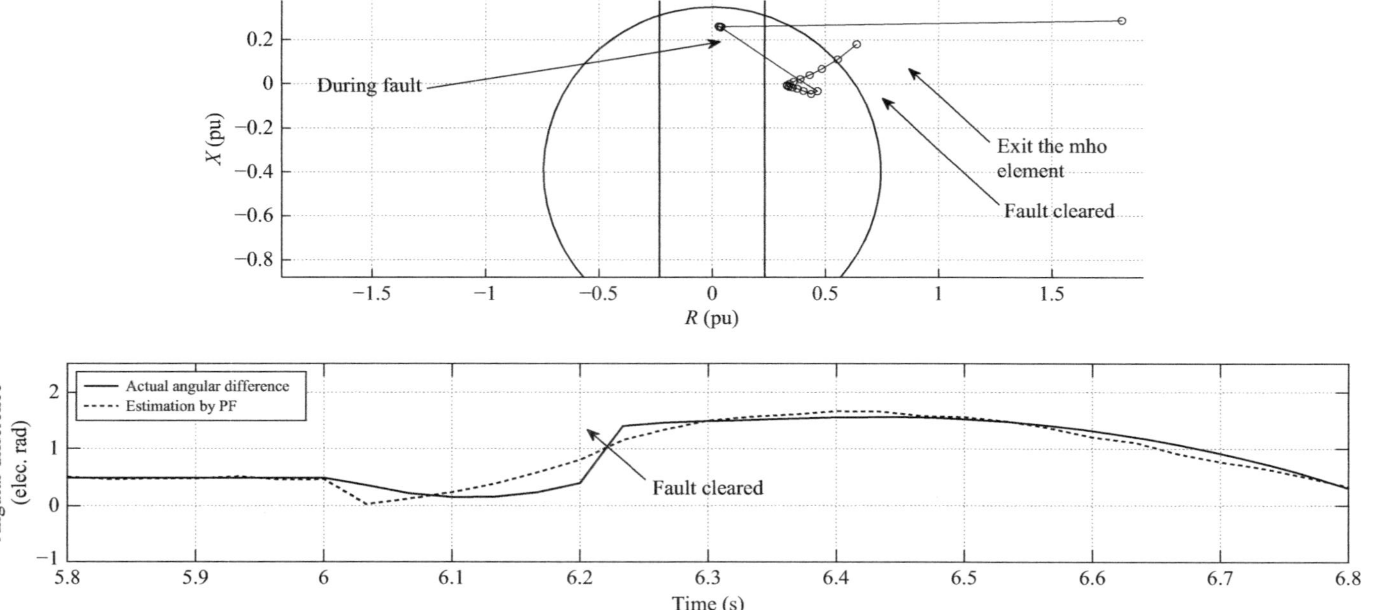

Figure 4.10 Impedance locus for the worst stable swing and estimated angular difference for G_8

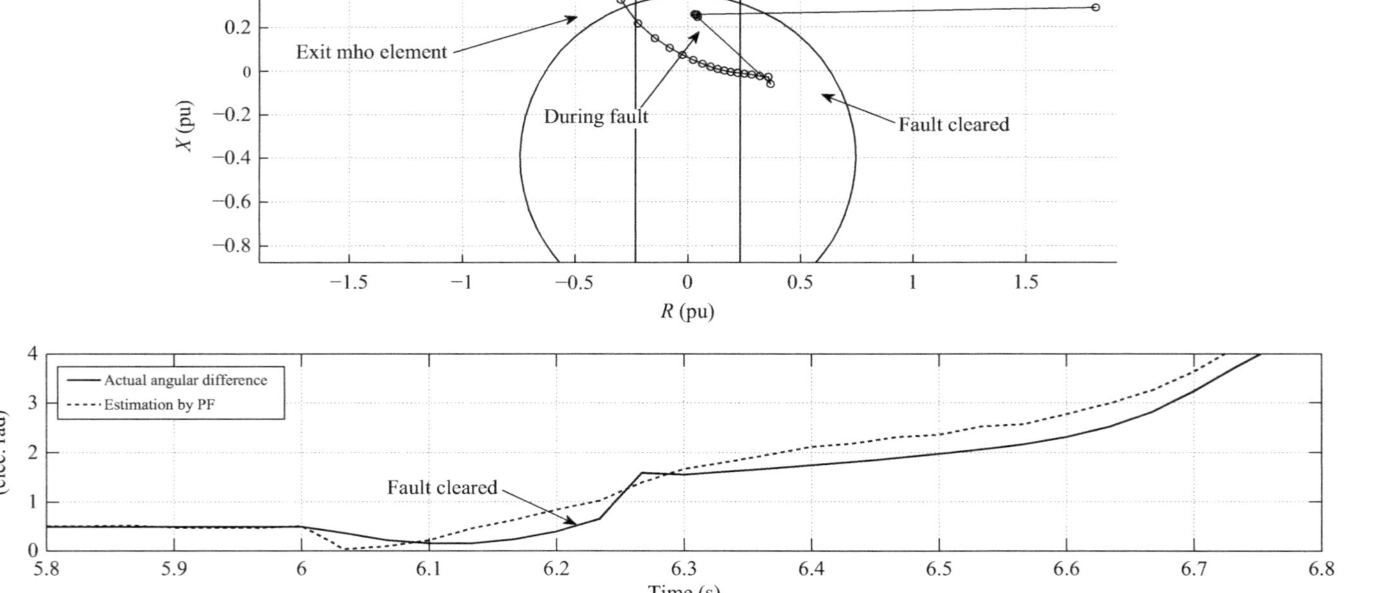

Figure 4.11 Impedance locus for the unstable swing and estimated angular difference for G_8

4.5 Conclusion

This chapter briefly describes the traditional visualization of power systems using SCADA and postulates the feasibility of adding another layer of situational awareness using PMU measurements. This layer would provide real-time detection and identification of major dynamic events taking place in power systems. Pattern creation and classification models that underpin this layer are described, and a case study is presented to illustrate the performance of some representative models on real and simulated disturbance data.

The chapter also assesses the feasibility of using PMU measurements for protection functions. Discussion includes the real-time constraints imposed by protection, measurements generated by different types of PMUs, and communication latencies in PMU data, finally identifying potential facets of power system protection that can benefit from PMU measurements. A new concept of supervisory protection to detect relay misoperations is explained. A case study describing synchrophasor-assisted generator OOS protection is presented.

References

[1] Dahal OP, Brahma SM, Cao H. Comprehensive Clustering of Disturbance Events Recorded by Phasor Measurement Units. IEEE Transactions on Power Delivery. 2014;29(3 (June)):1390–1397.

[2] Brown M, Biswal M, Brahma S, Ranade S, Cao H. Characterizing and Quantifying Noise in PMU data. In: 2016 IEEE Power and Energy Society General Meeting; 2016. p. 1–6.

[3] Biswal M, Brahma S, Cao H. Supervisory Protection and Automated Event Diagnosis using PMU data. IEEE Transactions on Power Delivery. 2016; 31(4 (Aug)):1855–1863.

[4] Ye L, Keogh EJ. Time Series Shapelets: A New Primitive for Data Mining. In: Proceedings of the ACM SIGKDD International Conference on Knowledge Discovery and Data Mining; 2009. p. 947–956. Available from: http://doi.acm.org/10.1145/1557019.1557122.

[5] Grabocka J, Schilling N, Wistuba M, Schmidt-Thieme L. Learning Time-Series Shapelets. In: SIGKDD; 2014. p. 392–401. Available from: http://doi.acm.org/10.1145/2623330.2623613.

[6] Keogh EJ, Rakthanmanon T. Fast Shapelets: A Scalable Algorithm for Discovering Time Series Shapelets. In: SDM; 2013. p. 668–676. Available from: http://dx.doi.org/10.1137/1.9781611972832.74.

[7] Chang K, Deka B, Hwu WW, Roth D. Efficient Pattern-Based Time Series Classification on GPU. In: ICDM; 2012. p. 131–140. Available from: http://dx.doi.org/10.1109/ICDM.2012.132.

[8] Lines J, Davis LM, Hills J, Bagnall A. A Shapelet Transform for Time Series Classification. In: SIGKDD; 2012. p. 289–297. Available from: http://doi.acm.org/10.1145/2339530.2339579.

[9] Mueen A, Keogh EJ, Young N. Logical-Shapelets: An Expressive Primitive for Time Series Classification. In: SIGKDD; 2011. p. 1154–1162. Available from: http://doi.acm.org/10.1145/2020408.2020587.

[10] Dahal OP, Cao H, Brahma S, Kavasseri R. Evaluating Performance of Classifiers for Supervisory Protection Using Disturbance Data from Phasor Measurement Units. In: 2014 IEEE PES Innovative Smart Grid Technologies Conference Europe (ISGT-Europe); 2014. p.1–6.

[11] Biswal M, Hao Y, Chen P, Brahma S, Cao H, DeLeon P. Signal Features for Classification of Power System Disturbances using PMU Data. In: Power Systems Computation Conference (PSCC), 2016; 2016. p. 1–7.

[12] Dahal OP, Brahma SM. Preliminary Work to Classify the Disturbance Events Recorded by Phasor Measurement Units. In: 2012 IEEE Power and Energy Society General Meeting; 2012. p. 1–8.

[13] O'Brien J, Deronja A, Apostolov A, *et al.* Use of Synchrophasor Measurements in Protective Relaying Applications. IEEE Power System Relaying Committee Report of Working Group C-14 of the System Protection Subcommittee; 2013.

[14] Cui Y, Kavasseri R. A Particle Filter for Dynamic State Estimation in Multi-Machine Systems with Detailed Models. IEEE Transactions on Power Systems. 2015;30(6 (Nov)):3377–3385.

[15] Pai MA. Energy Function Analysis for Power System Stability. Norwell, MA: Kluwer Academic Publishers; 1989.

[16] Kundur P. Power System Stability and Control. New York, NY: McGraw-Hill Professional; 1994.

[17] Cui Y, Kavasseri R, Brahma S. Dynamic State Estimation Assisted Posturing for Generator Out-of-Step Protection. In: 2016 IEEE Power and Energy Society General Meeting; 2016. p. 1–8.

Chapter 5

Using PMU measurements for enhanced power grid monitoring and protection

Xiangqing Jiao[1] *and Yuan Liao*[1]

5.1 Introduction

Power grid monitoring and protection play a pivotal role in ensuring safe and reliable system operation. Compared with conventional measurements from supervisory control and data acquisition system, phasor measurement units (PMUs) provide time-synchronized and direct measurements of phasors at a higher sampling frequency [1,2]. With increasing deployment of PMUs in power systems, monitoring and protection of power systems may be greatly improved. PMU measurements are capable of capturing system dynamics and thus can be used for dynamic model validation and calibration. PMU measurements can also be used to increase the efficiency of state estimation. IEEE Standard [1] addresses the definition of a synchronized phasor, time synchronization, application of timetags, method to verify measurement compliance with the standard and message formats for communication with a PMU. Standard [2] defines functional and performance requirements of a phasor data concentrator and provides guidance on concentrator testing.

This chapter discusses the opportunities that PMUs bring to enhance system monitoring and protection. Three topics are included: monitoring of power system operation, monitoring of power system model and wide-area fault location based on PMU measurements.

5.2 Monitoring of power system operation: state estimation

5.2.1 Background

State estimation serves to obtain the best estimate of the internal states of power grids based on redundant measurements. The states of a power system refer to voltage magnitudes and phase angles at every bus. State estimation is one of the most important functions in energy management systems (EMSs) because it provides inputs for

[1]Department of Electrical and Computer Engineering, University of Kentucky, Lexington, KY 40506, USA

Only with appropriate models that can truly replicate the system behaviour, can the model-based analysis, control, and planning be conducted with confidence. If there is a significant mismatch between prediction and actual system performance, unexpected power system instability, disturbances and large-scale outages may occur. Such inconsistency triggered serious power outages in history. An example is the 1996 Western Interconnection blackout [6]. The system model miscalculated the maximum transfer capabilities and expected that the system would still remain in a stable condition under a sequence of disturbances. However, serious violation of system stability occurred, which in turn caused system oscillation, breakup and the blackout. If the transfer limits could have been correctly calculated based on accurate models, such blackout might have been prevented. Another notable example is the 2003 US–Canada blackout that affected nearly 50 million people. It is revealed that the models used in the EMS software during that blackout underestimated reactive power load [7]. The model failed to accurately predict the severe instability. Therefore, the US–Canada Power System Outage Task Force recommended utilities to 'improve quality of system modelling data and data exchange practices' [8].

Moreover, future power systems will keep evolving. It is important to keep the system model accurate and updated, and the accuracy of the models needs to be regularly monitored to ensure the reliability of model-based analyses.

One approach is to use the PMUs to monitor the system models. In other words, the data collected by PMUs can be used to estimate model parameters and validate system models. For example, if PMUs are installed at two terminals of a transmission line, line parameters can then be estimated online based on PMU measurements. Generator model validation and tuning can also be realized using disturbances data collected by PMUs. Model validation refers to a procedure where simulation is performed based on the model to obtain simulated response to a disturbance, which is compared with measured response to a similar disturbance. If a reasonable matching is obtained, the model is said to be acceptable.

PMU data has a few distinguished advantages over conventional measurements collected by SCADA. First, the sampling rate of PMU is much higher than that of SCADA measurements, which facilitates observation of dynamic system phenomena and construction of dynamic models. Second, the PMU measurements are time-synchronized. PMUs installed at different locations of the system can provide measurements that correspond to the same event, which can be readily used to validate the system model. Without PMUs, it is difficult to correlate measurements captured at different locations and use them for model validation or any other analysis.

In the aspect of model validation, PMUs make online model monitoring possible. Previously, offline testing is the most common method to collect data for model validation and parameter estimation. For example, to obtain the model for a power plant, the plant needs to be taken offline so that various testing can be conducted. Short-circuit testing has also been used in load modelling [9]. Online model monitoring has a number of advantages over traditional testing. First, online monitoring provides prompt observation of models. Compared with conventional testing, PMU can use real-time responses to estimate parameters or validate model. For some advanced applications that are sensitive to model parameters, real-time parameter estimation is critical. For example, most fault location algorithms require the knowledge of transmission line parameters. However, the line parameters can change significantly due to line sag, conductor temperature, etc.[10] With the help of PMU-based online parameter estimation, the fault location result can be more accurate. Besides, as the mismatch between simulated and actual outputs can be identified promptly, severe accidents like outage and low-frequency oscillation may be avoided. Frequent model validation will boost system operators' confidence in the power system model. Second, online model monitoring is much more cost-effective. Compared with taking power system components offline to conduct testing, online monitoring allows the power system assets to continue operating without causing any additional costs. By using natural disturbance data, inconvenience for customers due to short-circuit testing and the demand of manpower to conduct testing can be omitted. Third, the measurement-based online monitoring reflects the actual component behaviour. The component's behaviour under off-line testing may be different with its online performance, considering impacts of other components in the system. Compared with testing of individual components, online monitoring gives a closer look at the components' online dynamic characteristics.

Furthermore, both steady-state and dynamic models can be validated by PMU measurements. Power system models can also be classified into steady-state model and dynamic model. Each of the components in the system can be represented by steady-state models. The combination of individual component models is a complete system model that represents the entire system's steady-state behaviour. Such model is known as power flow model, and it generally includes only positive-sequence quantities. PMU measurements during normal operation can be used to validate steady-state models. Dynamics property of the components can be represented by dynamic models. The response of a dynamic model depends on the historical inputs as well as on the present inputs. A dynamic model structure is usually expressed either in a form of ordinary differential equations, difference equations, partial differential equations, transfer functions or in other ways. PMUs can record not only steady-state behaviour but also dynamics response of the system. Hence, PMU measurements make dynamic model validation possible.

To demonstrate how to perform parameter estimation and model validation, the following part of this section presents the method of online transmission line parameter estimation and dynamic load model validation using PMU data.

5.3.2 Online transmission line parameter estimation

Transmission lines are essential components for delivering power from generation units to distribution substations. Transmission line parameters, including series resistance, series reactance and shunt susceptance, are important for analyses like protective relay setting, fault location and power flow studies. Precise line parameters can ensure the reliability and accuracy of these analyses. The parameters are usually calculated based on design data including conductor geometric parameters, conductor type, etc. However, actual values of these factors may not be the same as assumed, which can result in inaccurate estimation of line parameters. Moreover, the line conductor temperature varies depending on the current flow and weather conditions. Line length also changes due to line sag. Therefore, the actual line parameters are likely to differ from the calculated values. So, it is desirable to design effective methods for online estimation of transmission line parameters for improved accuracy.

This section presents an online optimal estimation method utilizing PMU measurements [10]. This method is capable of estimating positive-sequence line parameters, and filtering potential bad measurements. The studied transmission line is a transposed line between terminals P and Q. Distributed parameter line model is adopted to fully consider the shunt capacitance of the line. Assume PMUs are installed at both terminals, on which the steady state voltage and current phasors can be measured in real-time at different time instants during normal operation. The following notations are adopted: $V_{Pi}^{(1)}$ and $V_{Qi}^{(1)}$ are the ith phasor measurement of positive-sequence voltage at P and Q, respectively; $I_{Pi}^{(1)}$ and $I_{Qi}^{(1)}$ are the ith phasor measurement of positive-sequence current at P and Q, respectively; $Z_c^{(1)}$ is the positive-sequence characteristic impedance of the line; $\gamma^{(1)}$ is the positive-sequence propagation constant of the line; and l is the length of the line in miles or kilometres.

Based on the synchronized measurements obtained by PMUs at P and Q, the following two equations can be derived:

$$V_{Pi}^{(1)} - Z_c^{(1)} \sinh\left(\gamma^{(1)}l\right) + \sinh\left(\gamma^{(1)}l\right)\tanh\left(\frac{\gamma^{(1)}l}{2}\right)V_{Pi}^{(1)} - V_{Qi}^{(1)} = 0 \tag{5.9}$$

$$I_{Pi}^{(1)} - \tanh\left(\frac{\gamma^{(1)}l}{2}\right)\frac{V_{Pi}^{(1)}}{Z_c^{(1)}} + I_{Qi}^{(1)} - \tanh\left(\frac{\gamma^{(1)}l}{2}\right)\frac{V_{Qi}^{(1)}}{Z_c^{(1)}} = 0 \tag{5.10}$$

$$Z_c = \sqrt{\frac{z^{(1)}}{y^{(1)}}} \tag{5.11}$$

$$\gamma = \sqrt{z^{(1)}y^{(1)}} \tag{5.12}$$

where $z^{(1)}$and $y^{(1)}$ are the positive-sequence series impedance and shunt admittance of the line per mile or kilometre, respectively. $z^{(1)}$'s real part $r^{(1)}$ and imaginary part $a^{(1)}$, along with $y^{(1)}$'s imaginary part $b^{(1)}$ are the variables to be determined. For each set of the synchrophasor measurement, two complex equations can be obtained, which can be separated into four real equations. By making use of multiple measurement sets, possible measurement errors can be eliminated based on least square optimal estimation. Statistical approaches can be adopted to detect, identify and remove any possible bad measurements.

Assume N sets of measurements are available. The measurements are

$$\boldsymbol{M} = \left[V_{P1}, I_{P1}, V_{Q1}, I_{Q1}, \ldots, V_{PN}, I_{PN}, V_{QN}, I_{QN}\right] \tag{5.13}$$

Introduce $\boldsymbol{S}$ and $\boldsymbol{F}(\boldsymbol{X})$ as the measurement vector and function vector, respectively. They are related by

$$\boldsymbol{S} = \boldsymbol{F}(\boldsymbol{X}) + \boldsymbol{\mu} \tag{5.14}$$

where $\boldsymbol{\mu}$ is characterized by the following equation:

$$\boldsymbol{R} = E\left(\boldsymbol{\mu}\boldsymbol{\mu}^T\right) = \mathrm{diag}\left(\sigma_1^2, \ldots, \sigma_N^2\right) \tag{5.15}$$

Elements of $\boldsymbol{R}$ can be specified according to the accuracy of the metres, a smaller value of which indicates a more accurate measurement. The formation of measurement vector and function vector can be found in [11].

The optimal estimate of $\boldsymbol{X}$ is obtained by minimizing the cost function defined as

$$J = [\boldsymbol{S} - \boldsymbol{F}(\boldsymbol{X})]^T \boldsymbol{R}^{-1}[\boldsymbol{S} - \boldsymbol{F}(\boldsymbol{X})] \tag{5.16}$$

The chi-square test–based bad data identification method can be used. The expected value of the cost function is to be computed first, which is equal to the number of degrees of freedom k. Then, the estimated value of the cost function C_J is calculated. If $C_J \geq \chi^2_{k,\alpha}$, then the presence of bad data is suspected with probability of $1 - \alpha$, where the value of $\chi^2_{k,\alpha}$ can be calculated for a specific k and α based on chi-square distribution. If bad measurement is detected, the measurement corresponding to the largest standardized error will be identified as the bad measurement.

Such online estimation methodology is also applicable to some other transmission network topologies [10] and can also be adopted in fault location algorithms. Following is an example of locating fault occurred on a series compensated transmission line. Consider the transmission line between terminals P and Q to be a transposed line. The series compensation device is installed at the point of C. A fault may occur either to the left side or to the right side of the compensator.

Assume, there are N sets of measurements during normal operation, and one set of measurements during the fault. Then, the obtained measurement is

$$\boldsymbol{M} = \left[V_{P1}, I_{P1}, V_{Q1}, I_{Q1}, \ldots, V_{PN}, I_{PN}, V_{QN}, I_{QN}, V_{fP}, I_{fP}, V_{fQ}, I_{fQ}\right] \tag{5.17}$$

where V_{Pi}, I_{Pi}, V_{Qi} and I_{Qi} are the ith set of normal operation measurement, V_{fP}, I_{fP}, V_{fQ} and I_{fQ} are the during-fault measurement.

The unknown variable vector is defined as

$$\boldsymbol{X} = [x_1, x_2, \ldots, x_{8N}, x_{8N+1}, \ldots, x_{8N+8}, x_{8N+9}, x_{8N+10}, \ldots, x_{8N+18}]^T \tag{5.18}$$

where $x_{8i-7}, \ldots, x_{8i}$ $(i = 1, \ldots, N)$ are the variables required to represent the $4N$ pre-fault complex measurements; $x_{8N+1}, \ldots, x_{8N+8}$ are the variables required to represent the during-fault complex measurements; x_{8N+9} is the fault location in per unit; $x_{8N+10}, \ldots, x_{8N+18}$ are the variables required to represent the series impedance, series reactance, and shunt susceptance in positive-, negative- and zero-sequences.

The derivation of measurement vector $\boldsymbol{S}$ and function vector $\boldsymbol{F}(\boldsymbol{X})$ is described as follows.

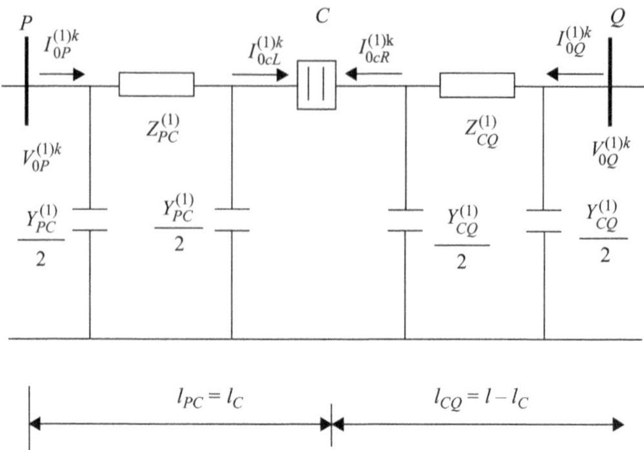

Figure 5.2 Positive-sequence equivalent π circuit of series compensated transmission line during normal operation

Based on the equivalent π circuit during normal operation shown in Figure 5.2, the following three equations can be establish for sections PC and CQ:

$$I_{0cL}^{(1)k} = -\frac{1}{Z_C^{(1)}} \sinh\left(\gamma^{(1)} l_{PC}\right) V_{0P}^{(1)k} + \cosh\left(\gamma^{(1)} l_{PC}\right) I_{0P}^{(1)k} \quad k = 1, \ldots, N \tag{5.19}$$

$$I_{0cR}^{(1)k} = -\frac{1}{Z_C^{(1)}} \sinh\left(\gamma^{(1)} l_{CQ}\right) V_{0Q}^{(1)k} + \cosh\left(\gamma^{(1)} l_{CQ}\right) I_{0Q}^{(1)k} \quad k = 1, \ldots, N \tag{5.20}$$

$$I_{0cL}^{(1)k} = -I_{0cR}^{(1)k} \quad k = 1, \ldots, N \tag{5.21}$$

where k is the index of set of pre-fault measurements; l is the distance between P and Q; l_C is the distance between P and C; $V_{0P}^{(1)k}, V_{0Q}^{(1)k}$ are the kth positive-sequence voltage phasor measurement at P and Q preceding fault, respectively; $I_{0P}^{(1)k}$ *and* $I_{0Q}^{(1)k}$ are the kth positive-sequence current phasor measurement at P and Q preceding fault, respectively; and $Z_c^{(i)}$ and $\gamma^{(i)}$ are the ith sequence characteristic impedance and propagation constant of the line, respectively, $i = 1$ in (5.19)–(5.21).

Substitution of (5.19) and (5.20) into (5.21) leads to:

$$\begin{aligned} f_k(X) = &\left[-\frac{1}{Z_C^{(1)}} \sinh\left(\gamma^{(1)} l_{PC}\right) V_{0P}^{(1)k} + \cosh\left(\gamma^{(1)} l_{PC}\right) I_{0P}^{(1)k}\right] \\ &+ \left[-\frac{1}{Z_C^{(1)}} \sinh\left(\gamma^{(1)} l_{CQ}\right) V_{0Q}^{(1)k} + \cosh\left(\gamma^{(1)} l_{CQ}\right) I_{0Q}^{(1)k}\right] = 0, \\ &k = 1, \ldots, N \end{aligned} \tag{5.22}$$

The first N functions in the function vector can be developed based on (5.22). The other functions are to be constructed based on the equivalent circuit during fault and the fault type.

If the fault is an AG fault, it holds that

$$V_f^{(1)} + V_f^{(2)} + V_f^{(0)} = 3R_f I_f^{(1)} \tag{5.23}$$

where $V_f^{(i)}$ and $I_f^{(i)}$ are the ith sequence fault voltage and current, respectively, and $i = 0, 1, 2$; R_f is the fault resistance. As R_f is a real number, $\left(V_f^{(1)} + V_f^{(2)} + V_f^{(0)}\right)$ and $I_f^{(1)}$ are in phase, which can be represented as follows:

$$f_{N+1}(X) = Im\left\{\frac{V_f^{(1)} + V_f^{(2)} + V_f^{(0)}}{I_f^{(1)}}\right\} = 0 \tag{5.24}$$

If the fault is a phase B to phase C fault, it holds that

$$V_f^{(1)} - V_f^{(2)} = 3R_f I_f^{(1)} \tag{5.25}$$

Thus, the function can be written as follows:

$$f_{N+1}(X) = Im\left\{\frac{V_f^{(1)} - V_f^{(2)}}{I_f^{(1)}}\right\} = 0 \tag{5.26}$$

If the fault is a phase B to phase C to ground fault, it holds that

$$V_f^{(0)} - V_f^{(1)} = 3R_f I_f^{(0)} \tag{5.27}$$

Therefore, the function can be written as follows:

$$f_{N+1}(X) = Im\left\{\frac{V_f^{(0)} - V_f^{(1)}}{I_f^{(0)}}\right\} = 0 \tag{5.28}$$

If the fault is a phase A to phase B to phase C fault, it holds that

$$V_f^{(1)} = R_f I_f^{(1)} \tag{5.29}$$

Hence, the function can be expressed by

$$f_{N+1}(X) = Im\left\{\frac{V_f^{(1)}}{I_f^{(1)}}\right\} = 0 \tag{5.30}$$

The $V_f^{(i)}$ and $I_f^{(i)}$ in (5.23), (5.25), (5.27) and (5.29) can be expressed as functions of the unknown fault location and line parameters based on analysis of the during-fault circuit. It notes that the series compensation device, which is installed at a fixed place along the transmission line, divides the line into two sections. As on which side the fault occurs is unknown to us, it is necessary to develop two subroutines addressing possible fault on either side.

- Subroutine 1: fault occurs to the left side of the compensator.

The equivalent circuit of *i*th sequence transmission line during the fault is shown in Figure 5.3. Based on section *PF* in Figure 5.3, the voltage at the fault point $V_f^{(i)}$ and current on the right-hand side of the fault point $I_{fR}^{(i)}$ can be calculated by the following equations:

$$V_f^{(i)} = \cos\text{h}\left(\gamma^{(i)} l_{PF}\right) V_P^{(i)} - Z_C^{(i)} \sin\text{h}\left(\gamma^{(i)} l_{PF}\right) I_P^{(i)} \tag{5.31}$$

$$I_{fR}^{(i)} = I_{fL}^{(i)} - I_f^{(i)} \tag{5.32}$$

where $V_P^{(i)}$ *and* $V_Q^{(i)}$ are the *i*th sequence voltage phasors at *P* and *Q* during fault, respectively, $i = 0,\ 1,\ 2$; $I_P^{(i)}$ *and* $I_Q^{(i)}$ are the *i*th sequence voltage phasors at *P* and *Q* during fault, respectively, $i = 0,\ 1,\ 2$, and the current on the left-hand side of the faulted point $I_{fL}^{(i)}$ can be obtained by the following equation:

$$I_{fL}^{(i)} = -\frac{1}{Z_C^{(i)}} \sin\text{h}\left(\gamma^{(i)} l_{PF}\right) V_P^{(i)} + \cos\text{h}\left(\gamma^{(i)} l_{PF}\right) I_P^{(i)} \tag{5.33}$$

Based on section *CQ* in Figure 5.3, the current on the left-hand side of the compensator can be calculated by the following equation:

$$I_{cL}^{(i)} = I_{cR}^{(i)} \tag{5.34}$$

where

$$I_{cR}^{(i)} = \frac{1}{Z_C^{(i)}} \sin\text{h}\left(\gamma^{(i)} l_{CQ}\right) V_Q^{(i)} + \cos\text{h}\left(\gamma^{(i)} l_{CQ}\right) I_Q^{(i)} \tag{5.35}$$

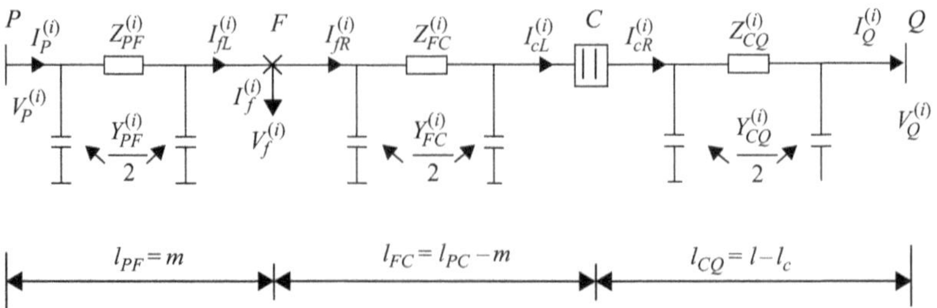

Figure 5.3 The equivalent π circuit of ith sequence transmission line during fault with fault to the left of the compensator

For section FC in Figure 5.3, it holds that

$$I_{cL}^{(i)} = -\frac{1}{Z_C^{(i)}} \sinh\left(\gamma^{(i)} l_{FC}\right) V_f^{(i)} + \cosh\left(\gamma^{(i)} l_{FC}\right) I_{fR}^{(i)} \tag{5.36}$$

By substituting (5.31), (5.32) and (5.34) into (5.36), the ith sequence fault current can be calculated by using the measurements at both ends:

$$I_f^{(i)} = \frac{\left[\left(1/Z_C^{(i)}\right)\sinh(\gamma^{(i)} l_{CQ}) V_Q^{(i)} + \cosh(\gamma^{(i)} l_{CQ}) I_Q^{(i)}\right] + \left(1/Z_C^{(i)}\right)\sinh(\gamma^{(i)} l_{FC}) V_f^{(i)}}{\cosh(\gamma^{(i)} l_{FC})}$$
$$+ \frac{1}{Z_C^{(i)}} \sinh\left(\gamma^{(i)} l_{PF}\right) V_P^{(i)} - \cosh\left(\gamma^{(i)} l_{PF}\right) I_P^{(i)} \tag{5.37}$$

- Subroutine 2: fault occurs to the right side the compensator.

Based on section FQ in Figure 5.4, the voltage at the fault point $V_f^{(i)}$ and the current on the left-hand side of the faulted point $I_{fL}^{(i)}$ can be calculated by the following equations:

$$V_f^{(i)} = \cosh\left(\gamma^{(i)} l_{FQ}\right) V_Q^{(i)} + Z_C^{(i)} \sinh\left(\gamma^{(i)} l_{FQ}\right) I_Q^{(i)} \tag{5.38}$$

$$I_{fL}^{(i)} = I_{fR}^{(i)} + I_f^{(i)} \tag{5.39}$$

where

$$I_{fR}^{(i)} = \frac{1}{Z_C^{(i)}} \sinh\left(\gamma^{(i)} l_{FQ}\right) V_Q^{(i)} + \cosh\left(\gamma^{(i)} l_{FQ}\right) I_Q^{(i)} \tag{5.40}$$

Based on section PC in Figure 5.4, the current on the right-hand side of the series compensator can be calculated by the following equation:

$$I_{cR}^{(i)} = I_{cL}^{(i)} \tag{5.41}$$

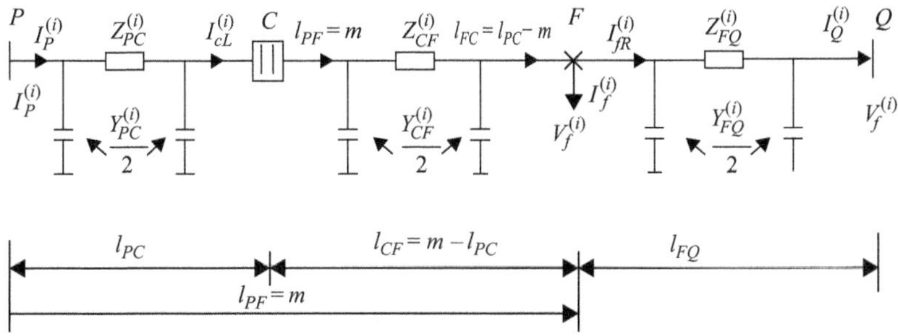

Figure 5.4 The equivalent π circuit of ith sequence transmission line during fault with fault to the right of the compensator

where

$$I_{cL}^{(i)} = -\frac{1}{Z_C^{(i)}} \sinh\left(\gamma^{(i)} l_{PC}\right) V_P^{(i)} + \cosh\left(\gamma^{(i)} l_{PC}\right) I_P^{(i)} \tag{5.42}$$

For section *CF* in Figure 5.4, it holds that

$$I_{cR}^{(i)} = \frac{1}{Z_C^{(i)}} \sinh\left(\gamma^{(i)} l_{CF}\right) V_f^{(i)} + \cosh\left(\gamma^{(i)} l_{CF}\right) I_{fL}^{(i)} \tag{5.43}$$

By substituting (5.38), (5.39) and (5.41) into (5.43), the *i*th sequence fault current can be calculated:

$$I_f^{(i)} = \frac{\left[-\left(1/Z_C^{(i)}\right)\sinh(\gamma^{(i)} l_{PC}) V_P^{(i)} + \cosh(\gamma^{(i)} l_{PC}) I_P^{(i)}\right] - \left(1/Z_C^{(i)}\right)\sinh(\gamma^{(i)} l_{CF}) V_f^{(i)}}{\cosh(\gamma^{(i)} l_{CF})}$$

$$-\frac{1}{Z_C^{(i)}} \sinh\left(\gamma^{(i)} l_{FQ}\right) V_Q^{(i)} - \cosh\left(\gamma^{(i)} l_{FQ}\right) I_Q^{(i)} \tag{5.44}$$

Therefore, the variables used in function vector $\boldsymbol{F}(\boldsymbol{X})$ are calculated, and $\boldsymbol{F}(\boldsymbol{X})$ is formulated as:

$$\boldsymbol{F}_{2i-1}(\boldsymbol{X}) = Re(f_i(\boldsymbol{X})), \quad i = 1, \ldots, N+1 \tag{5.45}$$

$$\boldsymbol{F}_{2i}(\boldsymbol{X}) = Im(f_i(\boldsymbol{X})), \quad i = 1, \ldots, N+1 \tag{5.46}$$

$$\boldsymbol{F}_{(2N+2)+(2i-1)}(\boldsymbol{X}) = x_{2i-1}, \quad i = 1, \ldots, 4N+4 \tag{5.47}$$

$$\boldsymbol{F}_{(2N+2)+2i}(\boldsymbol{X}) = x_{2i}, \quad i = 1, \ldots, 4N+4 \tag{5.48}$$

where *Re*(.) and *Im*(.) yield the real and imaginary part of their arguments, respectively.

The measurement vector $\boldsymbol{S}$ is formulated as follows:

$$\boldsymbol{S}_i = 0, \quad i = 1, \ldots, 2N+2 \tag{5.49}$$

$$\boldsymbol{S}_{(2N+2)+(2i-1)} = Re(\boldsymbol{M}_i), \quad i = 1, \ldots, 4N+4 \tag{5.50}$$

$$\boldsymbol{S}_{(2N+2)+2i} = Im(\boldsymbol{M}_i), \quad i = 1, \ldots, 4N+4 \tag{5.51}$$

where $\boldsymbol{M}$ is defined in (5.17).

The optimal estimate of $\boldsymbol{X}$ can be obtained by minimizing the cost function as defined in (5.16).

5.3.3 Load model monitoring

Load model monitoring is periodic validation of the load model used in simulation software such as the EMS software using both steady-state and disturbance data.

A load model is a set of equations that represents the relationship between voltage and/or frequency at a load bus and the real and reactive power consumed by load. Voltage dependency is the main concern, while frequency dependency is usually ignored. In this chapter, only voltage dependency is discussed.

Load model can be categorized as static and dynamic load models. A static load model is not dependent on time, and the model expresses a load's real and reactive powers at any time instant as functions of the bus voltage magnitude and/or frequency at the same instant. An example static load model is

$$P(t) = c[V(t)]^2 \tag{5.52}$$

where $P(t)$ is the power response at the time instant t, c is a constant coefficient, and $V(t)$ is the voltage magnitude at time instant t.

On the other hand, a dynamic load model expresses the relation, as functions of current and past voltage and/or frequency. An example of dynamic load model is

$$P(t) = c[V(t)]^2 + d[V(t-1)]^2 \tag{5.53}$$

where $P(t)$ is the power response at the time instant t, c and d are constant coefficients, $V(t)$ is the voltage magnitude at time instant t, and $V(t-1)$ is the voltage magnitude at time instant $t-1$.

When the load response reaches steady state, the steady-state power demand is a function of the steady-state voltage. This function defines the steady-state load characteristics known as voltage-dependent load models in load-flow studies. A sudden voltage change would cause an instantaneous power demand change. This change defines the dynamics/transient characteristic of the load. Most of the loads in a power system respond very quickly to the change in voltage. The steady state of the load response is achieved rapidly. Such cases can be analysed by using static models. However, when it comes to stability analysis, static load model is no longer adequate as load representation has critical effects in voltage stability.

Accurate load models are vital for many operation and analysis applications such as power-system stability analysis. The fact that power system instability is dynamic requires dynamics representation of the loads. If the load representation is not sufficiently accurate, significant mismatch will occur between the simulation results and the actual response of the load, which will affect the system stability assessment. In the past, power system instability usually occurs in weak systems. However, with increasing loads in the system, this problem has begun to appear in highly developed networks. Furthermore, power systems have been operated closer to their limits nowadays. As power system is more heavily stressed, reliable stability assessment is in greater demand than before, which in turn requires reliable load models.

For most utilities, load models have already been produced in the past. The accuracy of these load models, however, is not monitored based on actual measured power system data. In other words, these models are not regularly checked and validated through operation procedures. The studies based on these invalidated models may be unreliable. Besides, a lot of the existing models approximate loads as static models which neglect loads' dynamic properties. Usually, in the step of model conversion in the dynamic studies using software such as PSS/E, the real power of the load is represented as constant current load, and the reactive power of the load is represented as constant admittance load [7,12]. With growing penetration level of power electronics devices and motors, many dynamic phenomena cannot be replicated by existing models, such as air-conditioner stalling. Given that the dynamics characteristics of system load have a major impact on system stability, static load models are increasingly considered as inadequate for representing loads in dynamics studies. So, it is necessary to monitor the load models and calibrate them if needed.

PMU measurements enable on-line monitoring of load models. The load response captured during steady-state operation can be used to validate model's steady-state performance, and the load response captured during disturbances can be used to validate the dynamic property of the existing models. If serious mismatch is observed, then the model needs to be re-constructed.

5.3.3.1 Online load model validation

Online model verification and validation can be accomplished in the following steps:

1. Data collection: install PMU at load aggregation point, and obtain measurement data from PMUs.

   If field measurement devices have not been installed yet, monitoring devices should be first selected and installed. At medium and high voltage levels, the power system loads have to be aggregated to obtain manageable models that are appropriate for simulation and analysis. As a matter of fact, in the context of bulk power system studies, the term 'load' refers to an aggregated load of a feeder, substation or major delivery point on a system; and transmission and sub-transmission substations are usually the location of choice for load aggregation [13]. To achieve the most accurate estimation/validation result, the PMUs should be placed at the load aggregation points. In this way, the behaviour of other system components can be filtered. If PMUs are not installed at aggregation point, the collected data would reflect the combined performance of aggregated load and other asset that is between the load and the PMU. To use these data for validation, the models for the other components have to be included in the analysis.

   The quantities to be recorded are voltage and current phasors, based on which the real and reactive power can be calculated. The quantities should be recorded in two manners: continuous and exception-triggered. The continuous recording is used to validate the steady-state load model. The exception-triggered recording is used to validate dynamics load model; the data should cover the exceptional events such as voltage deviations and current deviations and should be recorded in a much higher sampling rate.
2. Event identification: for validation of dynamics property, the dynamic disturbances that are suitable for replication should be identified; for validation of steady-state responses, the steady-state normal operation data can be used.

   The extraction of steady-state data is straightforward: select a certain instant during steady state and record the measurements.

Disturbances such as system faults and loss of generation usually cause large voltage variations and can be used for dynamic model validation. In the context of online model monitoring, natural disturbance data rather than testing-based data are used. The use of natural disturbance data is advantageous from a technical and economical point of view.

In addition to the disturbance data, normal operation data could also be used in load modelling. First, the steady-state data can be used to validate model's steady-state response. Second, daily normal operation involves voltage variations as well, and these voltage variations can be used to validate model's dynamic response. Example voltage variations are switching on/off capacitors related to reactive power compensation, voltage step change due to tap changer operations, etc. The voltage variations during normal operation are relatively small, but may occur more frequently than large transients. These voltage variations may result in changing power consumed by the load. Such data can be employed to validate the dynamics property of the load. These data can also help to determine the load model structure.

3. Event replication and results comparison: use measured voltage as stimulus and obtain the response based on its model. Then, compare the simulated response with the actual response collected by PMUs and determine the model's validity.

   Suppose that, the voltage and current at the bus connecting the component to the external system are obtained. To validate the model of the component, the voltage measurements are used to calculate the current of the load. Then, the calculated currents are compared with the measured currents to determine whether the load model is acceptable, as shown in Figure 5.5.

   Model validity is confirmed if the simulated response to a disturbance reasonably matches the measured response to a similar disturbance. The predicted outputs are compared with the measured data (active and reactive power). If there are significant differences between the load's actual behaviour and the model's prediction, calibration of model is required. The differences are to be used to tune the model. More appropriate model structures will be selected, and parameters will be adjusted to obtain the best match between the simulated and measured responses. The revised model should be revalidated.

4. Generalization capability test: repeat validation procedure for various disturbances over a range of different conditions.

   Generalization capability of a model is the model's ability to reflect the load's behaviour under unseen conditions. A load model built based on one set of measurements may not be able to fit the load under other conditions. Therefore, the model's generalization capability needs to be examined as well. Assume measurements under different conditions are available, and these measurements are regarded as testing data. Use these testing data to the model as stimulus and obtain the model's validity under these conditions. Based on the model's performance, a rough estimation of generalization capability can be made.

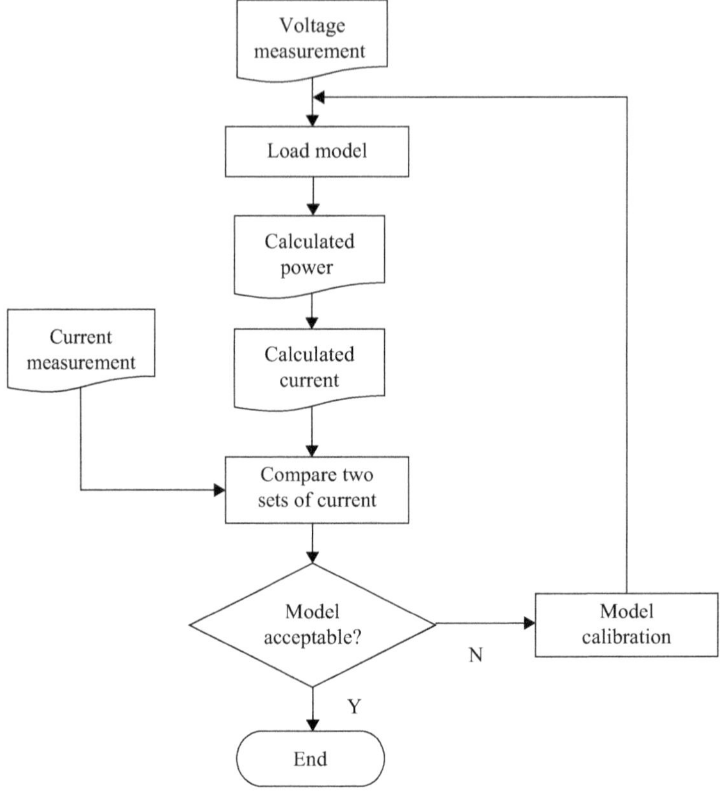

Figure 5.5 Flowchart of event replication and results comparison in load model validation

5.3.3.2 Discussion on load model validation procedure

Although a lot of work has been done on developing load model validation procedure, there are still improvements that can be made.

First, both 'verification' and 'validation' should be included in the procedure. Verification addresses whether the model implements the assumption correctly, and validation examines whether the assumption which have been made are reasonable. In the context of load modelling, the assumption refers to the load structure and model parameters. Therefore, verification means to check if the model is implemented correctly based on the assumed load type and model parameters. Then the validation procedure checks if the assumed type and parameters are correct.

Second, obtaining a confidence level about the model validity based on statistical analysis is more appropriate than a simple binary pass/fail conclusion. The accuracy and uncertainty of a model can be quantified by computing an index called 'accuracy index (AI)' and the respective 95 per cent confidence intervals (CIs). Significance test accepts or rejects a model based on the 95 per cent CI. Such statistical validation result will enable both the model developer and the model user to have quantitative support to decide whether the model is adequate for its intended purpose of use.

Assume in a recorded event, there are N time instants. At each time instant, an actual output value and a simulated output value is recorded. The AI that measures the closeness between the actual and simulated outputs can be calculated by the following equation:

$$\text{Accuracy index (AI)} = \left[\frac{X_{im}(\theta)}{X_{is}(\theta)} - 1\right]^2, \quad i = 1, \ldots, N \tag{5.54}$$

where X_{im} is the ith measured output, X_{is} is the ith simulated output, and N is the total number of time instants. Thus, the overall accuracy for this event can be quantified by

$$\sqrt{\frac{1}{N}\sum_{i=1}^{N}\left[\frac{X_{im}(\theta)}{X_{is}(\theta)} - 1\right]^2} \tag{5.55}$$

The uncertainty level of accuracy is to be expressed by an interval within which the true values of the AI are expected to lie with a specified level of confidence. To calculate the CI, first, the mean value of AI $\overline{\text{AI}}$, and the standard deviation for the mean value, σ should be determined.

Then, the margin of error is found by

$$\text{MOE} = \text{Critical value} \times \sigma \tag{5.56}$$

where the critical value for 95 per cent confidence level is 1.96 [14]. Other confidence levels can be chosen as well, and corresponding critical values should be calculated accordingly.

Finally, the CI is calculated as

$$\text{CI} = \overline{\text{AI}} \pm \text{MOE} \tag{5.57}$$

If the simulation model exactly represents the characteristics of the system or component under study, the simulated outputs will be the same as measured outputs, i.e. the difference between the simulated and measured output is zero. CI gives a plausible range of such difference. All values in the CI are plausible values, whereas values outside the interval are rejected as plausible values for the difference. Therefore, if the CI does not include zero, it can be concluded that the measured outputs are significantly different from the simulated outputs, and the load model is considered as an invalid model.

In addition to adopt 'verification & validation' and statistical judgement, the types of invalidity can be explored to help model calibration. There are a few invalidity types that could account for the mismatch between simulated response and the measured response:

- Invalidity Type I: external conditions in the simulation case are erroneous.
- Invalidity Type II: loads are not aggregated appropriately.

- Invalidity Type III: load model parameters are not accurate.
- Invalidity Type IV: the load structure and model parameters are correct, but the model does not implement the assumptions correctly.
- Invalidity Type V: the load model is not applicable to this particular case.

Based on different invalidity types, load model should be calibrated accordingly.

A few reasons could result in Invalidity Type I, such as bad models of other power system elements in the system and inaccurate estimation of power components from measurements. Invalidity Type I should be examined primarily if the model is considered not accurate. This type of invalidity should be excluded primarily.

If the load aggregation site is not appropriate, or the load model structure is not representative of the aggregated loads, the loads are said to be not aggregated well, which results in Invalidity Type II.

Invalidity Type III is straightforward: inaccurate model parameters would almost certainly result in inaccurate responses. From a practical point of view, the parameter estimation problem can be regarded as a curve-fitting problem. Nonlinear least squares based optimization algorithms can be used for parameter estimation.

Invalidity Type IV can be checked through the verification process. The computer code that implements the model requires thorough examination.

Invalidity Type V addresses the model's generalization capability, i.e. the model's ability to predict load's actual response of unseen events. The load characteristics are constantly changing over time. In extreme situations, such as those near the voltage collapse point, the load's characteristic is dramatically different from that during normal operation. Hence, it is acceptable if the model does not work for some particular situations. A dedicated model representing such kind of situation can be built and used whenever such situation occurs.

If a model needs calibration, the load model structure and parameters should be re-estimated.

5.3.4 System-level model monitoring

If sufficient synchronized measurements are available in a power grid, the validation of individual component models can be extended to system-level model monitoring. To validate the system model's steady-state performance, the results of power flow simulation are compared with measured voltage phasors and current phasors.

To validate the model's dynamics performance, simulated responses to a recorded event are compared with the measured responses. Before dynamics simulation can be run, the initial condition of the system model needs to be set. The initial condition should match the observed system condition immediately prior to the event. This task can be done by developing the power flow case. The power dispatch of generators, voltage set-point of generators, network topology, transformer tap position, status of switched reactive compensators, and load profiles should be set.

5.4 Wide-area fault location based on PMU measurements

5.4.1 Introduction

Electric power is delivered from generation stations to distribution substations through transmission lines. Transmission lines occasionally suffer from faults which are generally created by random and unpredictable events such as lightning, tree touch, animal contact, insulation breakdown due to animal waste and severe weather. When a fault occurs on a line, it is very important for the utilities to identify the fault location as quickly as possible to speed up maintenance work and improve the system reliability. If the fault location cannot be pinpointed quickly, system restoration cannot be performed and the line will be kept out of service, which will result in considerable economic losses. Therefore, accurate and fast transmission-line fault location plays a critical role in quick power system restoration.

Various algorithms for pinpointing faults have been proposed in the past. Algorithms for both single and parallel transmission lines are proposed to help utilities to perform quick restoration. Most of the existing methods require measurements at either one or both terminals of the line. However, such measurements may not always be available. A method that is able to accurately pinpoint the fault solely based on remote measurements is highly desirable. This session presents an accurate fault location methods based on wide-area sparse measurements [15]. The wide-area measurements are provided by PMUs, and these measurements may be captured from buses far away from the faulted transmission line. Even if the measurements are available from a single remote bus, the fault can still be reliably located. If measurements from multiple buses are available, optimal estimation theory is exploited to make the most of available measurements for enhanced estimation, and possible bad measurements can be filtered. The method is able to be applied to both transposed and untransposed transmission lines as long as the line parameters are known.

5.4.2 Fault location algorithm

The method is based on the bus impedance matrix concept [15] and is described as follows. To represent the fault, a fictitious bus r at the assumed fault location is added. The transfer impedances between the fictitious bus and the buses with measurement devices can be expressed as functions of the fault location. The driving point impedances of the faulted nodes, as well as the transfer impedances between the faulted nodes can also be represented. Based on short-circuit analysis theory, the during-fault voltage and current at the fictitious bus can be derived in terms of the according transfer impedance and driving point impedance. Consequently, they can be written as functions of the fault location. Hence, the reactive power consumed at the fictitious bus can be revealed in terms of fault location. By utilizing the fact that the fault resistance only consumes real power, the fault location will be found. Assume, the voltage measurements at bus k are available.

It can be shown that the reactive power consumed by the fault resistance can be computed as

$$Q = \text{Imag}\left\{-E_r\left[\left(Z_{kr}{}^T Z_{kr}\right)^{-1}\left(Z_{kr}{}^T \Delta E_k\right)\right]^*\right\} \tag{5.58}$$

where Imag{.} returns the imaginary part of its argument, and $(.)^*$ denotes the complex conjugate, E_r is the during-fault voltage at the fault bus, Z_{kr} contains the transfer impedances between the bus k and the fault bus, and ΔE_k is the superimposed voltage at the fault bus. It notes that E_r and Z_{kr} are functions of the fault location [15]. As the fault resistance only consumes real power, (5.58) should equal to zero. The fault location can then be solved by the Newton–Raphson technique. It is noted that the algorithm is independent of fault resistance.

If the faulted line is a double-circuit line, a fictitious bus containing six nodes is to be added. Suppose the line is between buses p and q. The shunt admittance and the series impedance of the line between p and the fault location are y_1 and z_1, respectively. The shunt admittance and the series impedance of the line between q and the fault location are y_2 and z_2, respectively. The transfer impedances between nodes on bus k and the faulted nodes can be derived as

$$Z_{k_i r} = \left(\frac{y_1}{2}+\frac{y_2}{2}+z_1^{-1}+z_2^{-1}\right)^{-1}\left(z_1^{-1}\left[Z_{k_i p}, Z_{k_i p}\right]^T + z_2^{-1}\left[Z_{k_i q}, Z_{k_i q}\right]^T\right),\quad i = 1,2,\ldots,6 \tag{5.59}$$

where $Z_{k_i p}$ contains the transfer impedances between the ith node on bus k and the nodes on bus p, and $Z_{k_i q}$ contains the transfer impedances between the ith node on bus k and the nodes on bus q.

The transfer impedances and the driving point impedances related to the faulted nodes are expressed as

$$Z_{rr_i} = \left(\frac{y_1}{2}+\frac{y_2}{2}+z_1^{-1}+z_2^{-1}\right)^{-1}\left(z_1^{-1}\left[Z_{pr_i}, Z_{pr_i}\right]^T + z_2^{-1}\left[Z_{qr_i}, Z_{qr_i}\right]^T + v_i\right),\quad i = 1,2,\ldots,6 \tag{5.60}$$

where Z_{pr_i} contains the transfer impedances between the bus p and the ith node on bus r, Z_{qr_i} contains the transfer impedances between the bus q and the ith node on bus r, and v is a six-by-six identity matrix whose ith column is denoted as v_i.

Compared to [15], this double-circuit line fault location method obviates the necessity of specifying which circuit of the two circuits is faulted.

In practical power systems, not all fault conditions remain unchanged during fault. An evolving fault refers to a fault where the faulted phase or phases change over time. Locating evolving faults is challenging due to the change in fault type after the fault inception. The presented fault location algorithm in this chapter, however, is immune to fault type. Therefore, to use the proposed algorithm to locate evolving faults, the need of detection of changing fault types is omitted. By applying a moving window to read the phasors from the beginning to the end of the fault data, continuous estimation of fault location can be achieved, regardless of the changing fault type. A window refers to the number of sample points analysed at a certain time instant. The running of the window is triggered by the fault occurrence and terminated at the end of the fault. A locus of estimation can be drawn as the window moves. As the transition of fault conditions, such as fault resistance and fault type, it does not affect the arguments in (5.58); the need of detecting and classification of evolving fault is no longer a necessity.

5.5 Conclusion

The unique advantages of PMUs have brought numerous opportunities to enhance power grid monitoring and protection. This chapter provides readers with a glance of possible improvements. In addition to the topics discussed above, many other applications can be enhanced by PMUs, such as frequency stability monitoring, power oscillations and voltage monitoring. In summary, PMU has potential to greatly improve system operators' ability to monitor the grid, and the real-time operation and control can be conducted with more confidence. In the future, increasingly, more PMU-based application will be fully deployed, tested and performed by utilities.

References

[1] IEEE Standard C37.118 2005, *IEEE Standard for Synchrophasors for Power Systems*, New York, NY, USA, 2006.

[2] IEEE Guide C37.244-2013, *Guide for Phasor Data Concentrator Requirements for Power System Protection. Control and Monitoring*, New York, NY, USA, 2013.

[3] Abur A., Gomez-Exposito A. *Power System State Estimation Theory and Implementation*. New York: Marcel Dekker, Inc.; 2004.

[4] Monticelli A. 'Electric power system state estimation'. *Proceedings of the IEEE*. 2000, vol. 88(2), pp. 262–282.

[5] Huang Y.-F., Werner S., Huang J., Kashyap N., Gupta V. 'State estimation in electric power grids: meeting new challenges presented by the requirements of the future grid'. *IEEE Signal Processing Magazine*. 2012, vol. 29(5), pp. 33–43.

[6] North American Synchrophasor Initiative (USA). *Model Validation Using Phasor Measurement Unit Data*. vol. 1. Richland: NASPI; 2015.

[7] North American Electric Reliability Corporation (USA). *Power System Model Validation*, vol. 1. Princeton: NERC; 2010.

[8] U.S.-Canada Power System Outage Task Force (USA and Canada). *Final Report on the August 14, 2003 Blackout in the United States and Canada: Causes and Recommendations.*, vol. 1. Washington: DOE; 2004.

[9] He R., Ma J., Hill D. 'Composite load modeling via measurement approach'. *IEEE Transaction on Power Systems*. 2006, vol. 21(2), pp. 663–672.

[10] Du Y., Liao Y. 'Online estimation of power transmission line parameters, temperature and sag'. *North American Power Symposium*; Boston, USA, Aug 2011. New York: IEEE; 2011. pp. 1–6.
[11] Liao Y., Kezunovic M. 'Optimal estimate of transmission line fault location considering measurement errors'. *IEEE Transaction on Power Delivery*. 2007, vol. 22(3), pp. 1335–1341.
[12] *PSS/E User Manual*. Raleigh: *Siemens Power Transmission & Distribution, Inc.*; 2007.
[13] Electric Power Research Institute (USA). *Measurement-Based Load Modeling*, vol. 1. Palo Alto: EPRI; 2006.
[14] Utts J.M., Heckard R.F. *Mind on Statistics*. 4th edn. Boston: Brooks/Cole, Cengage Learning; 2012.
[15] Jiao X., Liao Y. 'Accurate fault location for untransposed/transposed transmission lines using sparse wide area measurements'. *IEEE Transaction on Power Delivery*. 2016, vol. 31(4), pp. 1797–1805.

Chapter 6

Fault monitoring, detection, and correction using synchrophasor measurements in modern power systems

Anupam Mukherjee[1] *and Prakash Ranganathan*[1]

6.1 Clustering methods and openPDC interface

PMUs (phasor measurement units) with basic features can generate up to a least of 1.5 TB of data in a month. For complete observability and better situational awareness of the grid, there is a need to include multiple PMUs for a large-scale grid. Thus, it is evident that huge amount of data will get generated in the long run. In order to extract useful information from the huge data to control smart grid operations, a proper data handling and mining approach is required. This research focuses on the use of two common data clustering methods, namely density-based spatial clustering of applications with noise (DBSCAN) and k-means to power grid data sets. In addition, a novel, hybrid multitier clustering approach has been developed and integrated. This section describes the software setup to acquire and handle streaming datasets from PMUs through an openPDC software framework. The rest of sections are organized as (1) openPDC setup and data extraction process, (2) application of k-means clustering, (3) application of DBSCAN clustering, and (4) development of a novel method, multitier k-means clustering method.

6.1.1 openPDC setup and data extraction process

The openPDC is an open-source software framework that is maintained by the grid protection alliance [1–3]. It is a phasor data concentrator (PDC) software that allows utilities to connect multiple PMUs and extract time-synchronized phasor measurements. These data can then be archived and used at a later stage for any postevent analysis. The openPDC software also provides simulated PMU data that can be used for developing and testing streaming algorithms. This test dataset is actually a phasor data retrieved from a real PMU that is run in a cyclic order when it is used for testing purpose. The data are time-stamped with coordinated

[1]Electrical Engineering Department, University of North Dakota (UND), Grand Forks, ND 58201, USA

universal time (UTC) and archived as a "*.d" file in a source folder of the software. The data extraction stage of the data analysis process includes, collecting data from various PMUs, storing it in a PDC and access them through openPDC software. This process is essential for the later part of data analytics work. At this stage, data stored in various file formats are converted to a common ".csv" format. This data format serves as the source for the data analysis phase. For my research, streaming data from only one PMU is considered for investigating the clustering methods. It is a continuous stream of phasor data generated from the openPDC software. These data remain stored with its time-stamp information in the archive folder in a computer. The format of the stored data is ".d". These are shared object code, something like .obj files created by Visual Studio in windows [4]. The data analysis software codes are developed and executed in the MATLAB® environment [5]. The openPDC is a software created in C# language and developed in the Visual Studio environment. To bring clarity between the two data formats, there is a need for a common data format. It was realized that ".csv" would be the best fit for storing large amounts of data. A C#-based software code was developed to convert the ".d" files into ".csv" format. The code is present in Appendix A. The code developed will allow the user to select data based on input time periods (mm/dd/yyyy hh:mm:ss:ffff). Upon receiving time period, the code can be run to correct the formats. The rest of the process is automatic. Once the data get converted, the program would trigger the MATLAB code to run clustering algorithms on historical datasets. The process automatically forms clusters of good, bad (noise) and bordering data points. Figure 6.1 gives a diagrammatic representation of this process.

Upon clustering the data into different groups, an average value from the cluster is represented as a sample for decision-making. Depending on values of each cluster, alarms are generated to alert the grid operators to take appropriate actions. The entire process of data extraction to alarm generation has been summarized in Figure. 6.2.

6.1.2 *Application of* k*-means clustering*

k-means clustering is a kind of representative-based clustering method [6]. Given a *d*-dimensional space having *n* data points, a representative-based clustering method works toward grouping the *n* data points into *k* different clusters [6]. For each cluster, there is a point, usually the mean or average value, which represents the

Figure 6.1 Data extraction and conversion stage

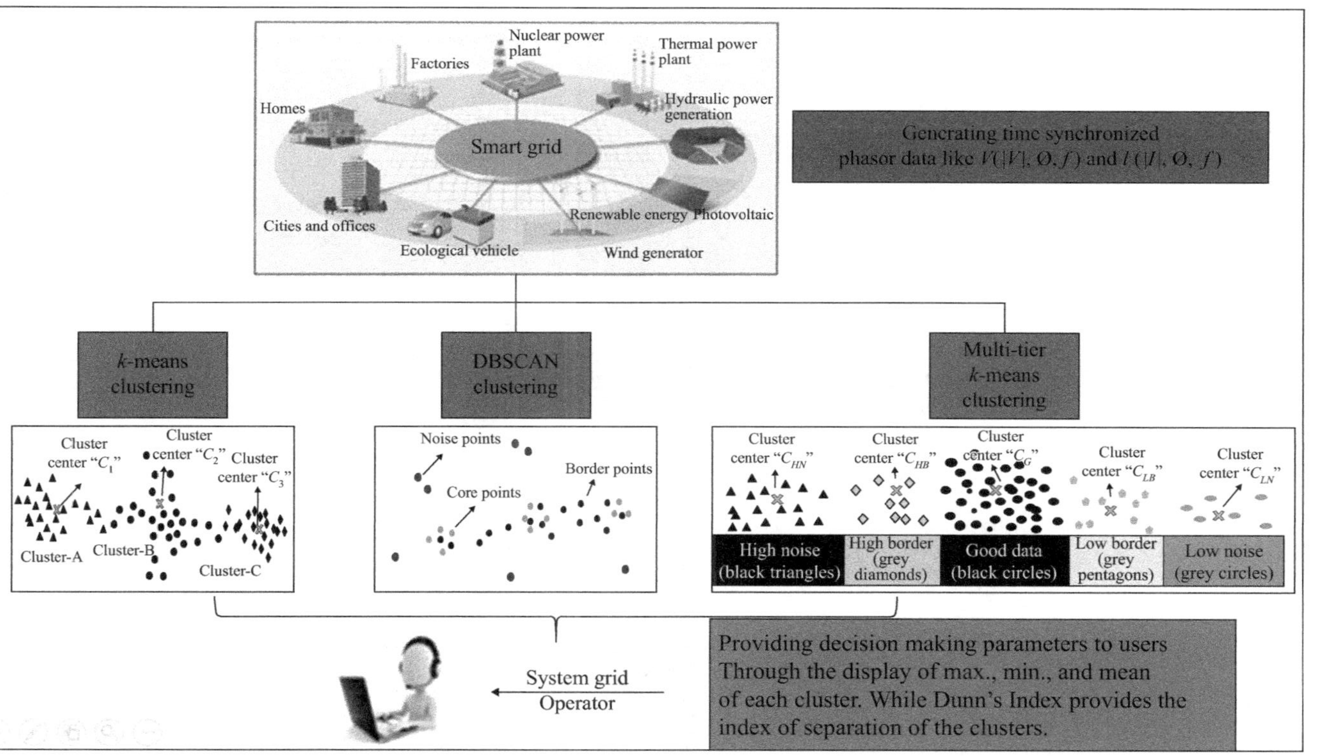

Figure 6.2 Data mining and visualization providing basis for decision-making

statistical information for the entire cluster. Thus, we can mathematically represent the clustering algorithm as follows:

If $d = \{x\}_{i=1}^{n}$, is the d-dimensional space, having k clusters, such that each cluster is represented as $C = \{C_1, C_2, \ldots, C_k\}$, then the centroid of each cluster is given by

$$\mu_i = \frac{1}{n_i} \sum_{x_j \in C_i} \{x_j\} \tag{6.1}$$

where $n_i = |C_i|$ is the number of points in the cluster C_i.

Using simple brute-force method or exhaustive algorithm, a good clustering method will be to simply generate all possible partitions of the n data points into k clusters, then calculate an optimization score for each of them and retain the one that performs the best with respect to other methods. The number of possibilities to partition n points into k nonempty and disjoint parts is given by

$$P(n, k) = \frac{1}{k!} \sum_{t=0}^{k} (-1)^t \binom{k}{t} (k - t)^n \tag{6.2}$$

k-means is a greedy iterative approach to find clustering that minimizes the sum of squared error (SSE), where the SSE is represented mathematically as [6]:

$$\text{SSE}(C) = \sum_{i=1}^{k} \sum_{x_j = C_i} ||x_j - \mu_i||^2 \tag{6.3}$$

where x_j is the actual data and μ_i is the mean value.

k-means initializes the cluster means by randomly generating k points in the data space. This is done by generating values randomly within the range for each dimension. Each iteration of k-means consists of the following two steps: (1) cluster assignment and (2) centroid update.

Given the k cluster means, in the cluster assignment step, each point $x_j \in d$ is assigned to the closest mean, which induces a clustering with each cluster C_i comprising points that are closer to μ_i than any other cluster mean. That is, each point x_j is assigned to cluster C_{j*}, where:

$$j^* = \arg \min_{i=1}^{k} \left\{ ||x_j - \mu_i||^2 \right\} \tag{6.4}$$

Given a set of clusters C_i, $i = 1, \ldots, k$, in the centroid update step, new average values are computed for each cluster from the point in C_i. The cluster assignment stage and centroid update steps are carried out iteratively until we reach a fixed point or local minima is reached. Practically speaking, it can be assumed that k-means has converged, if the centroids do not change from one iteration to the next.

For instance, the iterations can be stopped subject to the following condition being true:

$$\sum_{i=1}^{k} \left\|\mu_i^t - \mu_i^{t-1}\right\|^2 \leq \epsilon \tag{6.5}$$

where $\epsilon > 0$ and $t =$ the iteration number

The pseudocode for k-means is given in Figure 6.3.

Input: $X = \{x_1, x_2, \ldots, x_n\}$ (set of entities to be clustered)

k (number of clusters)

maxIters (limit of iterations)

Output: $C = \{c_1, c_2, \ldots, c_k\}$ (set of cluster centroids)

$L = \{1(e)|e = 1, 2, \ldots, n\}$ (set of cluster labels of E)

The k-means clustering algorithm was applied on the data collected from openPDC. The idea was to create a set of three clusters that will group the entire data into acceptable (good), nonacceptable (bad) and border (ok) values. The results obtained from the k-means clustering method are discussed in Section 6.3. Figure 6.4 gives a diagrammatic representation of cluster formation in k-means algorithm. Figure 6.4 shows the formation of three clusters based on user input. The centroid values act as user input. The algorithm starts with initial, random values of centroids, C_1, C_2, and C_3. With every iteration, the centroid values get updated, $C_i \rightarrow C_i^{'} \rightarrow C_i^{''} \rightarrow C_i^{'''}$, where $i \in (1, 2, 3)$. The centroid values are updated on the basis of threshold conditions. If the condition does not satisfy, then the centroid value is no longer updated, which basically implies that no other values from the

```
foreach ci ∈ C do
        ci = xj ∈ X (e.g. Random Selection)
end
foreach xi ∈ X do
        l(xi) = argminDistance(xi,cj) j ∈ {1,2,...,k}
end
If (changed = false) then
    Iter = 0;
end
repeat
        foreach ci ∈ C do
                Update Cluster (ci)
        end
        foreach xi ∈ X do
                minDist = argminDistance (xi,cj) where j ∈ {1,2,...,k}
                If (minDist ≠ l(xi) then
                   l(xi) = minDist;
                   changed = true;
                end
        end
        iter++;
until changed = true and iter ≤ maxIters;
```

Figure 6.3 Pseudocode for k-*means clustering*

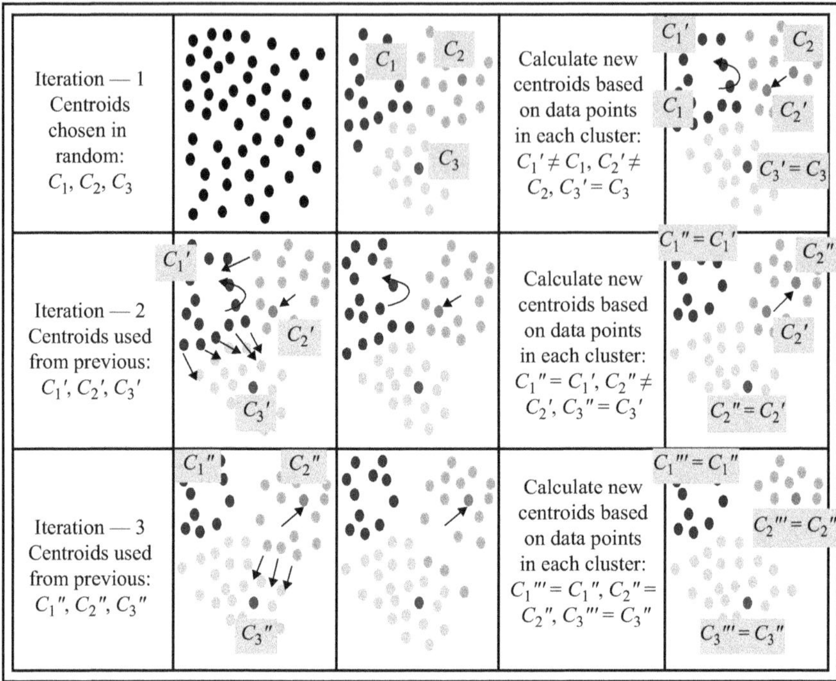

Figure 6.4 k-*means cluster formation*

input dataset can be fit into the cluster. In Figure 6.4, we can see the centroid $C_1^{''} = C_1^{'}$ for the cluster. This value remains unchanged after the first iteration as no other data points are included in the cluster for the rest of the k-means process.

6.1.3 Application of density-based spatial clustering of applications with noise clustering

DBSCAN of data points focuses on the density of points to determine the clusters instead of using only the distance between points [6]. A ball of radius ε is defined around a point $x \in R^d$, called the ε-neighborhood of x, as follows:

$$N_\varepsilon(x) = B_d(x, \varepsilon) = \{y \,|\, \delta(x, y) \leq \varepsilon\} \tag{6.6}$$

Here, $\delta(x, y)$ represents the distance between points x and y, which is usually assumed to be Euclidean distance, that is, $\delta(x, y) = ||x - y||_2$ [6]. Now, for any point $x \in d$, it is considered that x is a core point, if there are at least *minpts* points in its ε-neighborhood. That means, x is a core point if $|N_\varepsilon(x)| \geq minpts$, where *minpts* is a user-defined local density threshold. A border point is defined as one that does not meet the *minpts* threshold, that is it has $|N_\varepsilon(x)| < minpts$, but it belongs to the ε-neighborhood of some core point z, that is, $x \in N_\varepsilon(z)$. Finally, if a point is neither a core nor a border point, then it is called a noise point or an outlier.

It is considered that a point x is density reachable from y, if there exist a chain of points, $x_0, x_1, \ldots, x_i$, such that $x = x_0$ and $y = x_i$ and x_l is directly density reachable from x_{l-1} for all $l = 1, 2, \ldots, i$. In Figure 6.5, point r is density reachable from o, whereas point q is directly density reachable from m. The pseudocode for a DBSCAN clustering algorithm is given in Figure 6.6.

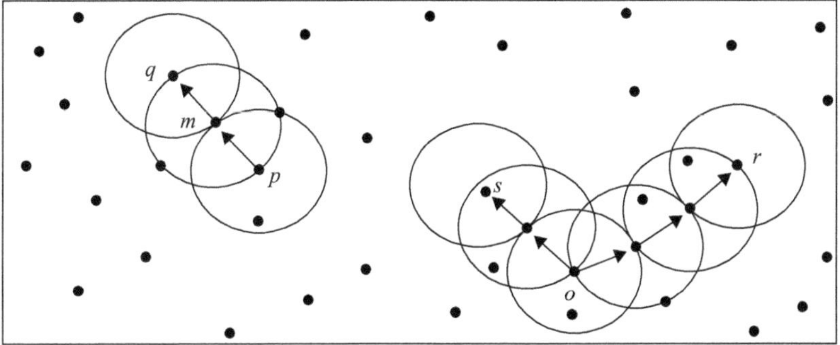

Figure 6.5 Growth of clusters based on their density [7]

```
DBSCAN (Input Set X, Pt, D)
        foreach xi in the input set do
                If (xi is not in any cluster) then
                  If (xi is a core point) then
                        Generate a new clusterID.
                        Label xi with clusterID.
                        expandCluster (xi, X, Pt, D, clusterID)
                  else
                        Label (xi, NOISE)
                  end
                end
        end
end
expandCluster (xi, X, Pt, D, clusterID)
        put xi in seed queue
        While the queue is not empty
                extract c from the queue (where c ∈ X and c ≠ xi)
                retrieve the neighborhood (eps) of c
                If (there are at least minPts neighbors) then
                  foreach neighbor n
                        If (n labeled NOISE) then
                          Label n with clusterID
                        elseIf (n is not labeled) then
                               Label n with clusterID
                               Put n in the queue
                        end
                  end
                end
        end
end
```

Figure 6.6 Pseudocode for DBSCAN clustering

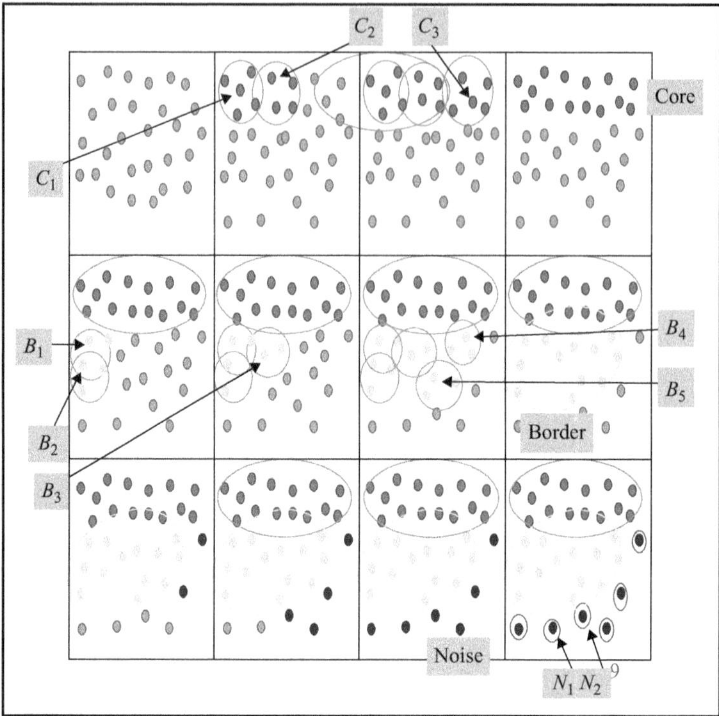

Figure 6.7 DBSCAN cluster formation

Input: $X = \{x_1, x_2, \ldots, x_n\}$ (set of entities to be clustered)

eps = minimum distance between two points to be clustered (D).

MinPts = minimum number of points that should be in a cluster to consider it as a border group (Pt).

Output: $L = \{1(x)|x = 1, 2, \ldots n\}$ (set of cluster labels of x)

Figure 6.7 gives a diagrammatic presentation of DBSCAN clustering. From Figure 6.7, it is evident how, that the clusters C_1, C_2, and C_3 that consists of core points (i.e., satisfies both conditions of minpts. and eps.), together form a core group of data points that have values close to each other, whereas the border group consists of points satisfying the border point properties (i.e., eps, not minpts.). Finally, the isolated points form the noise group.

6.1.4 Development of a novel method, multitier k*-means clustering method*

The multitier k-means algorithm can be described in two stages. The first stage is similar to k-means clustering, in which three clusters are formed based on dynamic data. In the second stage, the clusters are re-adjusted to a pre-defined threshold value set by utilities.

6.2 Need for a hybrid method

The purpose of introducing the concepts of data mining into power systems is to aid in taking critical decisions by exploring data. Thus, allowing the smart grid to be reliable and efficient in generating and delivering power. This research focuses on application of data-mining techniques to create a robust monitoring and alarm system. Thus, serving as a software interface can provide services like data mining, data visualization, intrusion detection and mitigation, alert services, topology capturing and state estimator, and forecasting. A diagrammatic representation of the entire software framework is presented in Figure 6.8.

The current research work deploys common clustering methods, namely *k*-means and DBSCAN on PMU data to study its grouping and its effects in capturing anomalies. A comparative analysis of this study has been presented in the results section, in Section 6.3. The results obtained demonstrated that there is a need for such clustering methods to accurately track the changes in load values on the transmission lines. These techniques can also detect any contingency scenarios and alert the system operators to take corrective measures to mitigate the problem. A hybrid data clustering (multitier *k*-means) method was developed to improve performance in capturing anomalies. The method has been discussed in the following section.

6.2.1 Development of multitier k*-means method*

There are several variations of *k*-means discussed in literatures, where the original *k*-means algorithm has been modified to suit the type of data [8–11]. They include *k*-medians, *k*-mediods, *k*-means ++, weighted *k*-means, fuzzy *c*-means, Kd-trees, and spherical *k*-means. *k*-means has been popular for its speed and accuracy of clustering data objects. Although DBSCAN and *k*-means clustering algorithms proved to accurately classify normal steady-state conditions (without faults), they failed to capture faults. It was mainly because these algorithms could not accurately identify the border lying data points, which are critical to generate alarm. Although DBSCAN could correctly track the shift in data points due to changing demand in the system, the computation time was too slow. On the other hand, *k*-means, although being one of the fastest clustering algorithms, could not correctly identify the good data points from the border lying ones, as far as the phasor data were concerned. Hence, there is a need for a method that capture both steady-state (normal) and abnormal conditions of the grid. This is carried out using multitier *k*-means method.

The multitier *k*-means algorithm can be described in two stages. The first stage is similar to *k*-means clustering, in which three clusters are formed. The algorithm begins with three initial clusters. Each of the three clusters represents normal, abnormal (faulty) data in low threshold, and abnormal (faulty) data in the high threshold ranges. The normal data indicate ideal steady-state scenario, that is, the power system is not under any stress. The high and low fault data clusters indicate that a contingency scenario (a line-to-ground fault) was recorded where magnitude of voltage (V) or current (I) or frequency (F) reached very large or very small values. The algorithm allows the user with two options. Option 1 allows user to choose the centroid values of the three clusters. Option 2 chooses centroid values automatically. In the case of automatic choice, the user is prompted to enter the

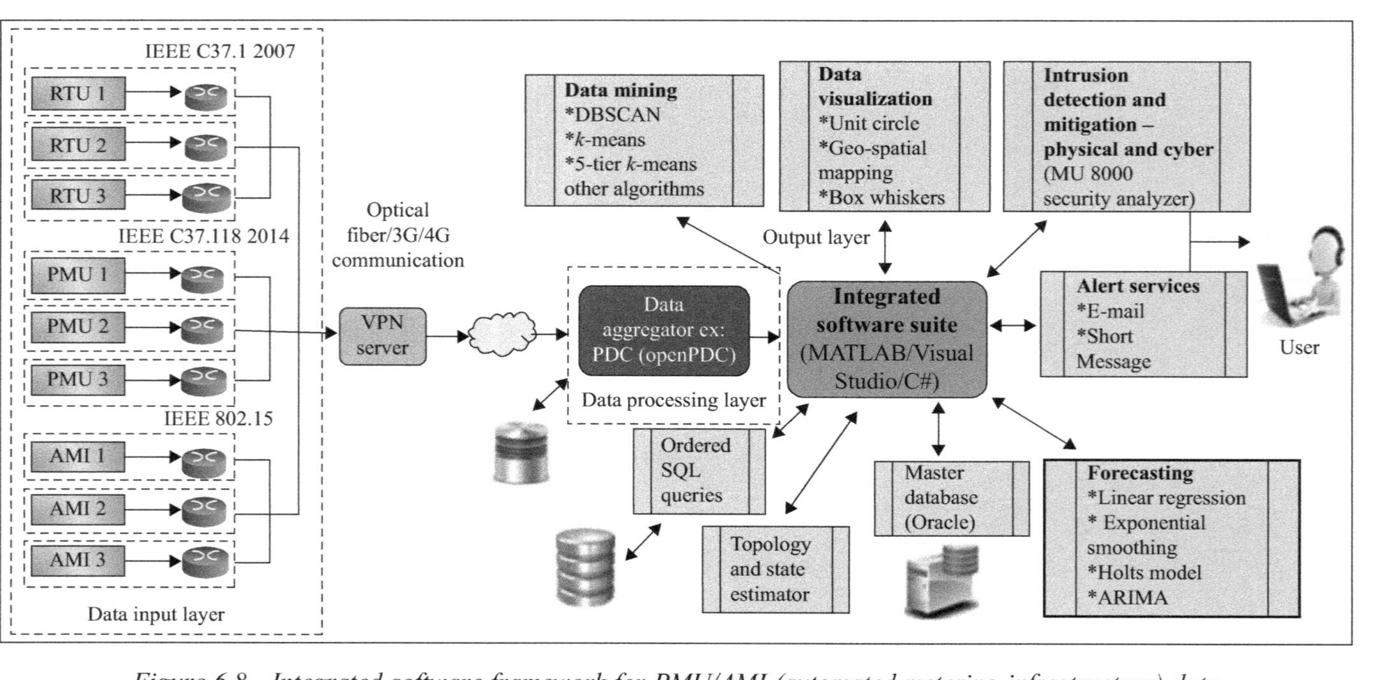

Figure 6.8 Integrated software framework for PMU/AMI (automated metering infrastructure) data

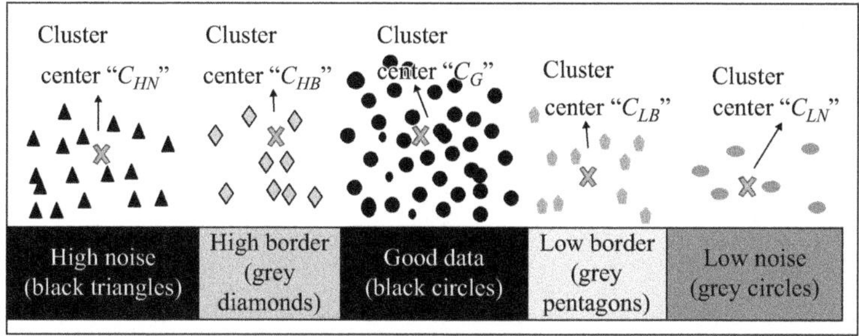

Figure 6.9 Multi-tier k*-means cluster categories*

ideal line parameters like voltage or current magnitude. The ideal value of the line parameter (current, voltage, or frequency) is taken as centroid of one of the clusters, and the other two cluster centroids are chosen at $\pm 1\%$ threshold of user input. Based on this input, the clusters are formed. There may or may not be three clusters, depending on the PMU data. Once the initial clusters are formed, the algorithm enters the second phase of execution. This second phase is developed with the goal of identifying the border regions (outliers) near the clusters. Based on specific threshold values (shown in (6.7)–(6.10)), the data points in the groups resulting from the initial stage of clustering are further categorized into border values. These are overlapping data between the three clusters from stage I, which have the potential to detect contingencies. The second stage also acts as a way of visualizing the data. Each resulting cluster is subjected to certain conditions. A color scheme has been developed, which categorizes data points as follows:

1. Good data—black circles
2. High-bordering data—grey diamonds
3. Low-bordering data—grey pentagons
4. High-noise (bad) data—black triangles
5. Low-noise (bad) data—grey circles

Figure 6.9 shows a diagrammatic representation of the color scheme.

The following variables indicate the centroid values of individual cluster.

1. CN_{good}—centroid value of a cluster with good data points during iteration N
2. CN_{low}—centroid value of a cluster with a low range of data points during iteration N
3. CN_{high}—centroid value of a cluster with a higher range of data points during iteration N.

The novel hybrid algorithm was applied to voltage magnitude, current magnitude, and frequency of the PMU datasets. The thresholds for each parameter, namely voltage magnitude, current magnitude, and frequency, are to be grouped into five clusters as per (6.7)–(6.10).

The algorithm can be mathematically represented as follows.

6.2.1.1 Stage I—Creation of 1–3 clusters based on the phasor data input

If $d = \{x\}_{i=1}^{n}$ is the d-dimensional space, having k clusters, such that each cluster is represented as $C = \{C_1, C_2, \ldots, C_k\}$, then the centroid of each cluster is given by

$$\mu_i = \frac{1}{n_i} \sum_{x_j \in C_i} \{x_j\} \tag{6.7}$$

where $n_i = |C_i|$ is the number of points in the cluster C_i. As the number of clusters are set at 3, $C = \{C_1, C_2, C_3\}$, using brute force, iterative method, the data points are clustered into any one of the three clusters. Each iteration results in new centroid values. The iterations continue until the centroids stop shifting to form a new centroid center points, that is, $|\mu_i - \mu'_i| = 0$. Here, μ'_i is the new centroid and $i = \{1, 2, 3\}$.

6.2.1.2 Stage II—Extending the clusters from stage I to 4 or 5 clusters based on any border data present

The clusters from stage I are further analyzed to detect the "outliers"—data points that are not exactly good data. These border clusters lie close enough to the upper or lower threshold values. They are to not be considered dangerous to the system. The density of these border points acts as an indicator for contingency scenarios. Each of the clusters is tested, if the data points lie within the following thresholds in the cluster:

$$\max.\{C_i\} \leq \left[1 + \frac{P}{100} * X_{\text{ideal}}\right] \quad \text{and} \quad \min.\{C_i\} \geq \left[1 - \frac{P}{100} * X_{\text{ideal}}\right] \tag{6.8}$$

If the maximum and the minimum of the cluster falls within the range $\left(\left[1 - \frac{P}{100} * V_{\text{ideal}}\right], \left[1 + \frac{P}{100} * X_{\text{ideal}}\right]\right)$, where P is a set percentage of points to be considered as good or bad, and X_{ideal} is the ideal voltage (V) or current (I) or frequency (F) of the line that is known or considered to be normal value:

$$\max.\{C_i\} > \left[1 + \frac{P}{100} * X_{\text{ideal}}\right] \quad \text{and} \quad \min.\{C_i\} \geq \left[1 - \frac{P}{100} * X_{\text{ideal}}\right] \tag{6.9}$$

where $X_{\text{ideal}} = \{V, I, F\}$.

If a certain number of points in a cluster are greater than large threshold values, then they are classified as high noise. This indicates that the system is under low load (demand) condition. The rest of the points in the cluster belong to the core group:

$$\max.\{C_i\} \leq \left[1 + \frac{P}{100} * X_{\text{ideal}}\right] \quad \text{and} \quad \min.\{C_i\} < \left[1 - \frac{P}{100} * X_{\text{ideal}}\right] \tag{6.10}$$

This condition suggests that data points that have values lower than the low threshold values are low border points. This indicates the presence of high load (demand) conditions:

$$\max.\{C_i\} > \left[1 + \frac{P}{100} * X_{\text{ideal}}\right] \quad \text{and} \quad \min.\{C_i\} < \left[1 - \frac{P}{100} * X_{\text{ideal}}\right] \tag{6.11}$$

Equation (6.10) illustrates the presence of core, high, and low border points in the cluster.

The multitier *k*-means method was developed with the knowledge that a *k*-means algorithm creates clusters based on their distance from a predefined centroid value. As the average value keeps updating, there is an inclination of the clusters to move further apart from each other. It is this property of the *k*-means algorithm that is used in stage I to isolate the noise cluster from the rest of its data. The second stage is used to differentiate the core and the border (high and low) points. Thus, if cluster extremities (max. and min.) do not match any of the conditions presented in (6.8)–(6.11), then it is categorized as noise cluster.

Figure 6.10 shows the pseudocode for multitier *k*-means:

Figure 6.11 illustrates two case scenarios that can be captured using multitier *k*-means. For the first case, it is evident that the three clusters with centroids C_1, C_2, and C_3 that were formed in stage I get further divided into five separate clusters indicating that the system is under a lot of stress. For the second case, however, only three clusters are formed indicating the system is under a high-load condition, which finally resulted in a fault. The results obtained from the application of the multitier *k*-means method are discussed in detail in Section 6.3.

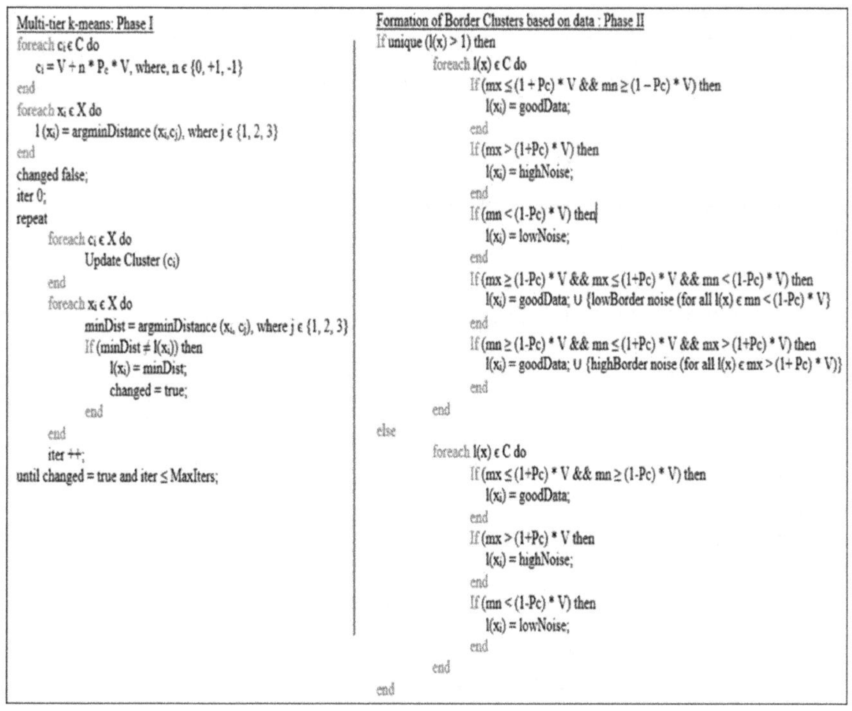

```
Multi-tier k-means: Phase I
foreach cᵢ ϵ C do
    cᵢ = V + n * Pₑ * V, where, n ϵ {0, +1, -1}
end
foreach xᵢ ϵ X do
    l (xᵢ) = argminDistance (xᵢ,cⱼ), where j ϵ {1, 2, 3}
end
changed false;
iter 0;
repeat
        foreach cᵢ ϵ X do
                Update Cluster (cᵢ)
        end
        foreach xᵢ ϵ X do
                minDist = argminDistance (xᵢ, cⱼ), where j ϵ {1, 2, 3}
                If (minDist ≠ l(xᵢ)) then
                    l(xᵢ) = minDist;
                    changed = true;
                end
        end
        iter ++;
until changed = true and iter ≤ MaxIters;
```

```
Formation of Border Clusters based on data : Phase II
If unique (l(x) > 1) then
        foreach l(x) ϵ C do
                If (mx ≤ (1 + Pc) * V && mn ≥ (1 – Pc) * V) then
                    l(xᵢ) = goodData;
                end
                If (mx > (1+Pc) * V) then
                    l(xᵢ) = highNoise;
                end
                If (mn < (1-Pc) * V) then
                    l(xᵢ) = lowNoise;
                end
                If (mx ≥ (1-Pc) * V && mx ≤ (1+Pc) * V && mn < (1-Pc) * V) then
                    l(xᵢ) = goodData; ∪ {lowBorder noise (for all l(x) ϵ mn < (1-Pc) * V}
                end
                If (mn ≥ (1-Pc) * V && mn ≤ (1+Pc) * V && mx > (1+Pc) * V) then
                    l(xᵢ) = goodData; ∪ {highBorder noise (for all l(x) ϵ mx > (1+ Pc) * V)}
                end
        end
else
        foreach l(x) ϵ C do
                If (mx ≤ (1+Pc) * V && mn ≥ (1-Pc) * V) then
                    l(xᵢ) = goodData;
                end
                If (mx > (1+Pc) * V then
                    l(xᵢ) = highNoise;
                end
                If (mn < (1-Pc) * V) then
                    l(xᵢ) = lowNoise;
                end
        end
end
```

Figure 6.10 Pseudocode of multitier k*-means (phases I and II)*

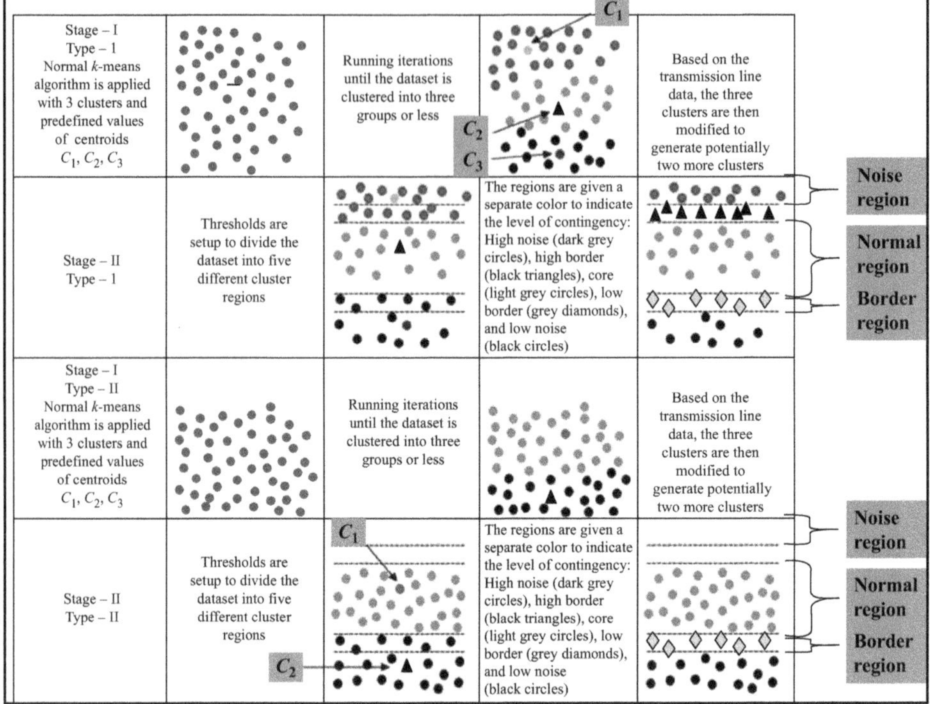

Figure 6.11 Cluster formation in multitier k-means

6.2.2 *Advantages of using proposed multitier* k*-means method*

There are three major advantages of using multitier *k*-means method. They are as follows:

1. The proposed approach dynamically forms a varying number of clusters. This could be anywhere from 1 to 5 clusters depending on the data thresholds and fault type. The need for such varying clusters in the approach is to clearly identify specific type of faults that may not be visible otherwise. For example, the addition of two new clusters; (1) low and (2) high border clusters indicate that the values are beginning to reach their end of thresholds. Detecting those border points will help operators to caution and arrive at proactive remedial actions to stabilize the grid.
2. Capable of clearly distinguishing the good, bad, and the noisy data regions based on thresholds.
3. Requires only one parameter (voltage or current) as input, initiates the algorithm.

Thus, the application of multitier *k*-means method cannot only track changes in load conditions but also aids in decision making of the system that enables operators to generate alarms. This will help avoid contingency scenarios that disrupt the normal functions of the power grid.

6.3 Results and discussions

A comparative analysis and investigation on various clustering has been discussed in this chapter. This chapter presents the results obtained from clustering algorithms (DBSCAN, *k*-means, and multitier *k*-means).

6.3.1 Results obtained from data clustering algorithms

Four different test cases have been formulated to evaluate the performance of existing clustering methods against multitier *k*-means. The cases under consideration are the following: (1) ideal or normal load condition, (2) heavy load condition, (3) light load condition, and (4) single line-to-ground fault condition. The methods evaluated are (1) DBSCAN, (2) *k*-means, and (3) multitier *k*-means. The performances of the clustering algorithms have been evaluated using Dunn's index (DI). This is a standard metric to evaluate cluster separation [12].

The parameters studied for clustering methods are voltage and current (magnitude and phase angle), and the frequency values from the openPDC software suite. It showcases the level of accuracy of the three algorithms during fault conditions. The four test cases are conducted for time-intervals of 5, 10, 15, and 20 min. The data include voltage, current, and frequency data from the synchrophasor sensors collected at a rate of 30 samples per second [13].

6.3.1.1 Test case 1: system under steady state (normal) conditions

Under normal conditions, the power system conditions are generally stable and its parameters lie within acceptable threshold values. Figure 6.12 shows clustering of voltage data from openPDC for a 15 min time period using DBSCAN. Figure 6.13 indicates the clustered voltage data from the proposed multitier *k*-means algorithms, and Figure 6.14 shows the clusters from *k*-means algorithm that is initialized to with a total of three clusters. These clusters are formed on the basis of the value of the data points, and they do not conform to any power system thresholds. Hence, we will see later in the chapter that although they have a desirable DI value, the clustering scheme fails to correctly classify the phasor data, in a manner that would be helpful for system operators to monitor the system. DBSCAN cluster the data into three clusters as core, border, and noise as shown in Figure 6.12. The core points are data points that are equal or nearer to the threshold value of ideal transmission line voltage (300 kV). They are encoded in dark grey color. The values that are slightly away from the nominal value of core points are defined as border points. Any further isolated points are classified as noise points. From power systems perspective, this is a miss classification, as the voltage data captured are still within the allowable range of $\pm5\%$ of ideal voltage value for the range of voltage considered [14]. Figure 6.13 represents the voltage data clustered by multitier *k*-means algorithm. The only input that is needed to be provided by the user is to initialize the value for ideal transmission line parameter, that is, a voltage value of 300 kV. It is evident from Figure 6.13 that the voltage data get correctly clustered into core points (dark grey) as the system is in a steady-state condition. Based

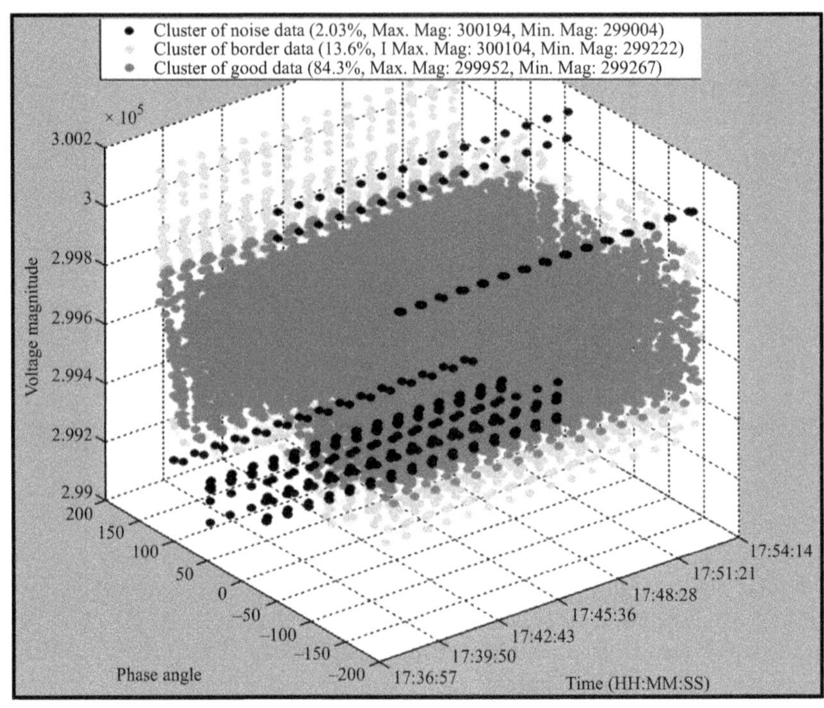

Figure 6.12 Steady-state voltage output—DBSCAN

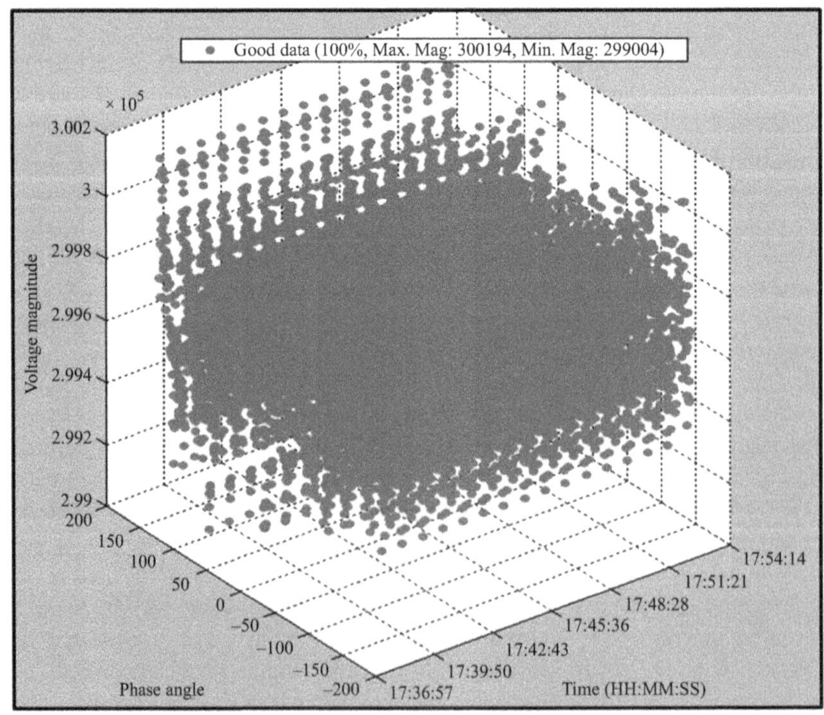

Figure 6.13 Steady-state voltage output—multitier k-means

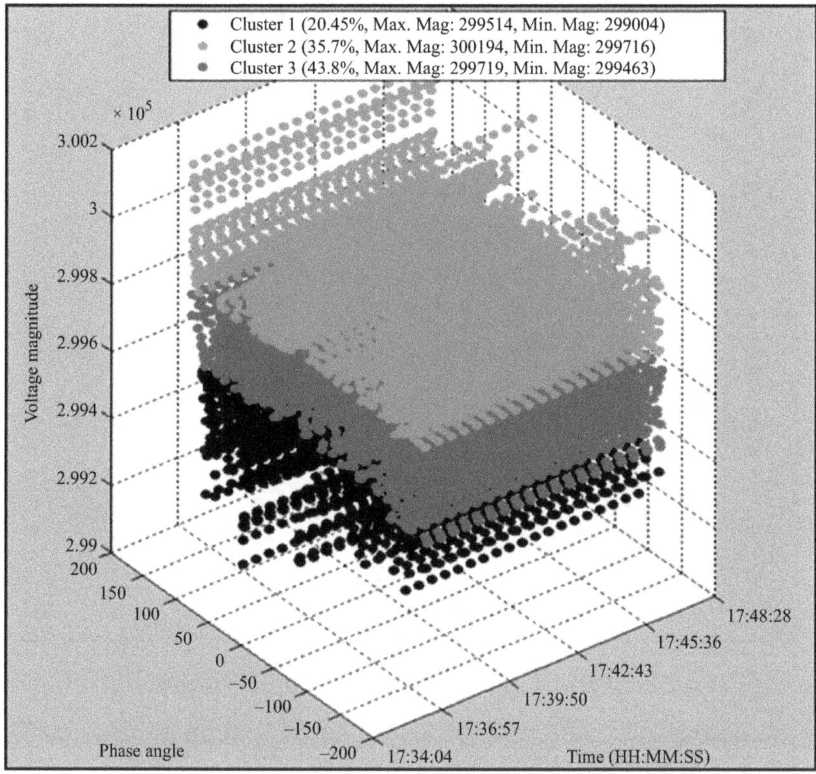

Figure 6.14 Steady-state voltage output—k-means

on its initial centroid values, the *k*-means algorithm continues to cluster all data points, in which the number of clusters was set to 3. This is done with the idea that the power data can have three possible clusters, the core group representing normal data, the border group consisting of voltage values just outside the threshold values, and finally the noise group consisting of voltage data having values further away from threshold values of border and core. Figure 6.14 shows the clustering output from *k*-means algorithm.

6.3.1.2 Test case 2: system under heavy load condition

During the heavy loaded condition, the demand increases, and thus voltage level starts to drop. Figures 6.15–6.17 represent the clustering outputs from DBSCAN, multitier *k*-means, and *k*-means algorithms. In Figure 6.15, operators can easily identify the densely populated areas (dark grey) and the sparsely populated data points (light grey). In comparison with the Figure 6.12, the shift in the density of core points from 299,800 to 299,500 V can be easily identified. Figure 6.16 represents the multitier *k*-means clustering of heavy load condition. As per the enforced conditions, the data points that are within tolerable limits are clustered as good cluster (dark grey), and other data points form a low-noise cluster (black), and

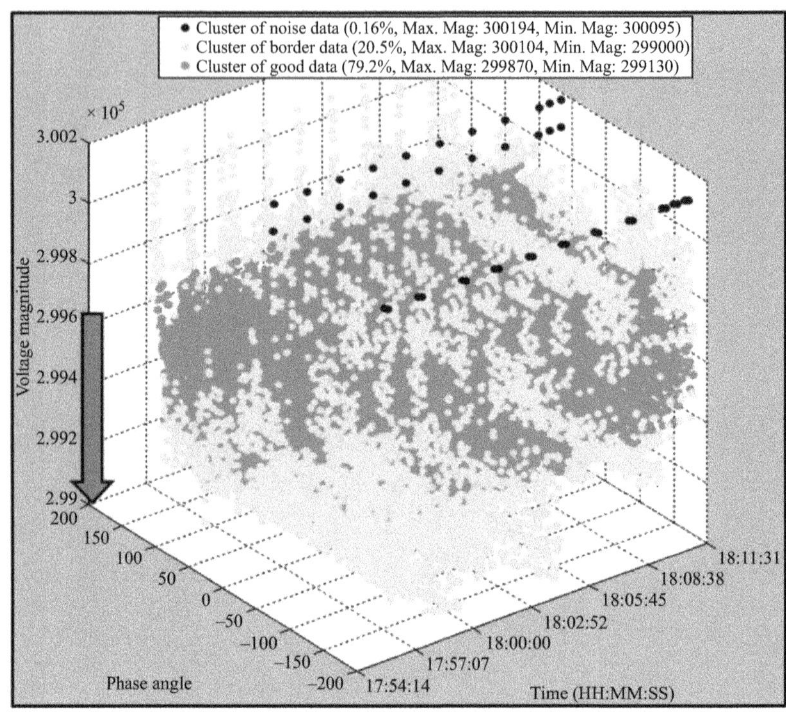

Figure 6.15 Heavy load condition—DBSCAN

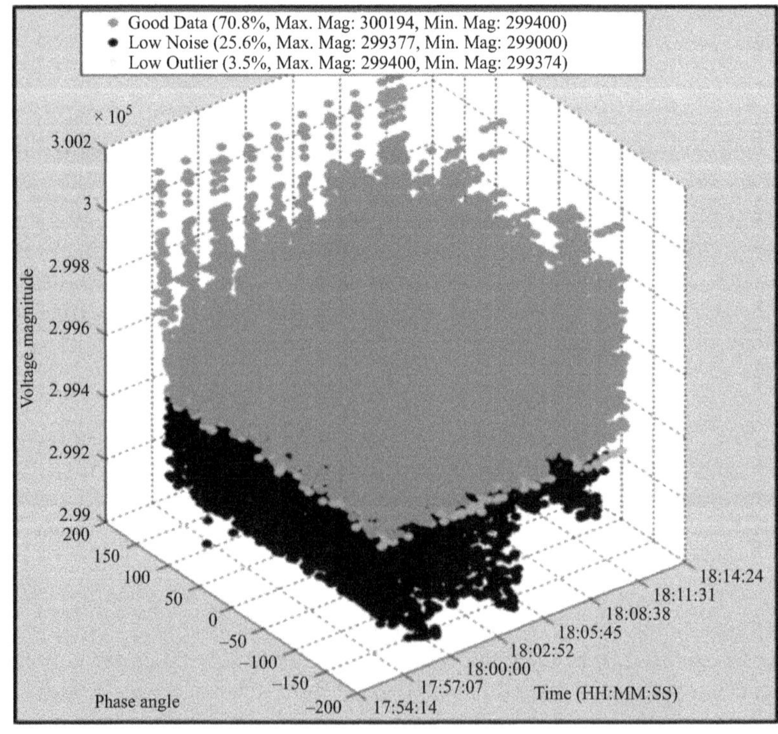

Figure 6.16 Heavy load condition—multitier k-means

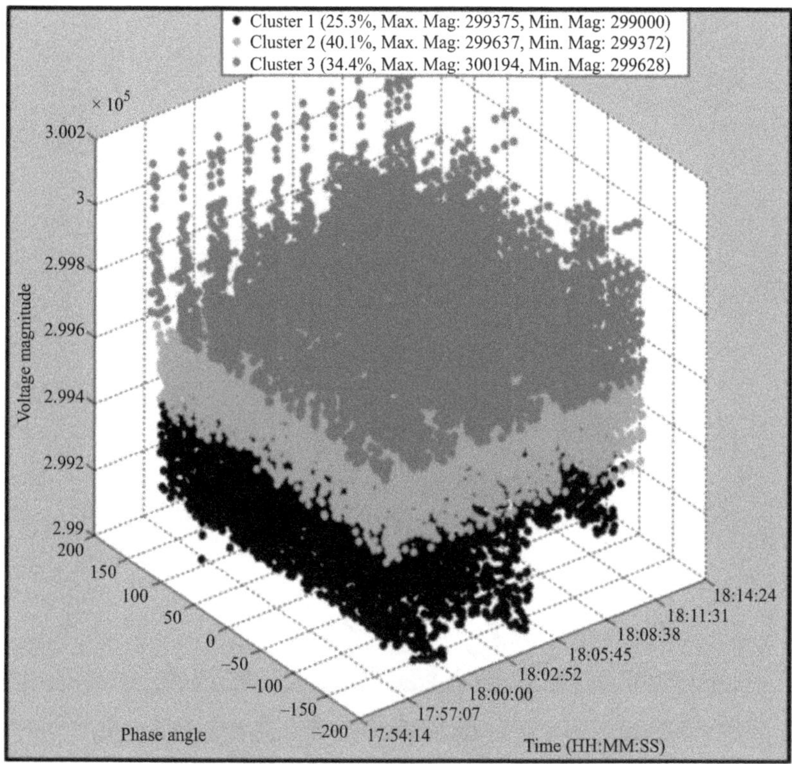

Figure 6.17 Heavy load condition—k-means

the data that do not fall into either of these two clusters are grouped as an outlier region (light grey). The shift in the data points due to heavy load conditions. This means that there is a reduction in the percentage of core group points and an increase in percentage of low outlier cluster points. Finally, Figure 6.17 represents the output from *k*-means algorithm to form three clusters. The *k*-means clustering method again forms three clusters but fails to clearly indicate that the density shift and thus cannot accurately capture the heavy load conditions.

6.3.1.3 Test case 3: system under light load conditions

Under light load conditions, there will be a slight increase in the voltage values due to low demand. Figure 6.18 shows the clustered output from DBSCAN. Under light load conditions, the voltage value increases to above normal. Figures 6.12 and 6.15 indicate the shift in the density of core points from 299,800 to 300,600 V. Figure 6.19 shows the data clustered by multitier *k*-means under light load conditions. As per the predefined centroid settings, the data are clustered into four clusters.

They are the good data (dark grey), the high-value outliers (light grey), the low-value outliers (medium black), and the high value noise data (black) clusters. The change in load conditions can again be captured by tracking the increase in

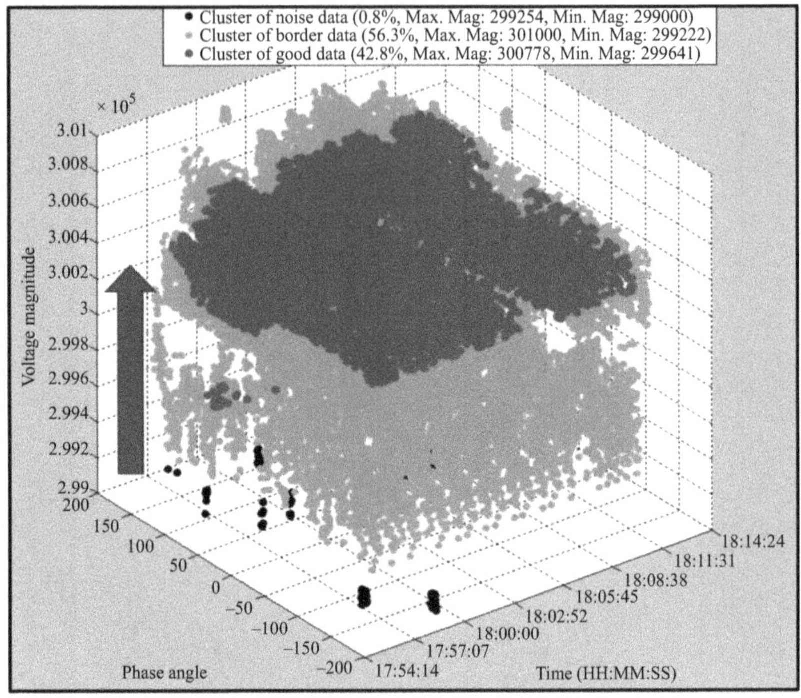

Figure 6.18 Light load condition—BSCAN

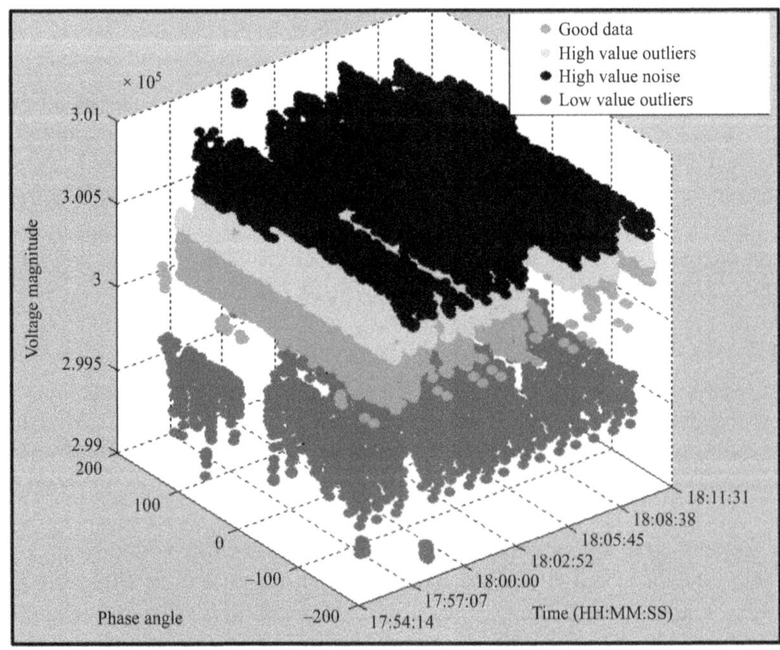

Figure 6.19 Light load condition—multitier k-means

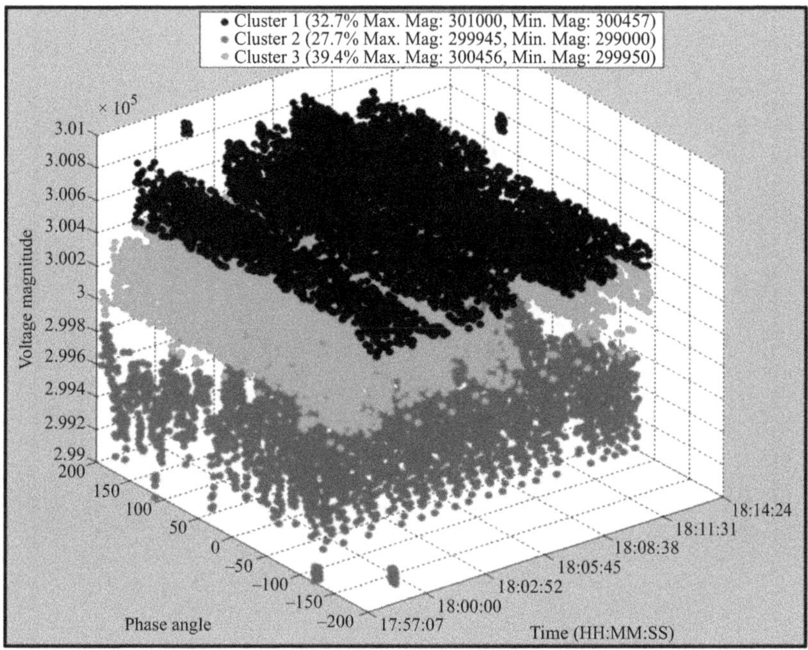

Figure 6.20 Light load condition—k-means

percentage of high outlier cluster points in Figure 6.19, with respect to that of Figure 6.13. Figure 6.20 is the output from the normal *k*-means algorithm to form three clusters for the light-loaded condition. It is again evident from Figure 6.20 that the *k*-means algorithm fails to capture the light load conditions.

6.3.1.4 Test case 4: system under single line to ground fault condition

The final test case is a single line to ground fault condition. This fault is captured using the proposed clustering methods. Figure 6.21 shows the DBSCAN-clustered output for the fault condition for voltage data. The DBSCAN clusters the fault conditions into noise and clearly separates the good and faulty points. However, it fails to clearly identify the beginning of the fault, that is, the capture the time when a contingency occurred, resulting in the drop in voltage below normal threshold. This makes it unsuitable to generate alarms. Figure 6.22 shows the same fault case clustered correctly using multitier *k*-means method. It does a good job in clustering the fault data into low outliers (light grey) and low noise (black) clusters. From Figure 6.22, the fault condition is clearly visible when the voltage level goes below a critical value (i.e., changes over from light grey to black). Thus, using this method, an alarm generated will alert the system operator through a prewarning message that there is a contingency scenario (the line-to-ground fault). Figure 6.23 shows the output from *k*-means algorithm. It clusters the given fault condition into three clusters irrespective of any threshold values.

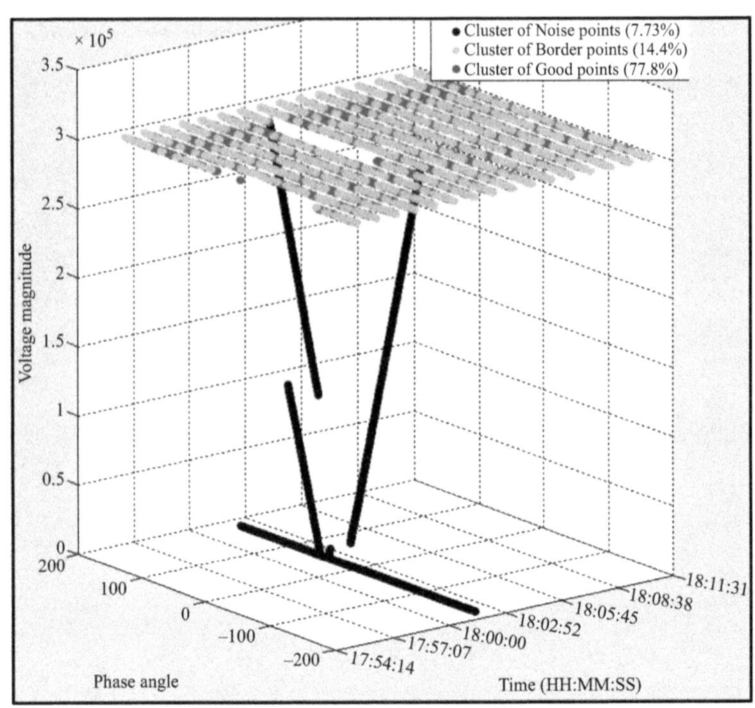

Figure 6.21 Line-to-ground fault—BSCAN

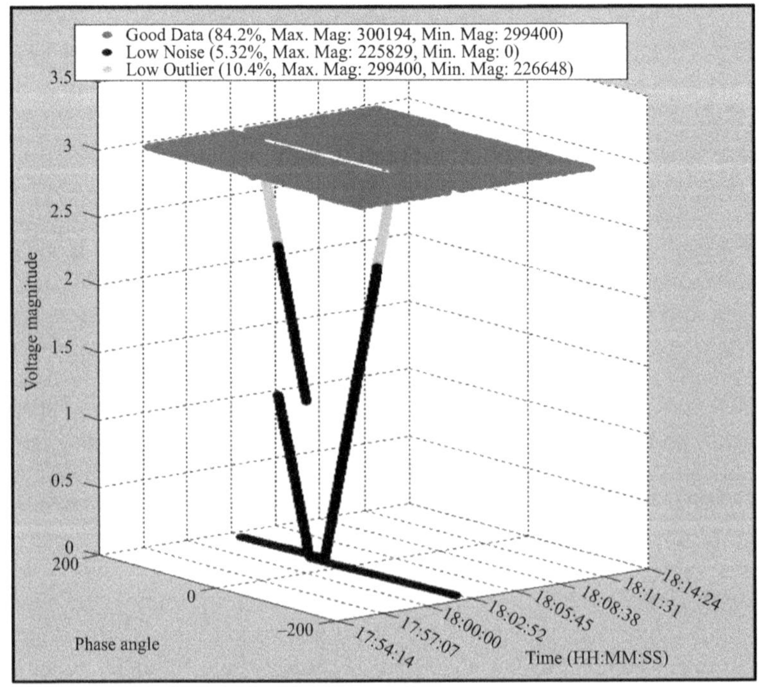

Figure 6.22 Line-to-ground fault—multitier k-means

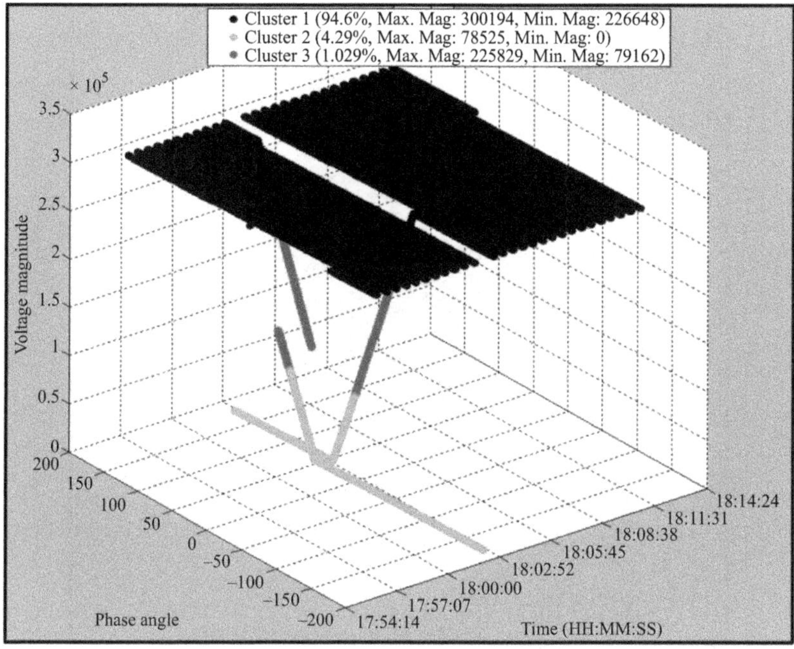

Figure 6.23 Line-to-ground fault—k-means

6.3.1.5 Test results with current (*I*) data

The clustering algorithms (DBSCAN, multitier *k*-means, and *k*-means) were applied to current (magnitude and phase angle) and frequency data as well. The results obtained were analogous to the voltage dataset. Figures 6.24–6.26 illustrate the results of these clustering methods on current parameter under normal, heavy load, and light load conditions. The ideal line current value is 500 A. The data were gathered over a period of 15 min.

From Figure 6.24, the multitier *k*-means captures accurately the normal state of the system and, thus generates only the core cluster. The percentage of good data points in the cluster is 100%.

Figure 6.25 illustrates how multitier *k*-means is able to capture the heavy load condition (when the current drawn from the system starts to increase). The percentage of the core data points drops from 100% to 91%, and the high outliers have 9% of the data.

Finally, Figure 6.26 captures the effect of light load conditions on the current data. The percentage of core points reduces, and the value shifts toward the lower outliers indicating that there is less demand than the ideal conditions.

6.3.1.6 Quantification of the visual data representation

This section discusses the distribution of voltage data captured in Figures 6.12–6.26.

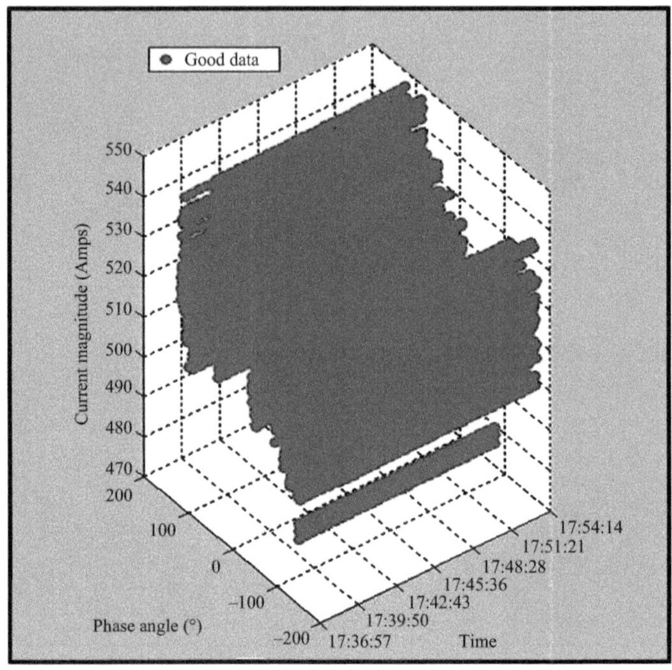

Figure 6.24 Steady-state condition multi-tier k-means

Under normal conditions, with different data sizes, Tables 6.1–6.3 provide the percentage distribution of cluster points in multitier *k*-means, DBSCAN, and *k*-means algorithms.

As the conditions change from normal values, the percentage of good points decrease from 89%. In a similar way, light load condition can be observed with high border points (16%) and the heavy load condition with low noise (24%).

From Table 6.2, the distribution of points is DBSCAN can be observed.

Table 6.3 shows the distribution of data points under different load conditions.

6.3.1.7 Computational time

The computational time is calculated for various phases in the data-mining process. A preprocessing time is required to interface with MATLAB environment and extract data from openPDC by C# application. Once the data are made available, the algorithms perform their operations of clustering. This run-time varies between the algorithms and with different data sizes. After the clustering process is completed, users visualize the clusters and decision samples in a MATLAB environment. The time it takes to form completed clusters is referred to as postprocessing time. The processing times for data sizes of various time-intervals 10, 15, 20, and 30 min are tabulated in Table 6.4. It is clear from Table 6.4 that *k*-means is fastest when compared to multitier and DBSCAN methods. The proposed multitier *k*-means does take extra seconds to cluster the data. The clustered information is very easy to

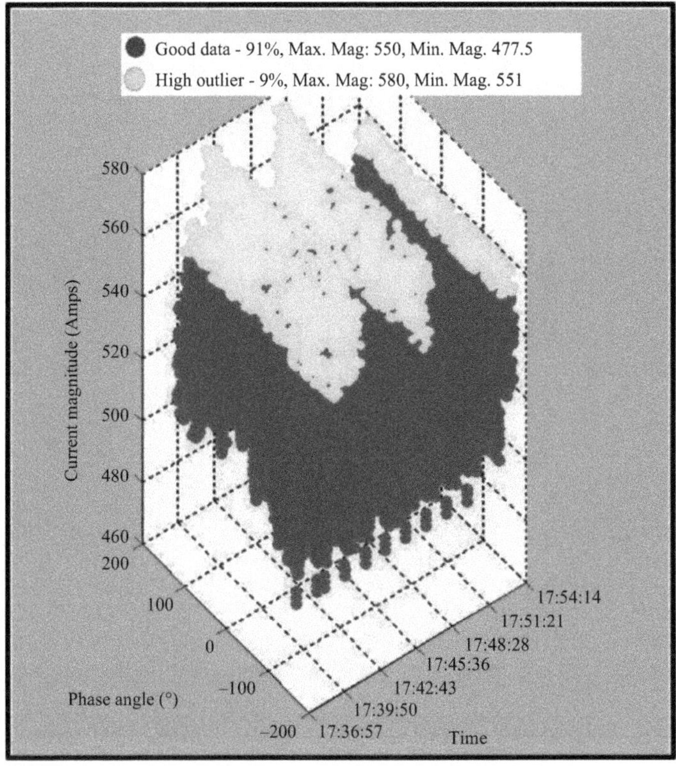

Figure 6.25 Heavy load condition multitier k-means

visualize and accurate in multitier *k*-means than DBSCAN and *k*-means. However, DBSCAN does take a longer time to process the data ranging from 10s of seconds to 100s of seconds depending on the time-interval. We believe that this is due to the fact that the number of iterations in DBSCAN is higher than other algorithms. All test cases have been conducted using the following PC configurations:

PC configuration tested:

1. Server configuration: AMD FX-9590 4.7 GHz 8-Core, 32 GB DDR3 RAM, 500 GB SSD
2. Software: Windows operating system, Microsoft Visual Studio 2012, openPDC 2.1, MySQL, SQLite, R, MATLAB/Simulink

From the case studies tested, each of the three algorithms has their advantages and disadvantages. Based on the feedback from utility operator's requirement, any of these algorithms can be selected at various stages for power system planning and operations. DBSCAN can be used to provide valuable understanding on the system during heavy load conditions. The multitier *k*-means can provide insight on the grid situation as well as cluster the data to show the load intensities. Although the

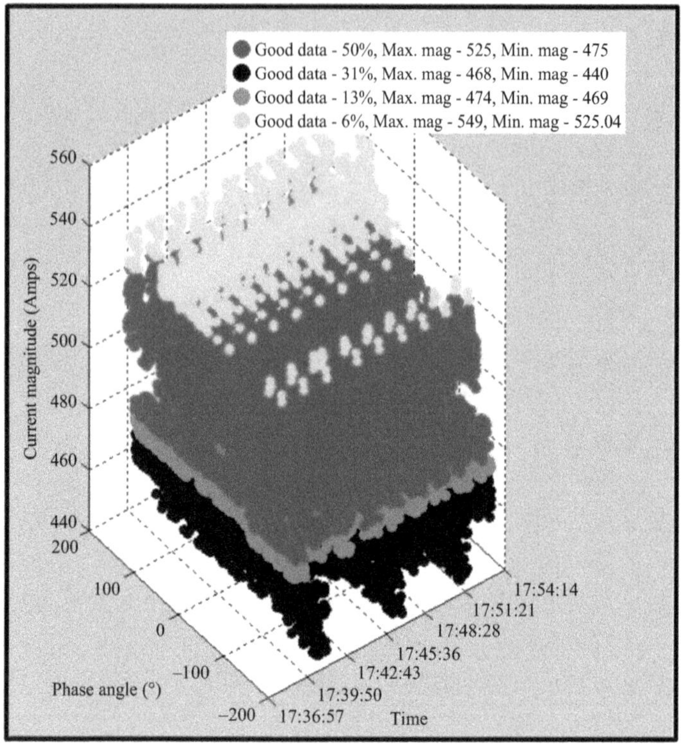

Figure 6.26 Heavy load condition multitier k-means

Table 6.1 Distribution of data points with multitier k-means

Load condition	Low noise	Low outliers	Good points	High outliers	High noise
Normal	0	10.53	89.47	0	0
Heavy	24.7	5.2	70.04	0	0
Light	0	3.3	79.76	16.94	0
Fault	5.32	10.4	84.2	0	0

Table 6.2 Distribution of data points with DBSCAN

Load condition	Noise points	Border points	Core points
Normal	0.5	6.3	93.2
Heavy	0.078	8.96	90.5
Light	0.8	56.3	42.8
Fault	7.73	14.4	77.8

Table 6.3 Distribution of data points with k-means

Load condition	Cluster 1	Cluster 2	Cluster 3
Normal (100%)	20.4	35.7	43.8
Heavy (100%)	25.3	40.1	34.4
Light (100%)	32.7	27.7	39.4
Fault (100%)	94.6	4.29	1.02

Table 6.4 Computation time for different data sizes

Algorithm	Time duration (min)	Preprocessing (s)	Processing (s)	Postprocessing (s)
DBSCAN	10	68.4	8.35	13.4
	15	126	26.04	30.6
	20	227.966	56.3988	66.7
	30	478.17	132.3	146.7
Multitier *k*-means	10	62.1	5.13	4.49
	15	123.8	8.2	6.71
	20	204.1	9.6	8.86
	30	429.9	15.2	13.1
k-Means	10	55.4	0.02	2.62
	15	113.7	0.03	3.86
	20	186.7	0.09	6.19
	30	414.8	0.14	7.49

Table 6.5 Visual fidelity of clustering algorithms

Visual fidelity under	*k*-means	DBSCAN	Multitier *k*-means
Heavy load condition	+ (MC ~ 50%)	−− (MC > 80%)	++ (MC < 20%)
Light load condition	++ (MC < 20%)	−− (MC > 80%)	+ (MC ~ 50%)
Normal condition	++ (MC < 20%)	−− (MC > 80%)	+ (MC ~ 50%)
Fault condition	−− (MC > 80%)	−− (MC > 80%)	++ (MC < 20%)

original *k*-means known to be the fastest algorithm in the literatures, it does not provide any useful information for power system data sets. Table 6.4 presents a breakup of the time taken to perform the experiments under various periods of collected data.

6.3.1.8 Visual fidelity

A survey of ten students were conducted to assess the visual appeal of the three clustering methods as well as their tendencies of misclassification (MC) and identify which method is effective for visualizing faults or anomalies under varying conditions. MC over here indicates the failure of a data clustering method to correctly place data points in the appropriate cluster. Table 6.5 indicates that proposed

Table 6.6 Dunn's indices for voltage and current data

Dunn's index	Time (min)	DBSCAN		Multitier *k*-means		*k*-Means	
		V	*C*	*V*	*C*	*V*	*C*
Heavy load	10	0.1894	0.0462	0.2069	0.0506	0.4350	101.3
	15	0.1922	0.0365	0.3714	0.0641	0.4297	178.9
	20	0.3136	0.0529	0.2004	0.0586	0.4241	56.8
	30	0.2167	0.0523	0.1901	0.0516	0.4543	59.2
Steady state	10	0.2236	0.326	0.3016	0.816	0.4203	75.9
	15	0.323	0.0425	0.3016	0.0818	0.2968	118.2
	20	0.1936	0.2443	0.3016	0.0818	0.2836	74.6
	30	0.1536	0.0904	0.3016	0.0818	0.4203	79.7
Light load	10	1.162	0.36	0.4035	0.04	0.4098	117.5
	15	0.1978	0.695	0.3714	0.0381	0.369	57.3
	20	1.158	0.852	0.3538	0.0372	0.405	59.9
	30	0.7375	0.09	1.1695	0.0516	0.4202	153.2

multitier *k*-means visually appeals better under heavy load conditions, and DBSCAN appeals good for light load conditions. A work is in progress to seek inputs on the visual fidelity of proposed clustering method from system operators.

6.3.1.9 Dunn's index

Dunn's index is used as a metric to evaluate clustering algorithms. The aim is to identify sets of clusters that are compact, with smaller variance values between members of the cluster, and are well separated. Here, the average values of different clusters are sufficiently far apart compared to values of individual cluster variance. Dunn's index can be mathematically represented:

$$\mathrm{DI}_m = \frac{\min_{1 \le i \le j \le m} \delta(C_i, C_j)}{\max_{1 \le k \le m} \Delta_k} \tag{6.12}$$

where C_i and C_j are the *i*th and *j*th centroids of *m* clusters formed from a clustering algorithm, and Δ_k is the distance between any two points within the *k*th cluster. Therefore, DI calculates the ratio of the minimum intracluster distance to the maximum intracluster distance [12].

The DI calculated in each experiment was used both as a performance metric to check the compactness of the clusters and also to track load conditions. In Table 6.6, the "time in min" column presents the period of data used for the calculations, and columns *V* and *C* indicate data calculated for voltage and current parameter, respectively. It was observed that under steady-state conditions, the DI for multitier *k*-means held the same value, regardless of the tested time period. This was further re-enforced in Figure 6.13, which showed that only core points were present in the overall data. Thus, any changes in the DI value (increase or decrease) would mean the formation of new clusters, which in turn indicated that there was presence of border or noise clusters. Table 6.6 suggests that the DI tends to increase under light load conditions and tends to decrease under heavy load conditions, for voltage data. Using this observation as a reference, the operators can capture whether a system is in steady state or transient condition.

6.4 Conclusion

A collection of clustering algorithms was presented for grouping the PMU datasets into good and bad data. The results presented provide promising results for situational awareness for the next generation power grid.

Acknowledgments

The authors would like to acknowledge the support from Division of Research & Economic Development at the University of North Dakota (UND-21418-4010-02294) and NSF award 1537565. Dr. Prakash Ranganathan acknowledges the support from his graduate students Anupam Mukherjee and Radhakrishnan Angamuthu Chinnathambi for their contributions.

References

[1] "Grid Protection Alliance." [Online]. 2016. Available: http://openpdc.codeplex.com/.

[2] "openPDC." [Online]. 2016. Available: http://gridprotectionalliance.org/pdf/openPDC_Overview.pdf.

[3] "Microsoft Developer Network." [Online]. Available: http://msdn.microsoft.com/en-us/library/.

[4] "D. File Extension." [Online]. 2016. http://fileinfo.com/extension/d.

[5] "MATLAB" [Online]. 2016. http://www.mathworks.com/products/matlab/.

[6] "Data mining and analysis" [Online]. 2016. http://www.cs.rpi.edu/~zaki/PaperDir/DMABOOK.pdf.

[7] "Recommended anomaly detection technique for simple, one-dimensional scenario?" [Online]. 2016. http://stackoverflow.com/questions/2303510/recommended-anomaly-detection-technique-for-simple-one-dimensional-scenario.

[8] R. Cordeiro de Amorim and B. Mirkin, "Minkowski metric, feature weighting and anomalous cluster initializing in *k*-means clustering," *Pattern Recognit.*, vol. 45, no. 3, pp. 1061–1075, 2012.

[9] I. S. Dhillon and D. S. Modha, "Concept decompositions for large sparse text data using clustering," *Mach. Learn.*, vol. 42, no. 1–2, pp. 143–175, 2001.

[10] T. Kanungo, D. M. Mount, N. S. Netanyahu, C. D. Piatko, R. Silverman, and A. Y. Wu, "An efficient *k*-means clustering algorithm: analysis and

implementation," *IEEE Trans. Pattern Anal. Mach. Intell.*, vol. 24, no. 7, pp. 881–892, 2002.

[11] J. C. Bezdek, R. Ehrlich, and W. Full, "FCM: the fuzzy *c*-means clustering algorithm," *Comput. Geosci.*, vol. 10, no. 2–3, pp. 191–203, 1984.

[12] "Dunn's cluster validity index as a contrast measure of VAT images" [Online]. 2008. http://www.ece.mtu.edu/~thavens/papers/ICPR_2008_Havens.pdf.

[13] R. Vallakati, A. Mukherjee, and P. Ranganathan, "A density based clustering scheme for situational awareness in a smart-grid", *2015 IEEE International Conference on Electro/Information Technology (EIT)*, Dekalb, IL, 2015, pp. 346–350.

[14] "Transmission 101" [Online]. 2004. http://pubs.naruc.org/pub/4AE08B08-2354-D714-518E-B2D936734DD4.

Chapter 7

Transmission line fault detection, classification, and localization in smart power grids using synchrophasor measurements

Pathirikkat Gopakumar[1], Maddikara Jaya Bharata Reddy[1] and Dusmanta Kumar Mohanta[2]

7.1 Introduction

Innovations in computational and communication technologies have been instrumental in the transition of conventional power grids to smart power grids (SPG) [1]. In SPG protection systems, one of the major research challenges is the development of real-time self-healing protection methodologies. A major constituent of self-healing technology is transmission line fault monitoring [1,2]. Intelligent methods play a vital role to precisely detect, classify, and localize the transmission line faults occurring anywhere in the grid. Conventional fault monitoring methodologies for transmission lines are limited to specific transmission line configurations, and these methods have necessarily to be incorporated in all transmission lines. The transfer of information from transmission lines to system protection center (SPC) necessitates proper communication channels, leading to heavy computational and communication burden. SPG protection system tides over this burden. This chapter discusses methodologies for transmission line fault detection, classification, and localization in SPGs using wide-area phasor measurement unit (PMU) measurements, which can overcome the computational and communication burden.

7.2 Real-time transmission line fault monitoring in power grids

Monitoring of faults in various power grid constituents, including transmission lines, is the backbone of any power grid. Transmission line faults in power grids are inevitable problems, especially in the case of overhead transmission systems. Faults are generally caused by random and unavoidable events like temporary contact due

[1]Department of Electrical and Electronics Engineering, National Institute of Technology (NIT), Tiruchirappalli, Tamilnadu 620015, India
[2]Department of Electrical and Electronics Engineering, Birla Institute of Technology, Mesra, Ranchi 835215, India

to trees or birds, lightning, thunderstorms, aging of insulators, overloading, short circuits, intended or unintended human actions, lack of maintenance and equipment failure, etc. Transmission line faults can be classified into asymmetric faults and symmetric faults [3]. Asymmetric faults include line-to-line (LL) faults, single line-to-ground (LG) faults, double line-to-ground (LLG) faults. Symmetric faults include line-to-line-to-line (LLL) faults, and line-to-line-to-line-to-ground (LLLG) faults. Faults may be temporary or permanent. Temporary faults are cleared after breaker reclosing, and permanent faults are not cleared even after several reclosing attempts [3,4]. To restore service after permanent faults, fast fault detection, classification, and localization with highest fidelity are indispensable [5]. Accurate fault detection, classification, and localization techniques facilitate the maintenance crew to reduce outage time and prevent the grid from entering into unstable state, restoring the line faster. Fault monitoring systems are becoming integral part of power grid real time monitoring system. Real-time monitoring enables reliable and stable operation of power grids under various operating conditions.

7.2.1 Emerging methodologies for fault monitoring

Real-time transmission line fault detection, classification, and localization methodologies present in literature can be broadly classified into those based on time domain and frequency domain. Each of them can be subdivided into conventional approach and artificial intelligence centered. These methodologies can be implemented using measurements acquired from remote telemetry units as well as PMUs. Categorization of methodologies based on method of execution is illustrated in Figure 7.1.

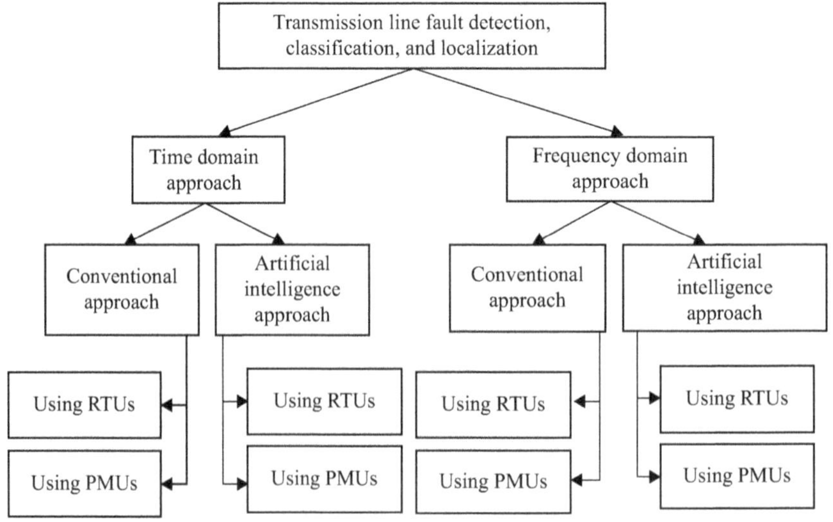

Figure 7.1 Classification of methodologies available in literature for transmission line fault detection, classification, and localization

Frequency domain–based methodologies analyze frequency components surviving in voltage or current waveform to detect, classify, and localize the fault that has occurred. Time domain–based methodologies, however, are based on magnitudes of voltage or current waveforms rather than frequency components present during the occurrence of fault. Many researches have proved that the frequency components present in the fault waveforms have hidden information relating to fault, and its type and location. Conventional approaches employed in both time domain and frequency domain are based on deterministic computations on well-defined models of the power systems and hence difficulties arise when the system undergoes dynamic changes in operating point. These difficulties can be overcome by the artificial intelligence approach, which deploy supervised learning methods. Fault monitoring presented here is based on frequency domain approach with the aid of artificial intelligence applied on features extracted from fault voltages and currents acquired using PMUs. The feature extraction is achieved by means of real time equivalent voltage phasor angle (EVPA) and equivalent current phasor angle (ECPA), as discussed in Section 7.3.

7.3 EVPA- and ECPA-based transmission line fault detection

Conventional approaches for fault detection are limited to specific transmission line configurations as described in the preceding section. Fault detection methods that can detect transmission line faults occurring in SPG can substantially contribute toward improved fault classification and localization. This section outlines a methodology for detecting transmission line fault occurred anywhere in the grid using PMU measurements at any one of the generator buses. The methodology presented here utilizes frequency domain analysis of EVPA and ECPA at any one of the generator buses [6]. Estimation of these angles and implementation of the presented transmission line fault detection methodology in power grids using frequency domain analysis has been elaborated in the following sections.

7.3.1 Real-time estimation of EVPA and ECPA

EVPA and ECPA are the angles made by three-phase equivalent voltage phasor and three-phase equivalent current phasor to the rotating reference frame (d^e-q^e) [7,8]. The d^e-q^e are orthogonal reference axes that rotate at a constant speed in counterclockwise direction. The speed of rotation is usually taken as equal to the power grid nominal frequency. Real-time estimations of EVPA and ECPA are achieved through mapping the three-phase synchronous voltage and current phasor measurements from PMU to a reference frame rotating at nominal frequency of the grid (d^e-q^e). The mapping of three-phase quantities to a rotating reference frame and achieving equivalent two-phase orthogonal quantities is commonly termed as

Park's transformation. Park's transformation of three-phase voltage and current waveforms can be expressed mathematically as in (7.1)–(7.4):

$$V_{\mathrm{ds}}^{e} = \frac{2}{3}\left\{\begin{array}{l} V_{\mathrm{RM}}\sin(\omega t)\cdot\sin(\omega_s t) + V_{\mathrm{YM}}\sin\left(\omega t - \frac{2\pi}{3}\right)\cdot\sin\left(\omega_s t - \frac{2\pi}{3}\right) \\ +V_{\mathrm{BM}}\sin\left(\omega t + \frac{2\pi}{3}\right)\cdot\sin\left(\omega_s t + \frac{2\pi}{3}\right) \end{array}\right\} \tag{7.1}$$

$$V_{\mathrm{qs}}^{e} = \frac{2}{3}\left\{\begin{array}{l} V_{\mathrm{RM}}\sin(\omega t)\cdot\cos(\omega_s t) + V_{\mathrm{YM}}\sin\left(\omega t - \frac{2\pi}{3}\right)\cdot\cos\left(\omega_s t - \frac{2\pi}{3}\right) \\ +V_{\mathrm{BM}}\sin\left(\omega t + \frac{2\pi}{3}\right)\cdot\cos\left(\omega_s t + \frac{2\pi}{3}\right) \end{array}\right\} \tag{7.2}$$

$$I_{\mathrm{ds}}^{e} = \frac{2}{3}\left\{\begin{array}{l} I_{\mathrm{RM}}\sin(\omega t - \Phi)\cdot\sin(\omega_s t) + I_{\mathrm{YM}}\sin\left(\omega t - \Phi - \frac{2\pi}{3}\right)\cdot\sin\left(\omega_s t - \frac{2\pi}{3}\right) \\ +I_{\mathrm{BM}}\sin\left(\omega t - \Phi + \frac{2\pi}{3}\right)\cdot\sin\left(\omega_s t + \frac{2\pi}{3}\right) \end{array}\right\} \tag{7.3}$$

$$I_{\mathrm{qs}}^{e} = \frac{2}{3}\left\{\begin{array}{l} I_{\mathrm{RM}}\sin(\omega t - \Phi)\cdot\cos(\omega_s t) + I_{\mathrm{YM}}\sin\left(\omega t - \Phi - \frac{2\pi}{3}\right)\cdot\cos\left(\omega_s t - \frac{2\pi}{3}\right) \\ +I_{\mathrm{BM}}\sin\left(\omega t - \Phi + \frac{2\pi}{3}\right)\cdot\cos\left(\omega_s t + \frac{2\pi}{3}\right) \end{array}\right\} \tag{7.4}$$

where V_{ds}^{e} and V_{qs}^{e} are the direct and quadrature axes voltages in rotating reference frame obtained after mapping. Similarly, I_{ds}^{e} and I_{qs}^{e} stand for currents. V_{RM}, V_{YM}, and V_{BM} are the peak values of phase voltages, which are equal during normal operating conditions. Similarly, I_{RM}, I_{YM}, and I_{BM} are peak values of three-phase line currents. All the aforementioned voltages are measured in volts (V) and currents in ampere (A). The angle Φ stands for power factor angle in radians between line voltage and line current. Frequencies ω and ω_s correspond to frequency (in radians per second) of three-phase voltages and rotating reference frame, respectively. A major advantage of this transformation for fault detection is that only two-phase data are required to be processed for analyzing three-phase system. Also, the power frequency components in the three-phase system become direct current (DC) components after the transformation and all other frequency components appear as

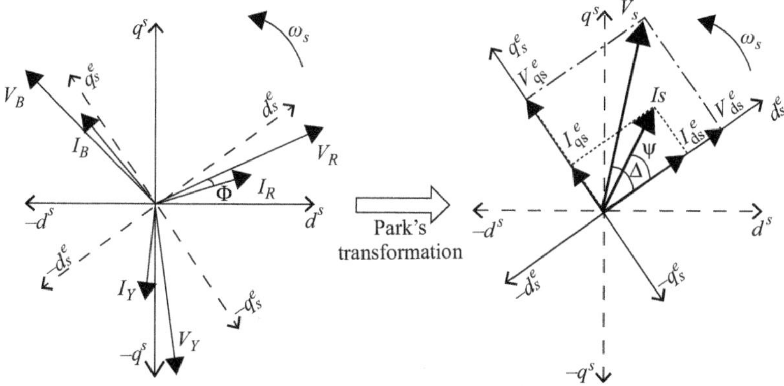

Figure 7.2 Phasor diagram of Park's transformation of three-phase voltages and currents [6]. © 2015 IET. Reprinted, with permission, from [6]

alternating components. This facilitates better perception of harmonic quantities in voltage and current phasors. The Park's transformation of three-phase quantities and the concepts of EVPA and ECPA have been illustrated with the help of phasor diagram as shown in Figure 7.2. In Figure 7.2, the axes d^s and q^s represent the stationary orthogonal reference frame. The axes d_e^s and q_e^s (or d^e and q^e) stand for an autonomous rotating reference frame as mentioned in preceding paragraph. This renders the rotating reference frame to be insensitive to variations in voltage or current phasors. The equivalent voltage phasor (V_s) and current phasor (I_s) form angles Δ and Ψ, respectively, with rotating axes d^e and qe. Parameters of Vs and Is are measured in Vs and As, whereas angles Δ and Ψ are in radians. These angles are termed EVPA and ECPA, respectively, which can be calculated as given in (7.5) and (7.6):

$$\Delta = \tan^{-1}\left(\frac{V_{\text{qs}}^{e}}{V_{\text{ds}}^{e}}\right) \text{ deg} \tag{7.5}$$

$$\Psi = \tan^{-1}\left(\frac{I_{\text{qs}}^{e}}{I_{\text{ds}}^{e}}\right) \text{ deg} \tag{7.6}$$

As the axes (d_e^s and q_e^s) are rotating at nominal grid frequency, the equivalent phase angles (Δ and Ψ) in radians have constant values during normal operating conditions. During transmission line faults of any type and at any location in the grid, these equivalent phase angles undergo variations with frequency components depending on the type of fault occurred (as described in the following sections). Frequency components present in the oscillations of equivalent phase angles (Δ and Ψ) are analyzed using N-point fast Fourier transform (FFT) over a time span which is equal to inverse of the nominal system frequency (*1/f*), as given in (7.7) and (7.8):

$$\chi_k^{\Delta} = \left(\frac{1}{N}\right)\sum_{n=0}^{N-1}\Delta(n)\, e^{-2\pi kn/N} \quad 0 < k < N-1 \tag{7.7}$$

$$\chi_k^{\Psi} = \left(\frac{1}{N}\right)\sum_{n=0}^{N-1}\Psi(n)\, e^{-2\pi kn/N} \quad 0 < k < N-1 \tag{7.8}$$

where χ_k^{Δ} and χ_k^{Δ} are FFT spectrum coefficients of EVPA and ECPA angular variations. The methodology utilizes frequency coefficients present in the variation of EVPA and ECPA for real-time detection of transmission line fault occurred anywhere in SPG. Frequency coefficients corresponding to 0 and 100 Hz in the FFT spectra of angular variations in EVPA and ECPA are deployed for fault detection in the methodology presented. The frequency coefficients corresponding to 0 and 100 Hz of EVPA variations are represented by α_{Δ} and β_{Δ}, respectively. The same for ECPA variation are represented by α_{Ψ} and β_{Ψ}, respectively. The spectral coefficients of EVPA variation corresponding to 0 (χ_0^{Δ}) and 100 Hz (χ_{100}^{Δ}) can be extracted by substituting for the variable k in (7.8) with values of 0 and 2. The resultant mathematical expressions are given by (7.9) and (7.10), respectively:

$$\chi_0^{\Delta} = \alpha_{\Delta} = \left(\frac{1}{N}\right)\sum_{n=0}^{N-1}\Delta(n) \tag{7.9}$$

$$\chi_{100}^{\Delta} = \beta_{\Delta} = \left(\frac{1}{N}\right)\sum_{n=0}^{N-1}\Delta(n)\, e^{-4\pi n/N} \tag{7.10}$$

where χ_0^{Δ} and χ_{100}^{Δ} are represented by α_{Δ} and β_{Δ}, respectively. Similarly, spectral coefficients of ECPA variation corresponding to 0 (χ_0^{Ψ}) and 100 Hz (χ_{100}^{Ψ}) are computed as given in (7.11) and (7.12), respectively:

$$\chi_0^{\Psi} = \alpha_{\Psi} = \left(\frac{1}{N}\right)\sum_{n=0}^{N-1}\Psi(n) \tag{7.11}$$

$$\chi_{100}^{\Psi} = \beta_{\Psi} = \left(\frac{1}{N}\right)\sum_{n=0}^{N-1}\Psi(n)\, e^{-4\pi n/N} \tag{7.12}$$

FFT coefficients χ_0^{Ψ} and χ_{100}^{Ψ} are represented by α_{Ψ} and β_{Ψ}, respectively. Frequency coefficients χ_0^{Δ}, χ_{100}^{Δ}, χ_0^{Ψ}, and χ_{100}^{Ψ} (α_{Δ}, β_{Δ}, α_{Ψ}, and β_{Ψ}, respectively) are decimals. The fault detection methodology, which utilizes these frequency coefficients in EVPA and ECPA, variations, has been elaborated in the following section.

7.3.2 Fault detection using EVPA and ECPA

Formulation of the presented fault detection methodology for SPG based on phase angle variations of EVPA and ECPA is illustrated here. EVPA and ECPA are

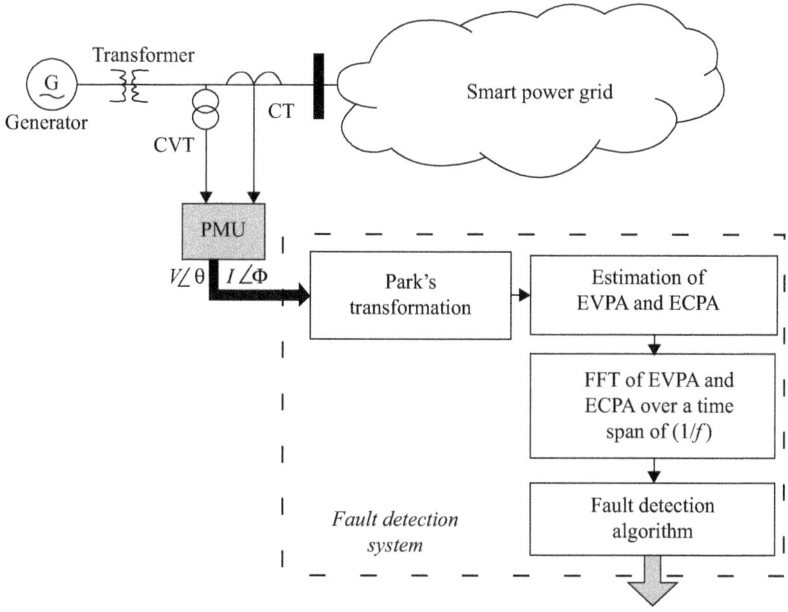

Figure 7.3 Schematic diagram of the fault detection scheme for SPG [6]. © 2015 IET. Reprinted, with permission, from [6]

estimated using PMU measurements at any one generator bus in the grid. The schematic diagram of the methodology for transmission line fault detection in SPG using PMU measurements acquired from one of the generator buses in the grid is shown in Figure 7.3. Capacitor voltage transformer and current transformer are used for acquiring real-time analog voltage, and current measurements for inputting to PMU. EVPA and ECPA at the generator bus are computed from PMU measurements using Park's transformation, as described in the preceding section. EVPA and ECPA waveforms are then analyzed through FFT over a time span of $(1/f)$ in order to estimate the embedded frequency coefficients. During normal operating conditions, EVPA and ECPA have constant values, and this is reflected as a peak value in FFT spectrum at 0 Hz. All other frequency coefficients remain at zero. However, during fault conditions, EVPA and ECPA undergo oscillations with frequency components depending on the type of the fault that has occurred.

Oscillations in EVPA and ECPA are the result of abrupt change in the total reactance seen by the generator due to alteration in the topology of the three-phase system during the occurrence of a fault. Also, the distortion in the line current (I_{line}) resulting from the fault, generates harmonic frequencies in the current. The harmonic currents, while flowing through various impedances of the system, generate harmonic voltages. The harmonic frequencies in voltages and currents, create variations in EVPA and ECPA. When the EVPA and ECPA variations are analyzed through FFT, other frequency components along with 0 Hz frequency can be seen in the spectrum. Furthermore, it has been observed from extensive case studies that frequency coefficients corresponding to 0 and 100 Hz in the spectrum of EVPA and ECPA have severe dependence on the type of fault occurred. This has been corroborated in the following section with case studies. The fault detection methodology for SPG utilizes coefficients corresponding to 0 and 100 Hz in oscillation of EVPA and ECPA for real-time detection of transmission line fault that has occurred anywhere in the grid. Presence of 100 Hz component in EVPA and ECPA variation has been taken as the measure for real-time fault detection. A case study conducted on IEEE-14 bus system is corroborated here for better understanding of the detection methodology. Mathematical modeling of the system and all simulation studies are conducted in MATLAB®/SIMULINK® platform [9,10]. EVPA and ECPA variation of G1 over time during normal operating conditions are shown in Figure 7.4. Corresponding frequency spectra of both angular variations are illustrated in Table 7.1. A sampling frequency of 10 kHz has been opted for frequency domain analysis with FFT applied over a time span which is equal to inverse of nominal system frequency (20 ms for 50 Hz system). The spectral coefficients present in the EVPA and ECPA variations during the normal operating conditions are tabulated in Table 7.1.

It can be observed from Figure 7.4(a) and (b) that EVPA and ECPA have constant values during normal operating conditions, and they are reflected as a high value for 0 Hz, whereas all other frequency coefficients remain at zero in frequency spectra as observed from Table 7.1. The impact of transmission line faults in EVPA and ECPA of IEEE-14 bus system is analyzed by applying various faults of LG (RG) fault, LL (RY) fault, LLG (RYG) fault, and LLL (RYB) fault one at a time in branch 4-5 at a location of 50 km from bus-4 with FR of 0 Ω and FIA of 0°. EVPA and ECPA variations for G1 during LG (RG) fault are shown in Figure 7.5. FFT of EVPA and ECPA has been applied over cycle data from the time of occurrence of fault. Spectral coefficients in the EVPA and ECPA variation are tabulated in Table 7.2 for better clarity. From the tables depicting spectral coefficients (Table 7.2), it can be observed that frequency components β_Δ and β_Ψ exist during transmission line faults, which are absent during normal operating conditions.

Similarly, angular variations for different faults, namely LL, LLG, and LLL faults are plotted in Figures 7.6–7.8, respectively. The spectral coefficients for each of these test cases have been tabulated in respective tables (Tables 7.3–7.5). In detail, results of various case studies are tabulated in Table 7.6.

From Table 7.6, it may be seen, the RG fault that has occurred on branch 2-4, 60 km from bus 2 of IEEE-14 bus system, has resulted in the frequency coefficients α_Δ, β_Δ, α_Ψ, and β_Ψ (decimals) to hold values of 0.17, 0.49, 0.04, and 0.02, respectively. As 100 Hz coefficients (β_Δ and β_Ψ) are greater than zero; the detection algorithm identifies it as a fault. Similarly, for RB and RYG faults, the β_Δ and β_Ψ coefficients are nonzero decimals. For RYB fault, at least one of the β coefficients has nonzero value. These results are in contradiction to the result obtained for no-fault condition, in which none of the β coefficients has nonzero value. Hence, the fault detection system can effectively detect the fault based on β coefficients of EVPA and ECPA variation at a generator bus. The results of the case study

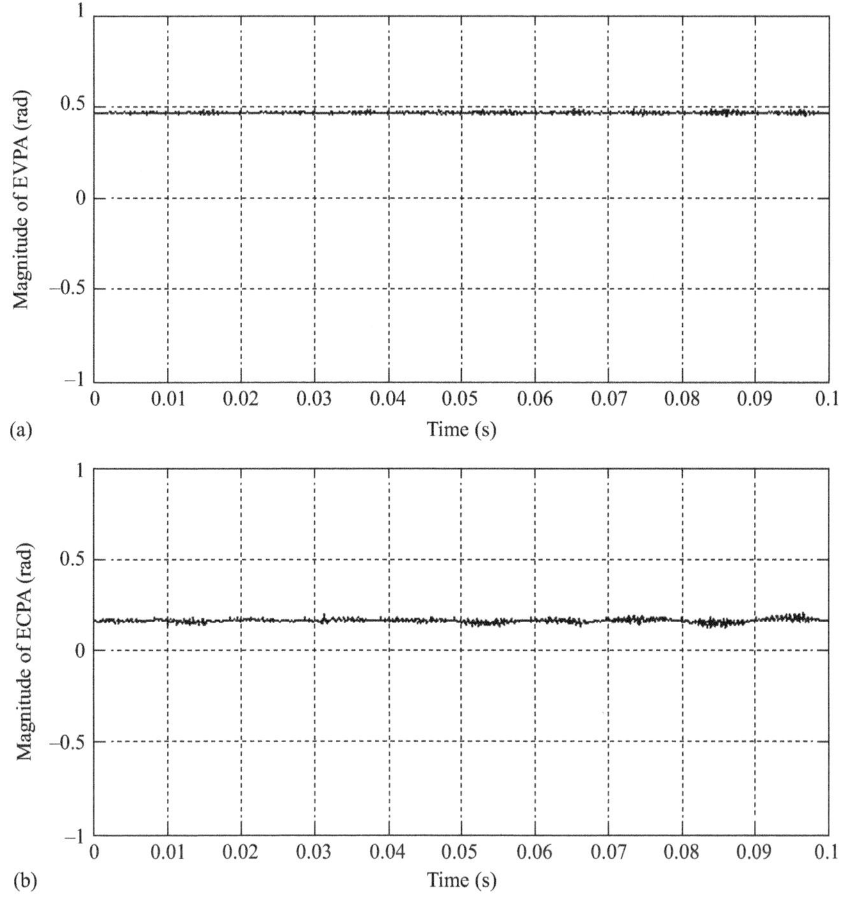

Figure 7.4 In IEEE-14 bus system, during normal operating conditions: (a) equivalent voltage phase angle variation, (b) equivalent current phase angle variation [6]. © 2015 IET. Reprinted, with permission, from [6]

Table 7.1 FFT spectral coefficients of EVPA and ECPA during no-fault condition in IEEE-14 bus system [6]. © 2015 IET. Reprinted, with permission, from [6]

Frequency→ Angles↓	0 Hz (α)	50 Hz	100 Hz (β)	150 Hz	200 Hz	250 Hz	300 Hz	350 Hz	400 Hz
Δ	**0.093**	0	**0**	0	0	0	0	0	0
Ψ	**0.032**	0	**0**	0	0	0	0	0	0

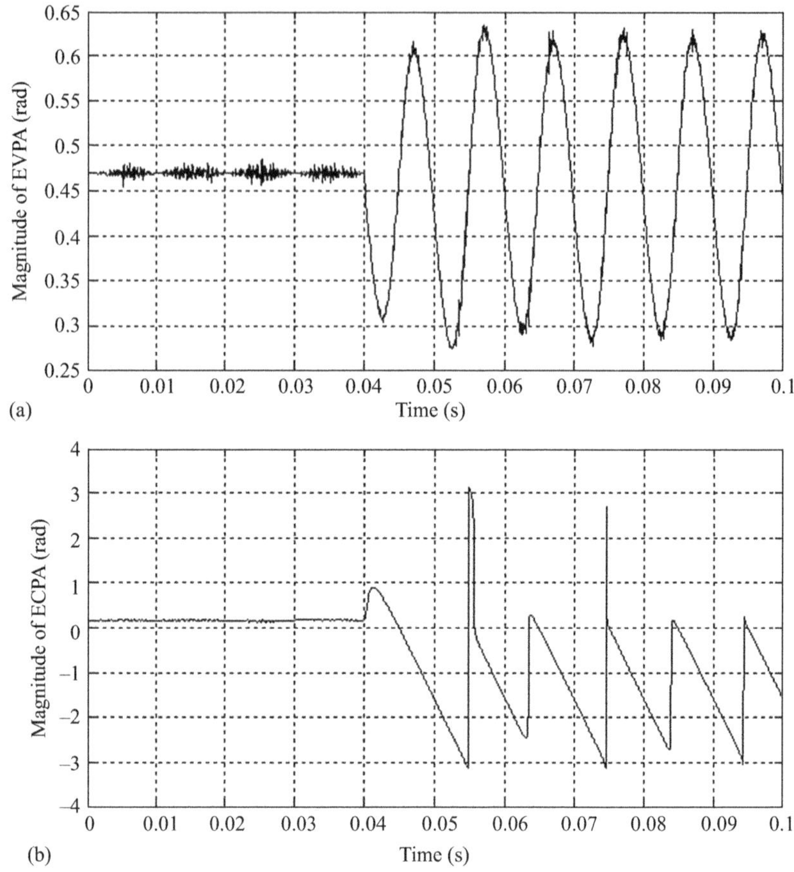

Figure 7.5 In IEEE-14 bus system, during LG Fault (RG) condition at branch 4-5, 50 km away from bus-4: (a) equivalent voltage phase angle variation, (b) equivalent current phase angle variation [6]. © 2015 IET. Reprinted, with permission, from [6]

Table 7.2 FFT spectral coefficients of EVPA and ECPA during LG (RG) fault condition in IEEE-14 bus system with FR of 0 Ω and FIA of 0° [6]. © 2015 IET. Reprinted, with permission, from [6]

Frequency→ Angles↓	**0 Hz (α)**	**50 Hz**	**100 Hz (β)**	**150 Hz**	**200 Hz**	**250 Hz**	**300 Hz**	**350 Hz**	**400 Hz**	**450 Hz**	**500 Hz**	**550 Hz**	**600 Hz**
Δ	**0.2**	0.07	**0.6**	0.01	0.07	0	0	0	0	0	0	0	0
Ψ	**0.12**	0.01	**0.04**	0.01	0.02	0.01	0.01	0.01	0.01	0.01	0.01	0.01	0.01

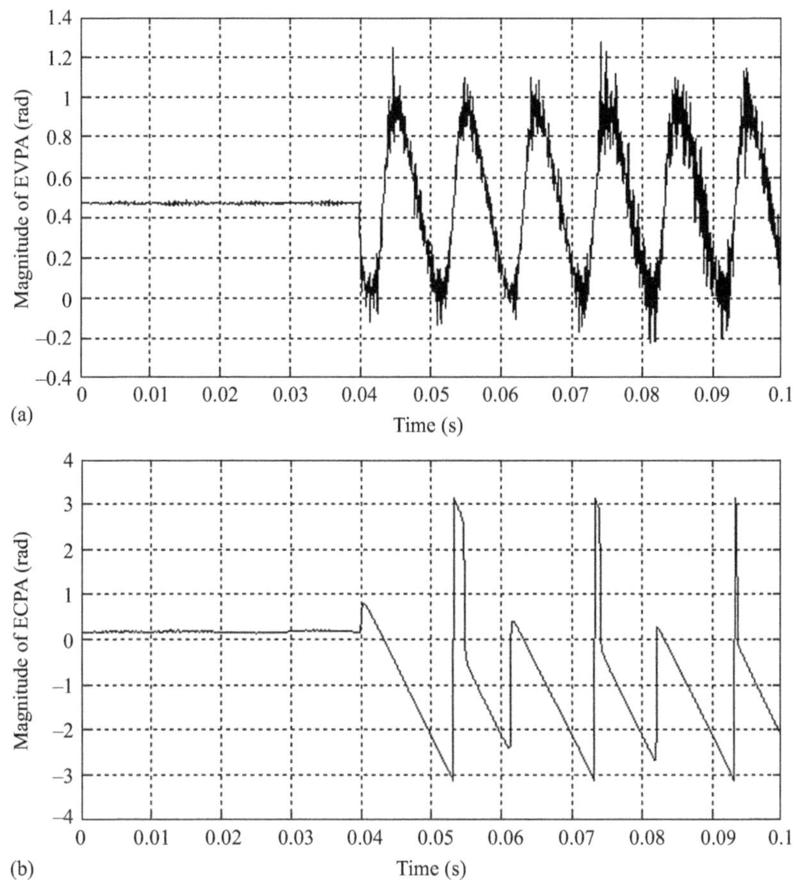

Figure 7.6 In IEEE-14 bus system, during LL-Fault condition at branch 4-5, 50 km away from bus-4: (a) EVPA variation, (b) ECPA variation [6]. © 2015 IET. Reprinted, with permission, from [6]

Table 7.3 FFT spectral coefficients of EVPA and ECPA during LL (RY) fault condition in IEEE-14 bus system with FR of 0 Ω and FIA of 0° [6]. © 2015 IET. Reprinted, with permission, from [6]

Frequency→ Angles↓	**0 Hz (α)**	**50 Hz**	**100 Hz (β)**	**150 Hz**	**200 Hz**	**250 Hz**	**300 Hz**	**350 Hz**	**400 Hz**	**450 Hz**	**500 Hz**	**550 Hz**	**600 Hz**
Δ	**0.1**	0	**1.6**	0	0.4	0	0.1	0	0	0	0	0	0
Ψ	**0.12**	0.01	**0.06**	0.01	0.02	0.02	0.02	0.02	0.01	0.01	0.01	0.01	0.01

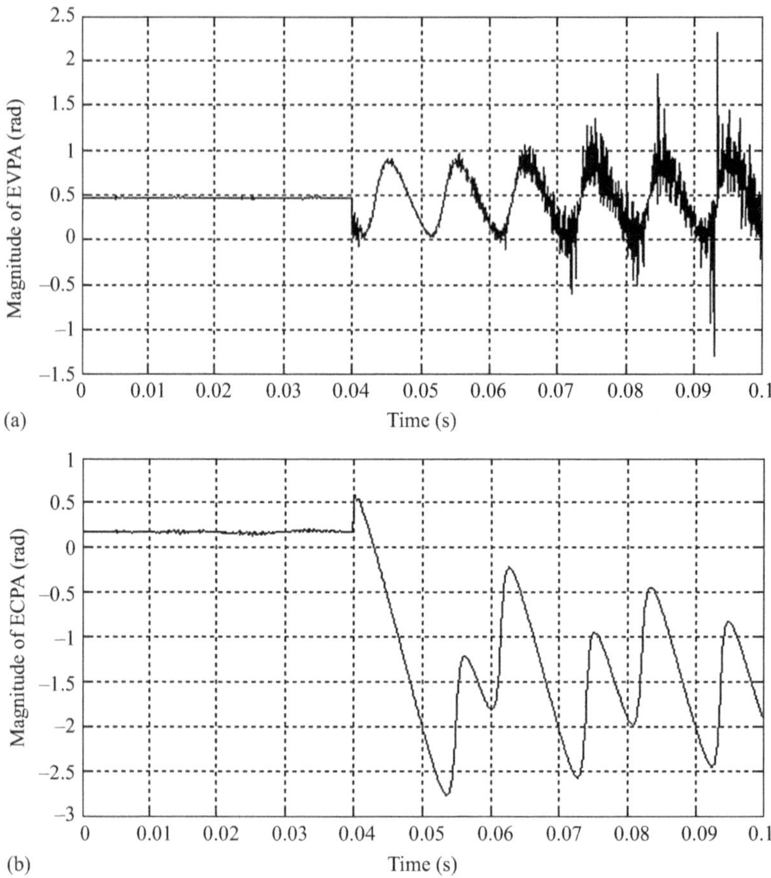

Figure 7.7 In IEEE-14 bus system, during LLG-Fault condition at branch 4-5, 50 km away from bus-4: (a) equivalent voltage phase angle variation, (b) equivalent current phase angle variation [6]. © 2015 IET. Reprinted, with permission, from [6]

Table 7.4 FFT Spectral coefficients of EVPA and ECPA during LLG (RYG) fault condition in IEEE-14 bus system with FR of 0 Ω and FIA of 0° [6]. © 2015 IET. Reprinted, with permission, from [6]

Frequency→ Angles↓	**0 Hz (α)**	**50 Hz**	**100 Hz (β)**	**150 Hz**	**200 Hz**	**250 Hz**	**300 Hz**	**350 Hz**	**400 Hz**	**450 Hz**	**500 Hz**	**550 Hz**	**600 Hz**
Δ	**0.1**	0.06	**1.7**	0.08	0.6	0.04	0.3	0.02	0.2	0.01	0.1	0.06	0.1
Ψ	**0.15**	0.03	**0.02**	0.01	0	0	0	0	0	0	0	0	0

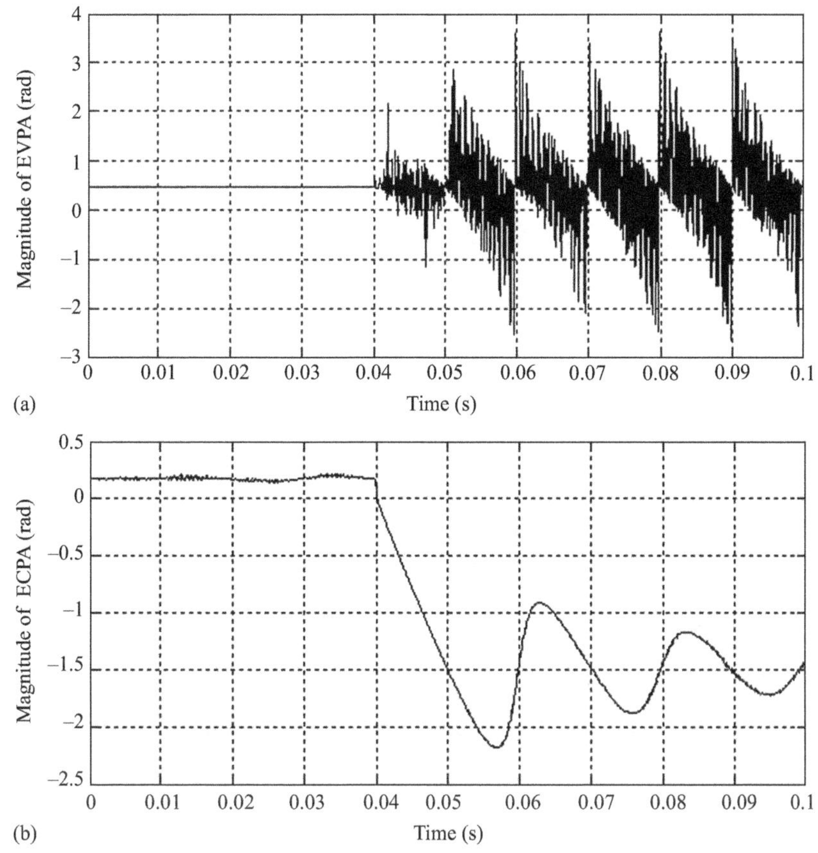

Figure 7.8 In IEEE-bus system, during LLL-Fault condition at branch 4-5, 50 km away from bus-4: (a) equivalent voltage phase angle variation, (b) equivalent current phase angle variation [6]. © 2015 IET. Reprinted, with permission, from [6]

Table 7.5 FFT spectral coefficients of EVPA and ECPA during LLL (RYB) fault condition in IEEE-14 bus system with FR of 0 Ω and FIA of 0° [6]. © 2015 IET. Reprinted, with permission, from [6]

Frequency→ Angles↓	**0 Hz (α)**	**50 Hz**	**100 Hz (β)**	**150 Hz**	**200 Hz**	**250 Hz**	**300 Hz**	**350 Hz**	**400 Hz**	**450 Hz**	**500 Hz**	**550 Hz**	**600 Hz**
Δ	**0.09**	0.37	**1.39**	0.17	0.6	0.01	0.39	0	0.36	0	0.2	0.13	0.23
Ψ	**0.16**	0.02	**0.01**	0	0	0	0	0	0	0	0	0	0

Table 7.6 Results of case studies conducted using IEEE-14 bus system for fault detection for different fault locations with FR of 0 Ω and FIA of 0° [6]. © 2015 IET. Reprinted, with permission, from [6]

Sl.	Fault location	Fault type created	α_Δ, β_Δ	α_Ψ, β_Ψ	Is $\beta_\Delta > 0 \| \beta_\Psi > 0$ (yes/no)	Fault detected?
1	Branch 2-4, 60 km from bus-2	RG	0.17, 0.49	0.04, 0.02	YES	YES
2	Branch 4-5, 50 km from bus-4	YG	0.20, 0.56	0.06, 0.02	YES	YES
3	Branch 1-2, 70 km from bus-1	BG	0.18, 0.44	0.04, 0.02	YES	YES
4	Branch 3-4, 20 km from bus-3	RY	0.05, 1.32	0.05, 0.03	YES	YES
5	Branch 13-14, 10 km from bus-13	YB	0.01, 0.9	0.06, 0.02	YES	YES
6	Branch 6-11, 40 km from bus-6	BR	0.02, 0.69	0.05, 0.02	YES	YES
7	Branch 12-13, 90 km from bus-12	RYG	0.11, 0.64	0.07, 0.01	YES	YES
8	Branch 7-9, 80 km from bus-7	YBG	0.14, 0.80	0.07, 0.01	YES	YES
9	Branch 4-9, 10 km from bus-4	BRG	0.13, 0.74	0.07, 0.01	YES	YES
10	Branch 9-10, 30 km from bus-6	RYB	0.11, 0.12	0.08, 0	YES	YES
11	Bus-3 (bus fault)	RG	0.22, 0.53	0.05, 0.02	YES	YES
12	Bus-4 (bus fault)	YG	0.24, 0.59	0.07, 0.02	YES	YES
13	Bus-5 (bus fault)	BG	0.20, 0.48	0.06, 0.02	YES	YES
14	Bus-6 (bus fault)	AB	0.07, 1.94	0.07, 0.04	YES	YES
15	Bus-7 (bus fault)	BC	0.04, 0.9	0.08, 0.02	YES	YES
16	Bus-8 (bus fault)	CA	0.05, 1.29	0.07, 0.02	YES	YES
17	Bus-9 (bus fault)	RYG	0.21, 0.82	0.09, 0.03	YES	YES
18	Bus-10 (bus fault)	YBG	0.23, 0.92	0.08, 0.02	YES	YES
19	Bus-11 (bus fault)	BRG	0.14, 0.88	0.09, 0.01	YES	YES
20	Bus-14 (bus fault)	RYB	0.15, 0.17	0.1, 0	YES	YES
21	No fault	–	0.49, 0	0.18, 0	NO	NO

depicted in this section validate that β coefficients (100 Hz) of EVPA and ECPA variations have impacted the occurrence of transmission line faults. Also, the fault detection methodology based on β_Δ and β_Ψ is shown to be efficient in detecting fault that occurring anywhere in the grid. FFT coefficients extracted from variations in EVPA and ECPA at a generator bus in the grid have been utilized for real-time transmission line fault detection in the complete power grid. The methodology does not require any data from other buses or PMUs. Fault detection is achieved locally at the generator bus, so that fault information alone can be sent to SPC.

The fault information can facilitate SPC to undertake wide-area control actions for maintaining stability after occurrence of the transmission line fault.

7.4 EVPA- and ECPA-based transmission line fault classification

After successful detection of a transmission line fault, classification of the type of fault that has occurred is significant in power system fault analysis. The conventional approach for classifying transmission line faults is limited to specific transmission line configurations as described in the preceding section. Transmission line fault classification methods that can classify faults occurring in SPGs can significantly contribute to enhanced fault localization and restoration. A fault classification methodology based on PMU measurements that can accurately classify any transmission line fault occurred anywhere in the grid is presented here. The methodology presented for fault classification in SPGs utilizes frequency domain analysis of EVPA and ECPA at any one of the generator buses. Estimation of these angles and method of frequency domain analysis using FFT have been elaborated in the preceding section. From the FFT results pertaining to various fault conditions in IEEE-14 bus system (Tables 7.2–7.6), it can be perceived that values of the spectral coefficients α_Δ, α_Ψ, β_Δ, and β_Ψ depend on the type of fault that has occurred. The hidden information regarding the type of fault can be extracted from spectral coefficients using the artificial intelligence approach. The fault classification approach from spectral coefficients of EVPA and ECPA variations (α_Δ, α_Ψ, β_Δ, and β_Ψ) using artificial intelligence technique is shown in Figure 7.9.

Spectral coefficients α_Δ, α_Ψ, β_Δ, and β_Ψ are fed to multiclass support vector machine (SVM) classifier (artificial intelligence module) for classification of the type of fault [11,12]. As phase information is lost during transformation of the three-phase phasors to two-phase quantities using Park's transformation, SVM classifier also necessitates normalized voltage coefficients (NVCs) (Γ_R, Γ_Y, and Γ_B) of three-phase voltages for faulty phase classification. NVC of each phase Γ_R, Γ_Y, and Γ_B are calculated from the FFT analysis of three-phase voltages as given in (7.13)–(7.15):

$$\Gamma_R = \frac{P_R}{\max\{P_R, P_Y, P_B\}} \tag{7.13}$$

$$\Gamma_Y = \frac{P_Y}{\max\{P_R, P_Y, P_B\}} \tag{7.14}$$

$$\Gamma_B = \frac{P_B}{\max\{P_R, P_Y, P_B\}} \tag{7.15}$$

where P_i is the FFT coefficient corresponding to nominal system frequency component of ith phase voltage. All the variables associated with (7.13)–(7.15) are decimals. NVC indicate the ratio of fundamental component of each phase with respect to maximum value of fundamental component in three phases. Hence with

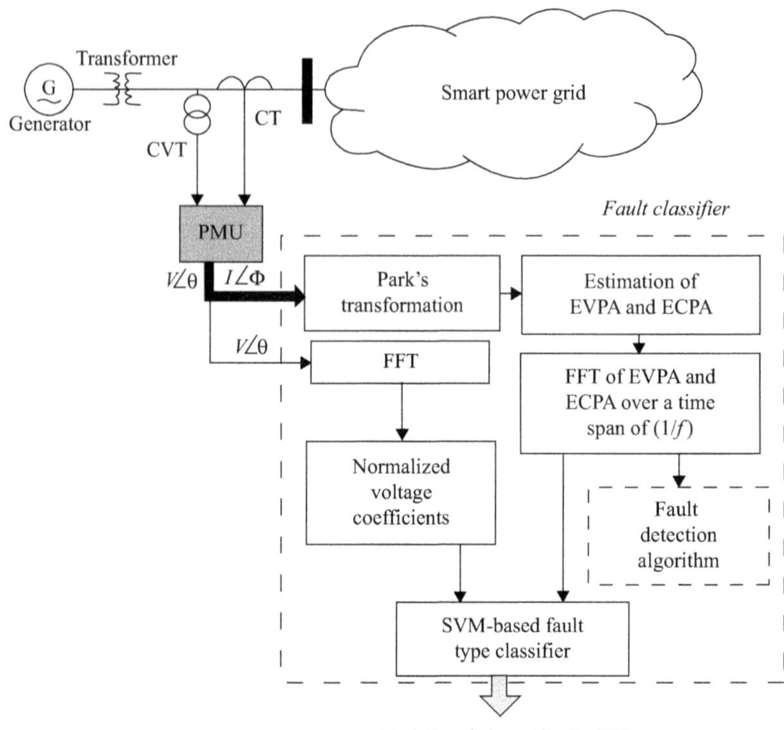

Figure 7.9 Functional block diagram of fault classification methodology presented for SPG [6]. © 2015 IET. Reprinted, with permission, from [6]

spectral coefficients, α_Δ, α_Ψ, β_Δ, and β_Ψ and NVCs, SVM classifier can classify the type of fault occurred with phases involved in the fault. One-step SVM classifier has been deployed in this research work for fault classification. The training data for the SVM module as a set of equivalent phase angle coefficients (α_Δ, β_Δ, α_Ψ, and β_Ψ) and NVCs are categorized with fault type or no-fault labels. The input data sets are generated for different fault locations, fault resistances, and inception angles. The labeled fault type data corresponding to each set of values in input data set constitute the output data set for SVM classifier. SVM classifier is trained with 1,300 data sets to afford enhanced boundary for fault classification. The methodology is corroborated here with a case study conducted in IEEE-14 bus system. Magnitude of the frequency components extracted through FFT analysis of EVPA and ECPA (measured from PMU installed at bus-1) along with NVCs at bus-1 has been utilized for real time fault classification. The same case studies conducted for different fault conditions in IEEE-14 bus system have been adopted for validating the fault classification methodology. It can be observed from Tables 7.2–7.6 that the magnitudes of FFT coefficients (α_Δ, α_Ψ, β_Δ, and β_Ψ) have impact on the type of

Table 7.7 Results of case studies conducted using IEEE-14 bus system for fault classification for different fault locations with FR of 0 Ω and FIA of 0° [6]. © 2015 IET. Reprinted, with permission, from [6]

Sl No.	Fault location	Fault type created	α_Δ, β_Δ	α_Ψ, β_Ψ	Γ_R, Γ_Y, Γ_B	Fault type identified by SVM classifier
1	Branch 2-4, 60 km from bus-2	RG	0.17, 0.49	0.04, 0.02	0.73, 1, 0.99	**RG**
2	Branch 4-5, 50 km from bus-4	YG	0.20, 0.56	0.06, 0.02	0.99, 0.68, 1	**YG**
3	Branch 1-2, 70 km from bus-1	BG	0.18, 0.44	0.04, 0.02	1, 0.99, 0.74	**BG**
4	Branch 3-4, 20 km from bus-3	RY	0.05, 1.32	0.05, 0.03	0.76, 0.77, 1	**RY**
5	Branch 13-14, 10 km from bus-13	YB	0.01, 0.9	0.06, 0.02	1, 0.84, 0.84	**YB**
6	Branch 6-11, 40 km from bus-6	BR	0.02, 0.69	0.05, 0.02	0.84, 1, 0.85	**BR**
7	Branch 12-13, 90 km from bus-12	RYG	0.11, 0.64	0.07, 0.01	0.79, 0.81, 1	**RYG**
8	Branch 7-9, 80 km from bus-7	YBG	0.14, 0.80	0.07, 0.01	1, 0.78, 0.72	**YBG**
9	Branch 4-9, 10 km from bus-4	BRG	0.13, 0.74	0.07, 0.01	0.77, 1, 0.73	**BRG**
10	Branch 9-10, 30 km from bus-6	RYB	0.11, 0.12	0.08, 0	1, 1, 1	**RYB**
11	Bus-3	RG	0.22, 0.53	0.05, 0.02	0.73, 1, 0.99	**RG**
12	Bus-4	YG	0.24, 0.59	0.07, 0.02	0.99, 0.72, 1	**YG**
13	Bus-5	BG	0.20, 0.48	0.06, 0.02	1, 0.99, 0.73	**BG**
14	Bus-6	RY	0.07, 1.94	0.07, 0.04	0.68, 0.67, 1	**RY**
15	Bus-7	YB	0.04, 0.9	0.08, 0.02	1, 0.66, 0.65	**YB**
16	Bus-8	BR	0.05, 1.29	0.07, 0.02	0.64, 1, 0.65	**BR**
17	Bus-9	RYG	0.21, 0.82	0.09, 0.03	0.69, 0.72, 1	**RYG**
18	Bus-10	YBG	0.23, 0.92	0.08, 0.02	1, 0.72, 0.72	**YBG**
19	Bus-11	BRG	0.14, 0.88	0.09, 0.01	0.71, 1, 0.73	**BRG**
20	Bus-14	RYB	0.15, 0.17	0.1, 0	1, 1, 1	**RYB**
21	No fault	–	0.49, 0	0.18, 0	1, 1, 1	**No-fault**

the fault. Hence the spectral coefficients (α_Δ, α_Ψ, β_Δ and β_Ψ) along with NVCs are fed to SVM classifier to classify the fault. It can be seen from Table 7.7 that all the faults that have occurred are successfully detected and classified by the SVM classifier. Also, it is obvious from Table 7.7 that the methodology can clearly discriminate fault conditions from no-fault condition.

Table 7.7 shows that the RG fault on branch 2-4, 60 km from bus-2 of IEEE-14 bus system has resulted in α_Δ, β_Δ, α_Ψ, β_Ψ, Γ_R, Γ_Y, and Γ_B to hold values of 0.17, 0.49, 0.04, 0.02, 0.73, 1, and 0.99, respectively. These values are given to SVM

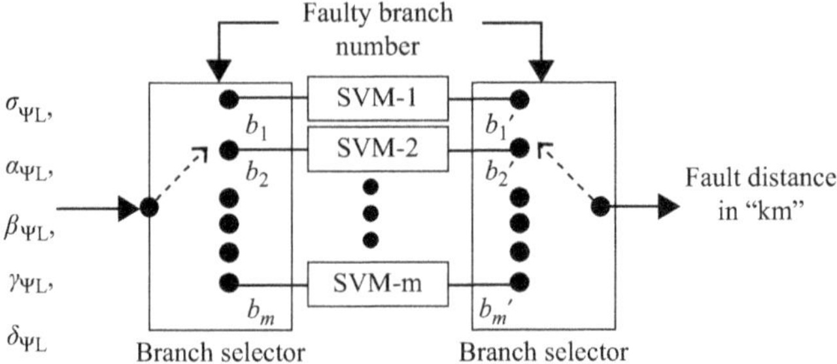

Figure 7.11 Functional block diagram of the fault localization using SVM classifier [16]. © 2015 IET. Reprinted, with permission, from [16]

because, the total impedance seen by the parent bus and the magnitude of the fault current changes with fault location. During normal operating conditions, frequency coefficient corresponding to only 0 Hz exists in the FFT spectrum of ECPA variation, and all other frequency coefficients remain at zero. However, during fault conditions, frequency coefficients corresponding to 0, 50, 100, 150, and 200 Hz are found to be existing in FFT spectrum of ECPA variation. These coefficients are represented by $\sigma_{\Psi L}$, $\alpha_{\Psi L}$, $\beta_{\Psi L}$, $\gamma_{\Psi L}$, and $\delta_{\Psi L}$, respectively. Fault location is estimated from the FFT coefficients with the help of multiclass SVM classifier. The excellent generalization properties and robustness in dealing with multidimensional data make multiclass SVM classifier to be an ideal tool for fault localizing in SPGs [12,13]. The functional block diagram of the fault localization using SVM classifier is shown in Figure 7.11. It consists of m number of SVM modules trained for different fault conditions (i.e., various FR, FIA, and fault locations) in each branch of the network, in which m is the number of branches in the network ($b_1, b_2, \ldots, b_m$). Branch selector is present at two sides of the SVM modules that operate synchronously with the faulty branch number.

Branch selector at the input side directs the input data ($\sigma_{\Psi L}$, $\alpha_{\Psi L}$, $\beta_{\Psi L}$, $\gamma_{\Psi L}$, and $\delta_{\Psi L}$) to that SVM module which is trained for the detected faulty branch. Branch selector at output side directs SVM module output (fault location distance) to output port. Faulty branch number, as an input, facilitates precise selection of SVM module in both input and output ports. From the FFT coefficients, SVM module estimates the distance of the fault location (in kilometers) in the faulty branch from parent bus. The methodology with all stages is corroborated here with a case study conducted in IEEE-14 bus system taking bus-1 (to which generator G1 is connected) as reference bus. EVPA deviation of all buses from that of G1 during various operating conditions is illustrated in following figures. Figure 7.12 displays the EVPA deviations during no-fault conditions. It can be seen from the figure that EVPA deviations are of constant value during these operating conditions. The constant value of EVPA deviations reflects as a nonzero value for 0-Hz component and 0 value for all other frequency components in FFT spectra. EVPA deviations during LG (RG) fault occurred in branch 3-4 connecting bus-3 and bus-4 at 25 km

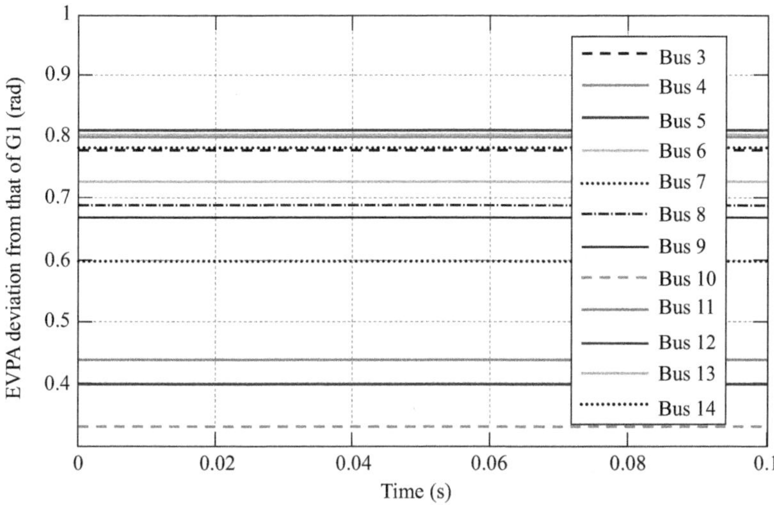

Figure 7.12 EVPA deviation from that of G1 during no-fault condition in IEEE-14 bus system [16]. © 2015 IET. Reprinted, with permission, from [16]

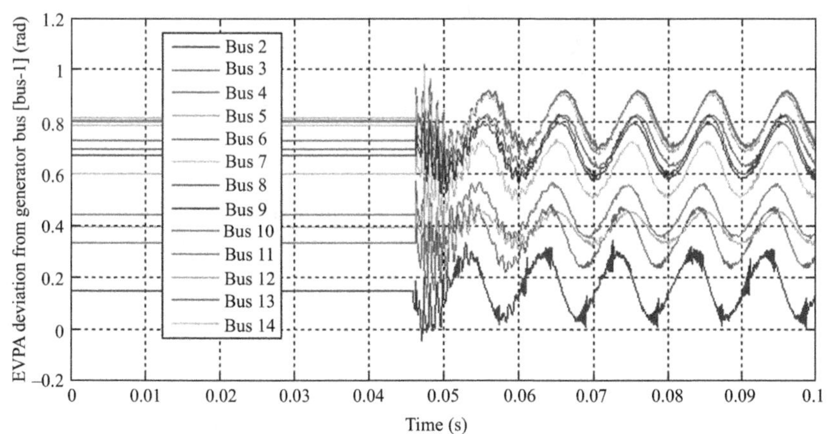

Figure 7.13 EVPA deviation from that of G1 in IEEE-14 bus system during LG-fault (RG) condition in branch 3-2 at 25 km away from bus-3 with FIA = 0° and FR = 0 Ω [16]. © 2015 IET. Reprinted, with permission, from [16]

from bus-3 with FR and FIA of 0 Ω and 0°, respectively, is shown in Figure 7.13. Fault has occurred at 0.04 s and hence till this time EVPA deviation has a constant value and afterward undergoes variations as a result of the fault. The magnitudes of FFT spectral coefficients $\alpha_{\Delta G}$, $\beta_{\Delta G}$, $\gamma_{\Delta G}$, and $\delta_{\Delta G}$ are shown in Figure 7.14.

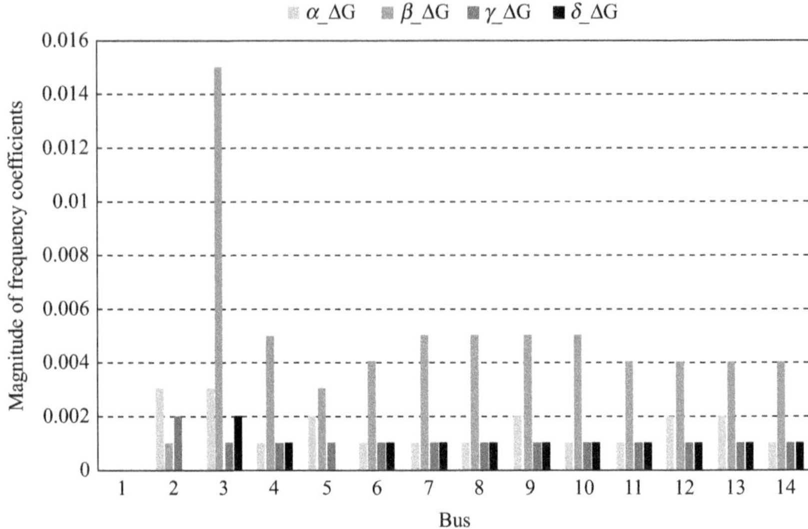

Figure 7.14 Magnitudes of FFT spectral coefficients $\alpha_{\Delta G}$, $\beta_{\Delta G}$, $\gamma_{\Delta G}$, and $\delta_{\Delta G}$ in EVPA deviation of each bus from that of G1 in IEEE-14 bus system during LG-fault (RG) condition in branch 3-2 at 25 km away from bus-3 with FIA = 0° and FR = 0 Ω

The value of $\sigma_{\Delta G}$ is not included as it has much higher magnitude compared to other coefficients. Also it has minimal impact on parent bus identification. Since bus-1 (the bus to which G1 is connected) is taken as reference, EVPA deviation is zero and hence the FFT coefficients for this bus are also zeros. EVPA variations during the occurrence of LL fault on branch 11-6 at 60 km from bus-11 with FR of 25 Ω and FIA of 50° is shown in Figure 7.15. Similar to LG fault, up to 0.04 s EVPA is of constant value and thereafter experiences variation. The magnitudes of FFT spectral coefficients $\alpha_{\Delta G}$, $\beta_{\Delta G}$, $\gamma_{\Delta G}$, and $\delta_{\Delta G}$ are shown in Figure 7.16. Similar behavior can be observed for LLG and LLL faults, as given in Figures 7.17 and 7.19. Corresponding FFT spectral coefficients $\alpha_{\Delta G}$, $\beta_{\Delta G}$, $\gamma_{\Delta G}$, and $\delta_{\Delta G}$ are shown in Figures 7.18 and 7.20. In contradiction to no-fault case, frequency coefficients other than σ_{Δ} exist in fault cases. This condition is utilized for real-time detection of transmission line faults. It is obvious from the FFT spectra shown in Figures 7.14, 7.16, 7.18, and 7.20 that the bus to which faulty branch is connected will have the highest FFT coefficient of $\beta_{\Delta G}$.

After identification of the parent bus, EVPA deviations of all the connected buses from the parent bus are estimated, and FFT has been performed on this deviation. The deviation of EVPA between parent bus and connected buses are shown in Figures 7.21–7.24, respectively, for each fault condition. For better understanding, spectral coefficients extracted from FFT spectra of EVPA deviations between buses for all four fault cases are tabulated in Table 7.8. It is clear

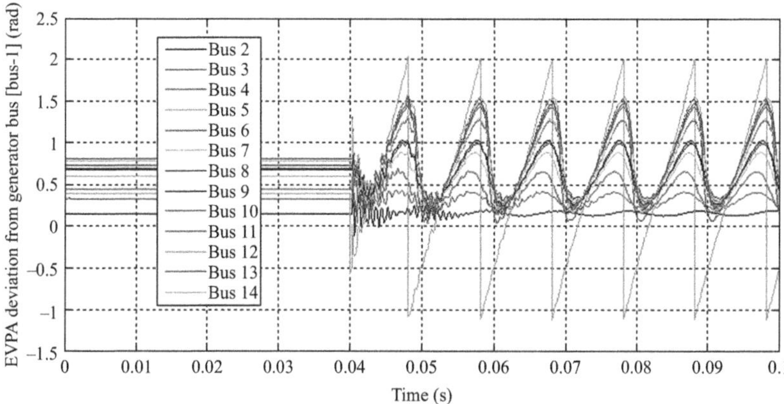

Figure 7.15 *EVPA deviation of each bus from that of G1 in IEEE-14 bus system during LL-fault (RY) condition in branch 11-6 at 60 km away from bus-11 with FIA = 50° and FR = 25 Ω [16]. © 2015 IET. Reprinted, with permission, from [16]*

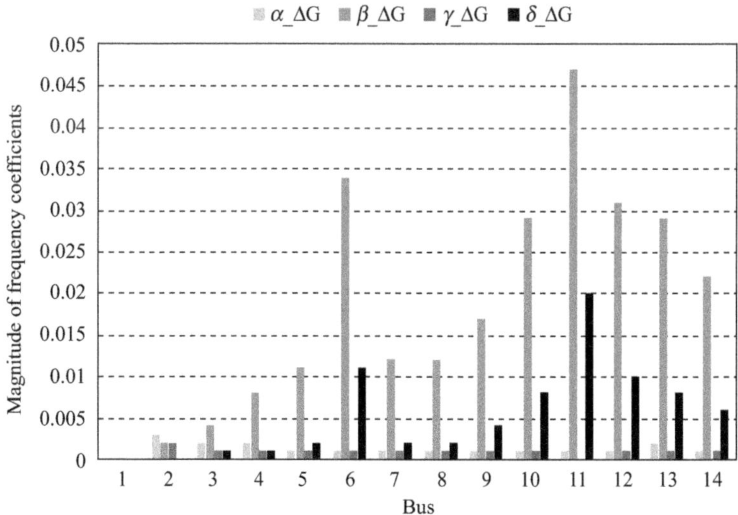

Figure 7.16 *Magnitudes of FFT spectral coefficients $\alpha_{\Delta G}$, $\beta_{\Delta G}$, $\gamma_{\Delta G}$, and $\delta_{\Delta G}$ in EVPA deviation of each bus from that of G1 in IEEE-14 bus system during LL-fault (RY) condition in branch 11-6 at 60 km away from bus-11 with FIA = 50° and FR = 25 Ω*

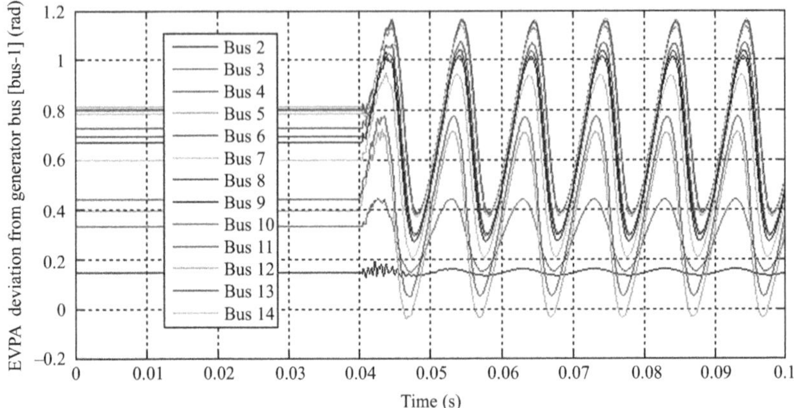

Figure 7.17 EVPA deviation of each bus from that of G1 in IEEE-14 bus system during LLG-fault (RYG) condition in branch 5-4 at 70 km away from bus-5 with FIA = 90° and FR = 60 Ω [16]. © 2015 IET. Reprinted, with permission, from [16]

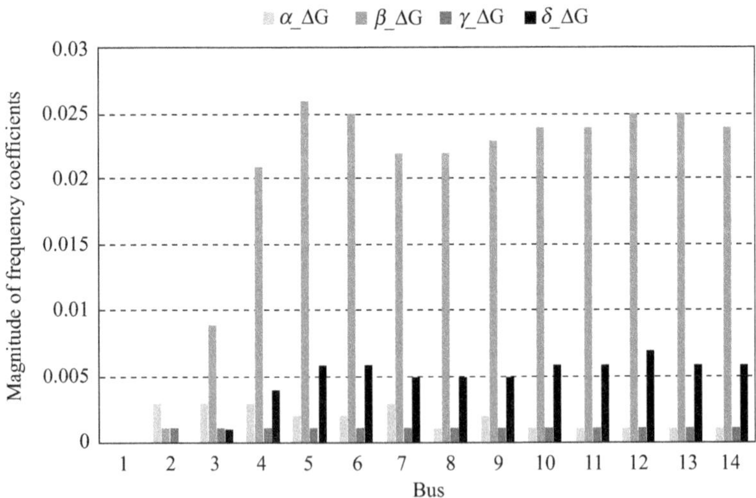

Figure 7.18 Magnitudes of FFT spectral coefficients $\alpha_{\Delta G}$, $\beta_{\Delta G}$, $\gamma_{\Delta G}$, and $\delta_{\Delta G}$ in EVPA deviation of each bus from that of G1 in IEEE-14 bus system during LLG-fault (RYG) condition in branch 5-4 at 70 km away from bus-5 with FIA = 90° and FR = 60 Ω [16]

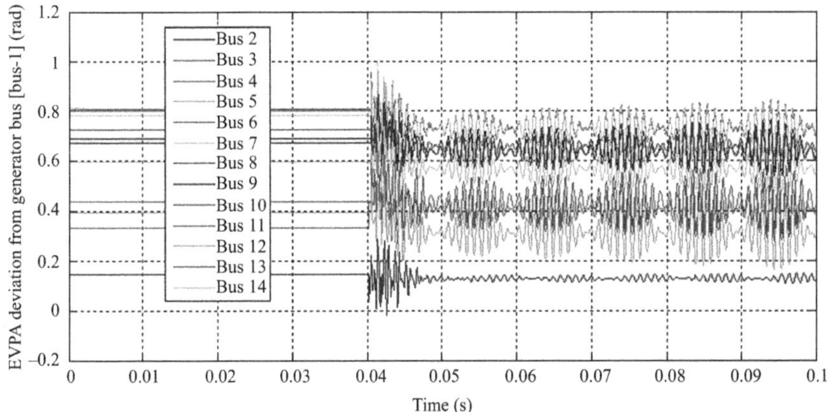

Figure 7.19 EVPA deviation of each bus from that of G1 in IEEE-14 bus system during LLL-fault (RYB) condition in branch 12-13 at 80 km away from bus-12 with FIA = 120° and FR = 75 Ω [16]. © 2015 IET. Reprinted, with permission, from [16]

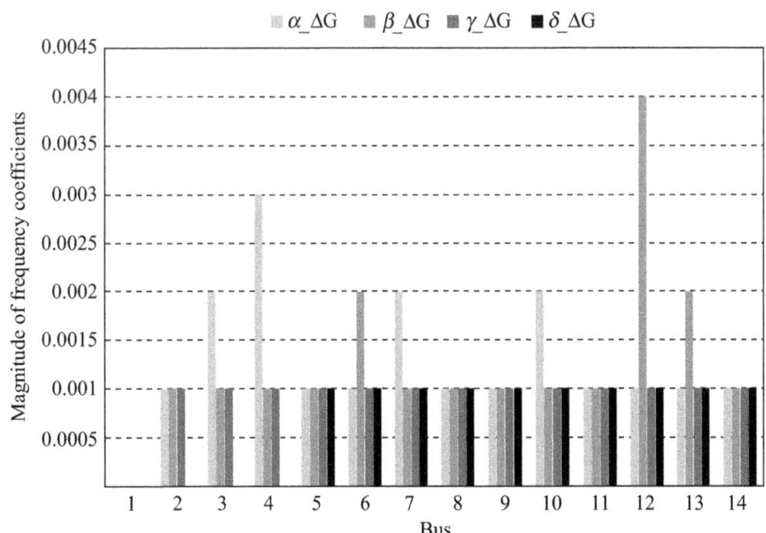

Figure 7.20 Magnitudes of FFT spectral coefficients $\alpha_{\Delta G}$, $\beta_{\Delta G}$, $\gamma_{\Delta G}$, and $\delta_{\Delta G}$ in EVPA deviation of each bus from that of G1 in IEEE-14 bus system during LLL-fault (RYB) condition in branch 12-13 at 80 km away from bus-12 with FIA = 120° and FR = 75 Ω [16]

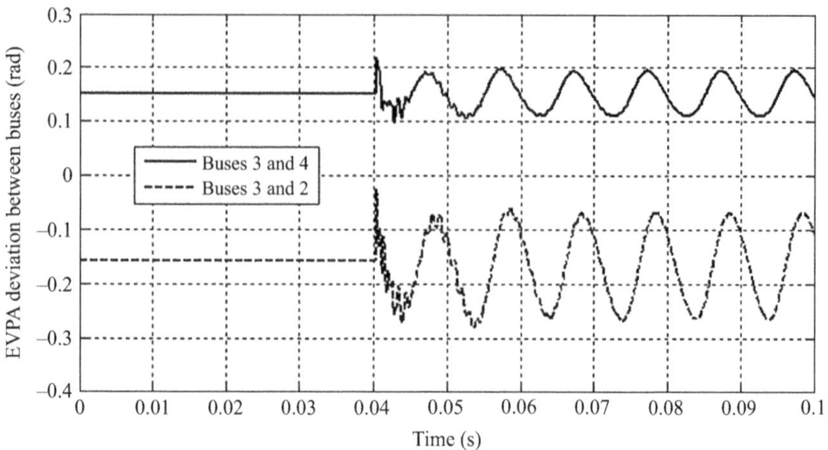

Figure 7.21 EVPA deviation between bus-3 and connected buses during LG fault (RG) in IEEE-14 bus system [16]. © 2015 IET. Reprinted, with permission, from [16]

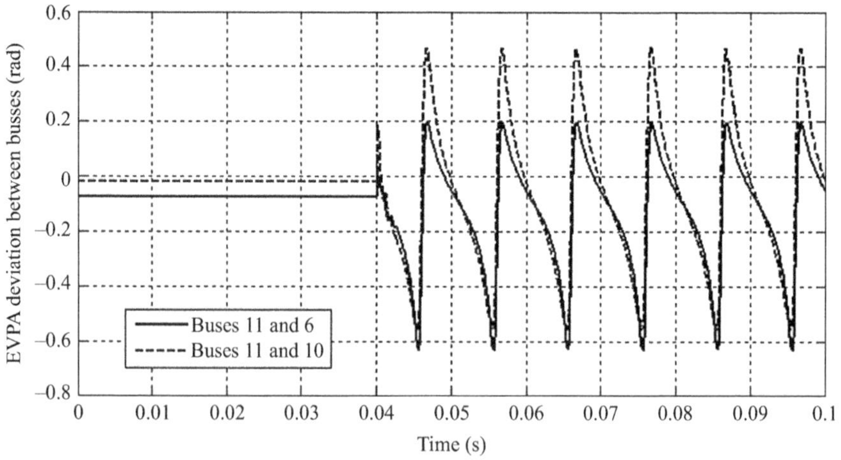

Figure 7.22 EVPA deviation between bus-11 and connected buses during LL fault (RY) in IEEE-14 bus system [16]. © 2015 IET. Reprinted, with permission, from [16]

from Table 7.8 that FFT coefficient corresponding to 100 Hz ($\beta_{\Delta\Delta}$) of EVPA deviation between parent bus and that bus connected through faulty branch will have least value compared to that of other connected branches. For example, during LG fault condition in branch 3-4, the parent bus (bus-3) has two branches connected to it, namely branches 3-4 and 3-2. In branch 3-2, the FFT coefficients of

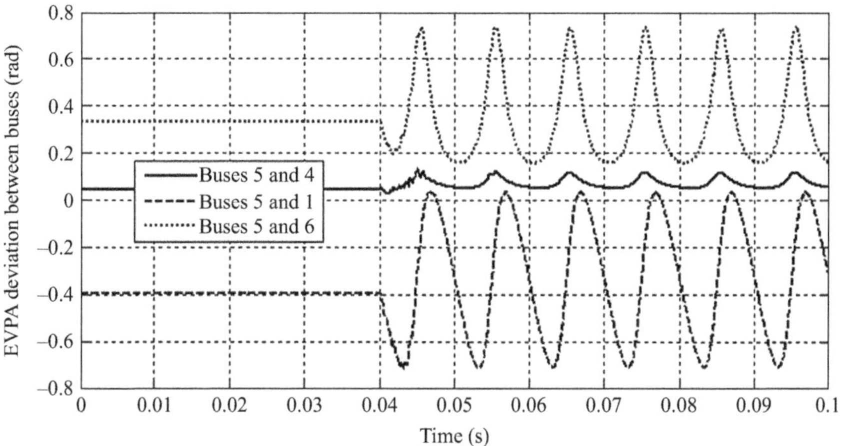

Figure 7.23 EVPA deviation between bus-5 and connected buses during LLG fault (RYG) in IEEE-14 bus system [16]. © 2015 IET. Reprinted, with permission, from [16]

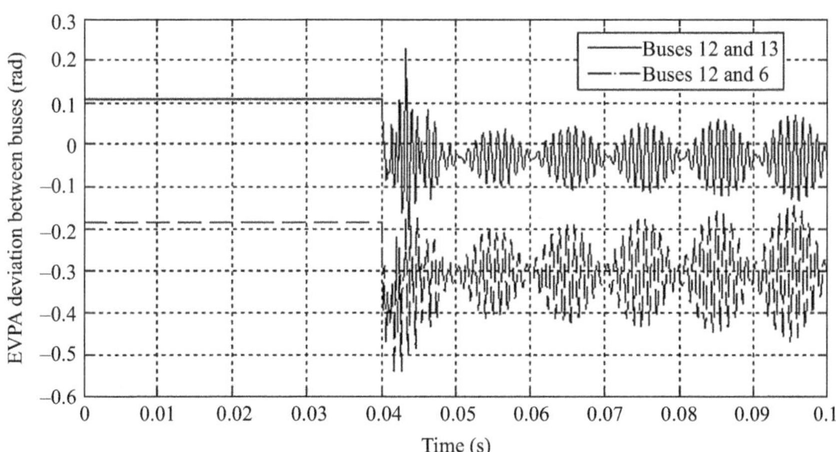

Figure 7.24 EVPA deviation between bus-12 and connected buses during LLL (RYB) fault in IEEE-14 bus system [16]. © 2015 IET. Reprinted, with permission, from [16]

ECPA variation ($\sigma_{\Delta\Delta}$, $\alpha_{\Delta\Delta}$, $\beta_{\Delta\Delta}$, $\gamma_{\Delta\Delta}$, and $\delta_{\Delta\Delta}$) are given by 0.040, 0.000, 0.005, 0.000, and 0.000, respectively. The same for branch 3-4 are given by 0.030, 0.000, 0.002, 0.000, and 0.000, respectively. It is clear that $\beta_{\Delta\Delta}$ of branch 3-4 (0.002) is less than that of branch 3-2 (0.005). This indicates that the branch 3-4 is faulty. The main reason for lower value of $\beta_{\Delta\Delta}$is that the buses connecting faulty branch will

Table 7.8 FFT coefficients of EVPA deviations between parent bus and connected buses in IEEE-14 bus system [16]. © 2015 IET. Reprinted, with permission, from [16]

Fault details (type, branch, distance, FR, FIA)	**Parent and connected buses**	$\sigma_{\Delta\Delta}$	$\alpha_{\Delta\Delta}$	$\beta_{\Delta\Delta}$	$\gamma_{\Delta\Delta}$	$\delta_{\Delta\Delta}$	**Identified faulty branch**
LG, (**3-4**), 25 km,	3,2	0.040	0.000	0.005	0.000	0.000	Branch 3-4
0 Ω, 0°	**3,4**	0.030	0.000	**0.002**	0.000	0.000	
LL, (**11-6**), 60 km,	**11,6**	0.022	0.000	**0.013**	0.000	0.009	Branch 11-6
25 Ω, 30°	11,10	0.010	0.000	0.017	0.000	0.011	
LLG, (**5-4**), 70 km,	5,6	0.067	0.000	0.015	0.000	0.006	Branch 5-4
60 Ω, 50°	**5,4**	0.012	0.000	**0.002**	0.000	0.000	
	5,1	0.072	0.000	0.022	0.000	0.004	
LLL, (**12-13**), 80 km,	12,6	0.052	0.000	0.001	0.000	0.000	Branch
100 Ω, 75°	**12,13**	0.005	0.000	**0.000**	0.000	0.000	12-13

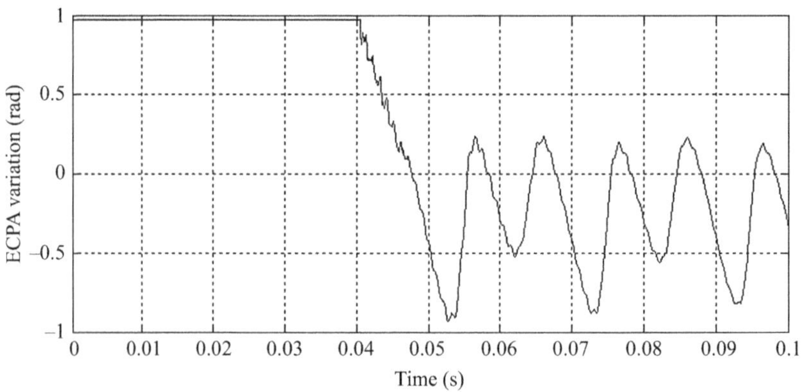

Figure 7.25 ECPA variation in branch 3-4 during LG fault (RG) in IEEE-14 bus system [16]. © 2015 IET. Reprinted, with permission, from [16]

have similar ECPA variation and hence getting canceled each other during subtraction of the same. Hence value of $\beta_{\Delta\Delta}$ can be utilized for faulty branch discrimination. The value of $\beta_{\Delta\Delta}$ for faulty branch in each test case is highlighted in Table 7.8 for better clarity.

Location of the fault has been identified through FFT analysis of ECPA variation in the faulty branch current. ECPA variation with respect to fault distance for each fault condition considered in the case study is shown in Figures 7.25–7.28, respectively.

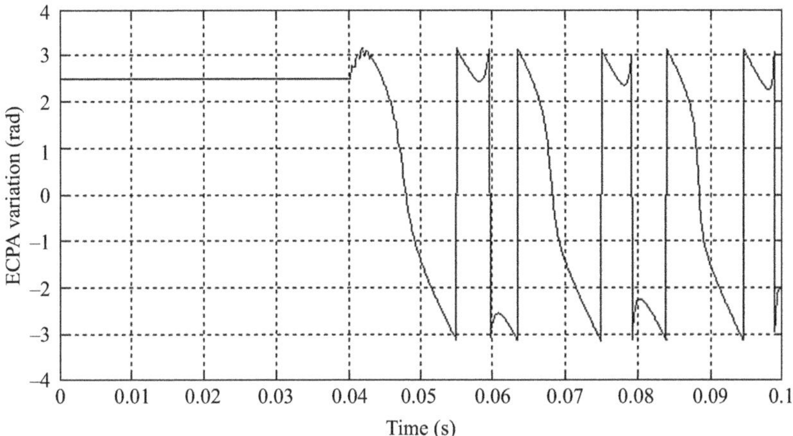

Figure 7.26 ECPA variation in branch 11-6 during LL fault (RY) in IEEE-14 bus system [16]. © 2015 IET. Reprinted, with permission, from [16]

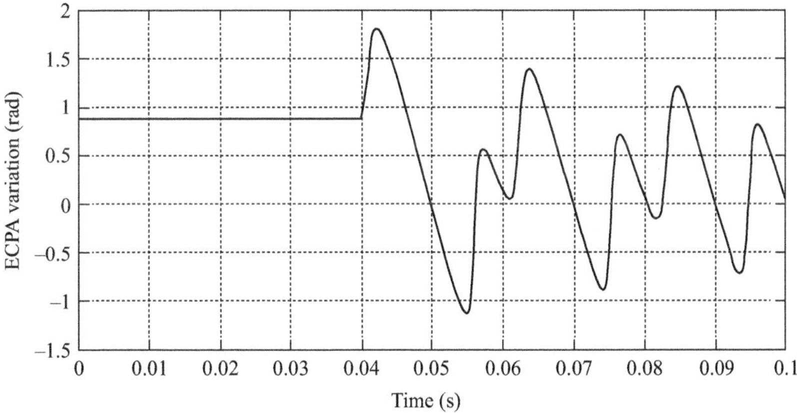

Figure 7.27 ECPA variation in branch 5-4 during LLG (RYG) fault in IEEE-14 bus system [16]. © 2015 IET. Reprinted, with permission, from [16]

FFT coefficients $\sigma_{\Psi L}$, $\alpha_{\Psi L}$, $\beta_{\Psi L}$, $\gamma_{\Psi L}$, and $\delta_{\Psi L}$ in FFT spectra are utilized for estimation of fault location. Variations in FFT coefficients $\sigma_{\Psi L}$, $\alpha_{\Psi L}$, $\beta_{\Psi L}$, $\gamma_{\Psi L}$, and $\delta_{\Psi L}$ with fault distance are shown in Figures 7.29–7.32. Fault conditions considered are the same as that of previous cases.

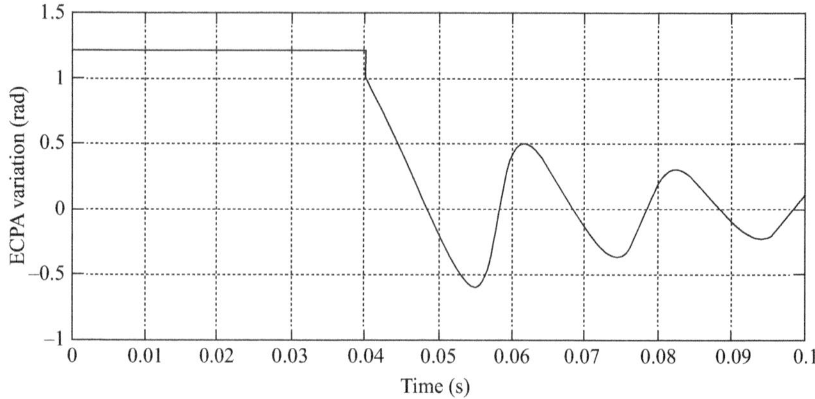

Figure 7.28 ECPA variation in branch 12-13 during LLL fault (RYG) in IEEE-14 bus system [16]. © 2015 IET. Reprinted, with permission, from [16]

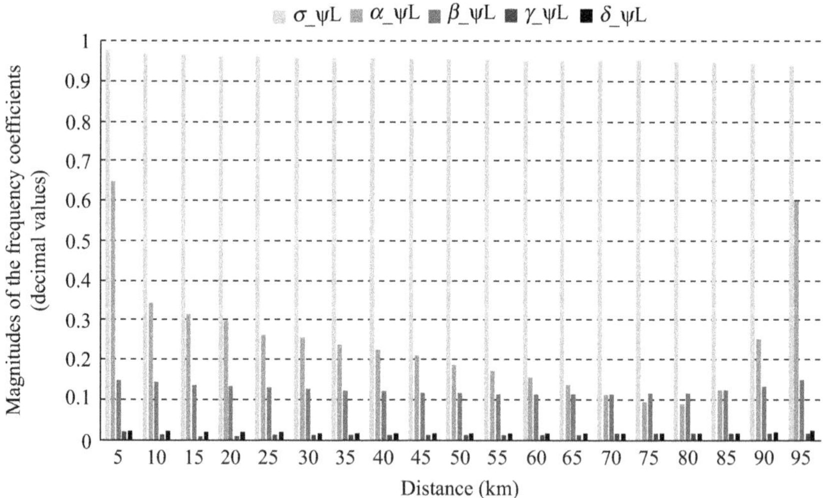

Figure 7.29 Magnitudes of FFT coefficients of ECPA variation with fault location distance from bus-3 of IEEE-14 bus system during LG fault in branch 3-4 with FR of 0 Ω and FIA of 0°

Figures 7.29–7.32 clearly indicate the dependency of the FFT coefficients of ECPA variation on location of the fault from the parent bus. To extract the fault location information from these coefficients, multiclass SVM has been deployed [14,15]. The results of the fault localization are tabulated in Table 7.9 that clearly

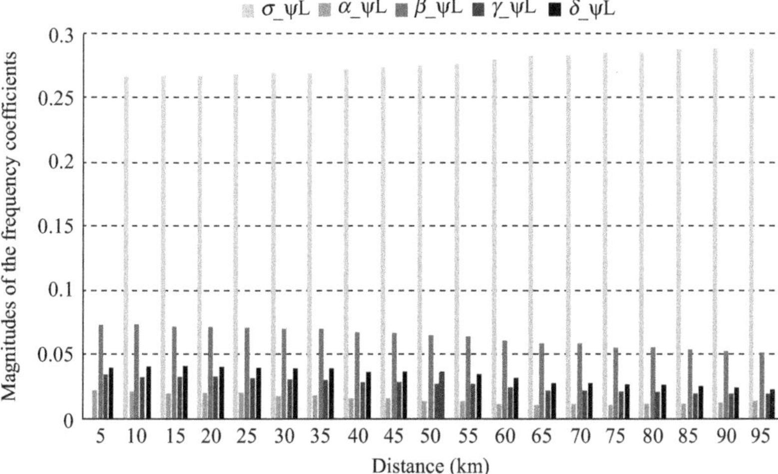

Figure 7.30 Magnitudes of FFT coefficients of ECPA variation with fault location distance from bus-3 of IEEE-14 bus system during LL fault in branch 11-6 with FR of 25 Ω and FIA of 30°

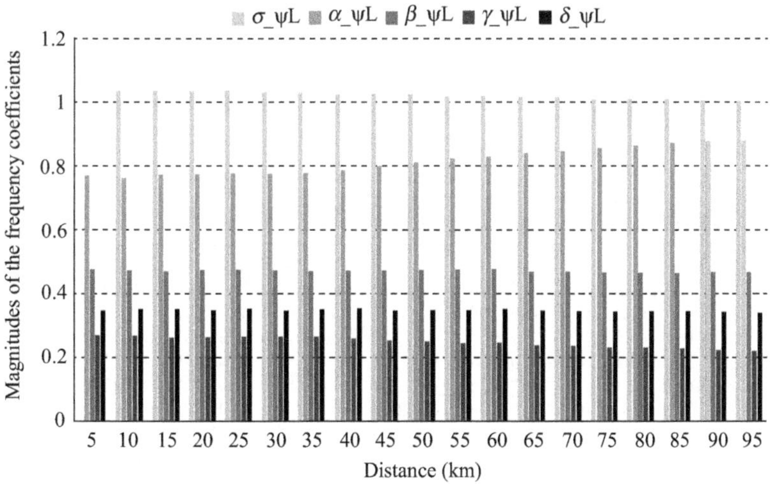

Figure 7.31 Magnitudes of FFT coefficients of ECPA variation with fault location distance from bus-3 of IEEE-14 bus system during LLG fault in branch (5-4) with FR of 60 Ω and FIA of 50°

shows that SVM can accurately predict the fault locations from FFT coefficients of ECPA variation. Time required for fault localization varies with the topology of the network and fault location. For IEEE-14 bus system, it takes approximately 4 s for localizing the LG fault condition considered in branch 3-4.

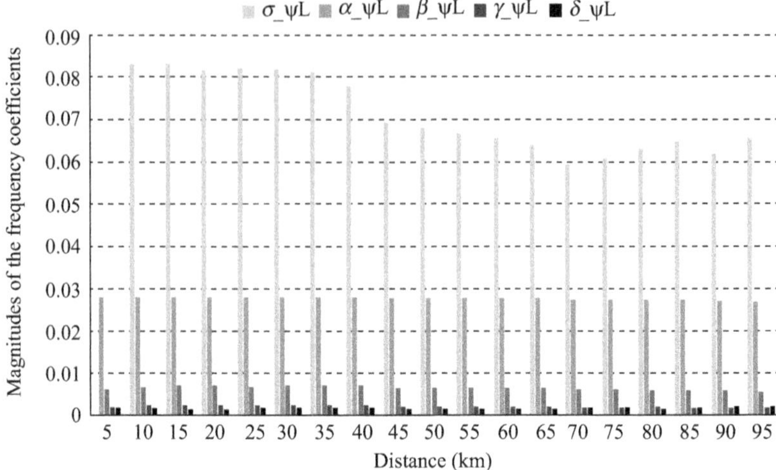

Figure 7.32 Magnitudes of FFT coefficients of ECPA variation with fault location distance from bus-3 of IEEE-14 bus system during LLL fault in branch 12-13 with FR of 100 Ω and FIA of 75°

Table 7.9 Results of case studies conducted in IEEE-14 bus system for fault localization [16]. © 2015 IET. Reprinted, with permission, from [16]

Fault type	Fault condition (fault distance, FR, FIA)	FFT coefficients of ECPA variation					Location estimation by SVM
		$\sigma_{\Psi L}$	$\alpha_{\Psi L}$	$\beta_{\Psi L}$	$\gamma_{\Psi L}$	$\delta_{\Psi L}$	
LG 3-4	5 km, 5 Ω, 0°	0.94841	0.63938	0.14307	0.01750	0.02132	5
	20 km, 20 Ω, 10°	0.93244	0.29495	0.12808	0.00884	0.01756	20
	50 km, 40 Ω, 70°	0.92409	0.18313	0.11144	0.01244	0.01473	50
	70 km, 60 Ω, 90°	0.92074	0.11052	0.10926	0.01366	0.01449	70
	95 km, 80 Ω, 120°	0.91019	0.59281	0.14559	0.01427	0.02261	95
LL 11-6	5 km, 5 Ω, 0°	0.29913	0.02554	0.07483	0.03514	0.04218	5
	20 km, 20 Ω, 10°	0.27336	0.02083	0.07296	0.03287	0.04148	20
	50 km, 40 Ω, 70°	0.26619	0.01408	0.06301	0.02734	0.03497	50
	70 km, 60 Ω, 90°	0.27400	0.01138	0.05635	0.02260	0.02743	70
	95 km, 80 Ω, 120°	0.27971	0.01357	0.05003	0.01963	0.02246	95
LLG 5-4	5 km, 5 Ω, 0°	1.15475	0.84425	0.52265	0.27088	0.35241	5
	20 km, 20 Ω, 10°	1.14431	0.86803	0.52240	0.26182	0.35515	20
	50 km, 40 Ω, 70°	1.12711	0.93216	0.51579	0.23808	0.34905	50
	70 km, 60 Ω, 90°	1.00696	0.87695	0.46278	0.22361	0.33963	70
	95 km, 80 Ω, 120°	0.97508	0.88529	0.44904	0.21598	0.33187	95
LLL 12-13	5 km, 5 Ω, 0°	0.10073	0.03060	0.00655	0.00195	0.00161	5
	20 km, 20 Ω, 10°	0.08634	0.03062	0.00758	0.00244	0.00156	20
	50 km, 40 Ω, 70°	0.06592	0.02989	0.00664	0.00180	0.00156	50
	70 km, 60 Ω, 90°	0.06336	0.02956	0.00569	0.00168	0.00187	70
	95 km, 80 Ω, 120°	0.08030	0.02915	0.00570	0.00206	0.00233	95

7.6 Conclusion

Transmission line fault monitoring using PMU measurement is an integral part of SPG monitoring system. Present fault monitoring system is limited to specific transmission line configurations and hence, necessarily to be incorporated in all transmission lines. Such system can lead to heavy computational and communication burden. To overcome such issues, methodology for detecting, classifying, and localizing transmission line fault anywhere within the grid has been presented in this chapter. The method utilizes PMU measurements of voltage and current phasors at any one of the generator buses. EVPA and ECPA at the generator bus are estimated from the phasor measurements. Detection of the transmission line fault is achieved using FFT spectral analysis of EVPA and ECPA at any one of the generator buses. Multiclass SVM has been adopted for classifying the type of fault using the above-mentioned FFT spectral coefficients. Once transmission line fault is detected, transmission line fault localization methodology identifies the parent bus through FFT spectral analysis of EVPA deviation of buses from that of any single generator bus. After identifying the parent bus, the faulty branch is discriminated from all the branches connected to the parent bus through FFT spectral analysis of EVPA deviation between the parent bus and all other connected buses. Fault location distance (in km) from the parent bus in faulty branch is estimated on the basis of FFT spectral coefficients of ECPA variation in the faulty branch using multiclass SVM. Presented methodologies have been corroborated with case studies conducted in standard IEEE-14 bus system.

References

[1] Bose A. 'Smart transmission grid applications and their supporting infrastructure'. *IEEE Transactions on Smart Grid.* 2010; 1 (1): 11–19.

[2] Fang X., Misra S., Xue G., Yang D. 'Smart grid—the new and improved power grid: a survey'. *IEEE Communications Surveys & Tutorials.* 2011; 14: 944–980.

[3] Yusuff A.A., Jimoh A.A., Munda J.L. 'Determinant-based feature extraction for fault detection and classification for power transmission lines'. *IET Generation, Transmission & Distribution.* 2011; 5 (12): 1259–1267.

[4] Manassero G., Senger E.C., Nakagomi R.M., Pellini E.L., Neves Rodrifgues E.C. 'Fault-location system for multiterminal transmission lines'. *IEEE Transactions on Power Delivery.* 2010; 25 (3): 1418–1426.

[5] Zhang Y., Zhang Q., Song W., Yu Y., Li X. 'Transmission line fault location for double phase-to-earth on non-direct-ground neutral system'. *IEEE Transactions on Power Delivery.* 2000; 15 (2): 520–530.

[6] Gopakumar P., Jaya Bharata Reddy M., Mohanta D.K. 'Adaptive fault identification and classification methodology for smart power grids using synchronous phasor angle measurements'. *IET Generation, Transmission and Distribution.* 2015; 9 (2): 133–145.

[7] Lopes F.V., Santos W.C., Fernandes D., Neves W.L.A., Souza B.A. 'An adaptive fault location method for smart distribution and transmission grids'. *Proceedings IEEE PES Conference on Innovative Smart Grid Technologies (ISGT Latin America)*, 2011: 1–7.

[8] Ferraz R.G., Iurinic L.U., Filomena A.D., Bretas A.S. 'Park's transformation analytical approach of transient signal analysis for power systems'. *Proceedings of North American Power Symposium (NAPS)*, 2012: 1–6.

[9] Hashiesh F., Mostafa H.E., Khatib A.R., Helal I., Mansour M.M. 'An intelligent wide area synchrophasor based system for predicting and mitigating transient instabilities'. *IEEE Transactions on Smart Grid*. 2012; 3 (2): 645–652.

[10] Jaya Bharata Reddy M., Mohanta D.K. 'Adaptive-neuro-fuzzy inference system approach for transmission line fault classification and location incorporating effects of power swings'. *IET Generation, Transmission & Distribution*. 2008; 2 (2): 235–244.

[11] Kalyani S., Shanti Swarup K. 'Classification and assessment of power system security using multiclass SVM'. *IEEE Transactions on Systems, Man, and Cybernetics—Part C: Applications and Reviews*. 2011; 41 (5): 753–758.

[12] Ravikumar B., Thukaram D, Khincha H.P. 'Comparison of multiclass SVM classification methods to use in a supportive system for distance relay coordination'. *IEEE Transactions on Power Delivery*. 2010; 25 (3): 1296–1305.

[13] Fan Z., Kulkarni P., Gormus S., *et al*. 'Smart grid communications: overview of research challenges, solutions, and standardization activities'. *IEEE Communications Surveys & Tutorials*, 2012; 15: 21–38.

[14] Chang C.C., Lin C.J. 'LIBSVM: a library for support vector machines'. *ACM Transactions on Intelligent Systems and Technology*, 2011; 2: 27:1–27:27; Software available from: http://www.csie.ntu.edu.tw/~cjlin/libsvm.

[15] Janik P., Lobos T. 'Automated classification of power-quality disturbances using SVM and RBF networks'. *IEEE Transactions on Power Delivery*. 2006; 21 (3): 1663–1669.

[16] Gopakumar P., Jaya Bharata Reddy M., Mohanta D.K. 'Transmission line fault detection and localization methodology using phasor measurement unit measurements'. *IET Generation Transmission and Distribution*. 2015; 9 (11): 1033–1042.

Chapter 8

PMU-based vulnerability assessment of power systems

Guanqun Wang[1], *Chen-Ching Liu*[1], *Evangelos Farantatos*[2] *and Navin Bhatt*[2]

8.1 Introduction

8.1.1 Cascaded events and vulnerability assessment

Since the 1960s, the scale and complexity of interconnected power grids have increased to such an extent that the vulnerability of power systems with respect to large-scale blackouts has increased significantly. For example, on November 9, 1965, the blackout in Northeast U.S. caused a loss of 21 GW load and left 30 million people without power for up to 13 h [1]. The 2003 blackout in North America affected more than 50 million people, costing about $3 billion economic losses and contributed to 11 reported fatalities [2]. On July 30 and 31, 2012, two consecutive blackouts took place in India. The second one is the largest power outage that has ever happened in the world. A total of 600 million people were affected, nearly 10% of the world's population. An estimated 48 GW of generating capacity was lost during the outage [3,4]. Table 8.1 lists some major blackouts in the world in recent years.

Analysis of these large-scale blackouts indicates that these outages were triggered by a cascading sequence of events [8]. Generally, these sequences of events follow a common process [8,9]. First, the cascaded events are initiated by a single or multiple events such as line outages or loss of generators. Due to relay actions, one initiating contingency may trigger another, leading to successive tripping of transmission lines and generators if the contingencies are not cleared in time. The subsequent events can cause power-flow rerouting, line overloading, voltage problems, and reactive power deficiency, which further weaken the system. Frequency instability, voltage instability, and rotor-angle instability can take place during this process, leading to an uncontrolled system partitioning, and eventually a widespread blackout.

[1]Electrical Engineering and Computer Science Department, Washington State University, Pullman, WA 99164-2752, USA
[2]Electric Power Research Institute (EPRI), Palo Alto, CA 94303-0813, USA

Table 8.1 Some major blackouts in recent years [3–7]

Date	Location	Effects
2012-7-30 and 7-31	India	Affected more than 600 million people. 48 GW of generating capacity was taken off
2011-9-8	United States and Mexico	2.7 million people were affected, including 1.4 million customers in San Diego County and 1.1 million in Mexico
2009-11-10	Brazil and Paraguay	60 million people were affected and 28.8 GW of load was lost
2006-11-04	West Europe	Eight countries and approximately 50 million people were affected. 17 GW of load was lost. The blackout was restored after 1.5 h
2003-09-28	Italy	180 GW of load was lost, 57 million people were affected. The blackout was restored after 20 h
2003-09-23	North Europe	1800 MW of load was lost, 5 million people were affected. The blackout was restored after 6.5 h
2003-08-14	United States and Canada	One province and eight states were out of power, and more than 50 million people were affected. The blackout last for up to 29 h, cost the loss of 61.8 GW of power and economic loss of $6 billion

As a result, it is important to identify the vulnerable patterns of power systems. That is, events that may trigger other events. Yamashita *et al.* [8] report the patterns of cascaded events as line tripping due to overloading, generator tripping due to overexcitation, line tripping due to loss of synchronism, generator tripping due to abnormal voltage, and frequency system condition and underfrequency/voltage load shedding. In [10], it is stated that zone 3 relay operation serves as a basic pattern for cascaded events. A fuzzy inference system is developed to identify contingencies that are likely to trigger cascaded zone 3 relay operations. Furthermore, the concept of strategic power infrastructure defense systems is proposed in [11,12], which involves real-time sensing and communication, failure analysis, vulnerability assessment, and self-healing control features. This system is a real-time, wide-area, adaptive protection, and control system that depends on the power, communication, and computer infrastructures. It is intended to prevent catastrophic failures that may lead to large-scale power outages.

Rotor-angle instability that can further cause frequency instability and voltage instability is one of the main vulnerable patterns of power systems. It is important to predict rotor-angle instability before the system loses synchronization. For transient stability, the power system dynamics are divided into three stages, represented by three sets of nonlinear dynamic equations:

$$\text{Pre-fault} \quad \begin{cases} \dot{x} = f_0(x, y, u) \\ 0 = g_0(x, y, u) \end{cases} \quad 0 < t \le t_0^- \tag{8.1}$$

$$\text{Fault-on} \quad \begin{cases} \dot{x} = f_F(x, y, u) \\ 0 = g_F(x, y, u) \end{cases} \quad t_0^+ \leq t \leq t_{cl}^- \tag{8.2}$$

$$\text{Post-fault} \quad \begin{cases} \dot{x} = f_{\text{PF}}(x, y, u) \\ 0 = g_{\text{PF}}(x, y, u) \end{cases} \quad t_{cl}^+ \leq t \tag{8.3}$$

where x represents state variables that describe the dynamics of generators, speed governors and excitation systems, etc. The symbol y represents algebraic variables such as node voltage amplitudes and angles, and u represents control variables, such as tripping and restoration of generators and load. The time instants t_0 and t_{cl} represent the time at which a fault occurs and is cleared, respectively. After severe disturbances, the rotor-angle speed of generators will oscillate due to imbalance between mechanical power inputs and electrical power outputs of generators. If the system does not have the ability to eliminate such oscillations, it will lose synchronism and become unstable. As a result, it is important to monitor vulnerable patterns and assess transient stability of power system online so that controls can be taken to prevent the system from entering an emergency operating state.

There are two major approaches to transient stability analysis for power systems. One is time-domain simulation methods [13,14], which is based on numerical integration of system dynamic equations. With an accurate component model, this method can provide details of the dynamic process for a large-scale system. However, the high computation burden and lack of a stability index limits the online applicability of this method. The other one is the energy function-based direct methods, such as relevant unstable equilibrium point method [15], controlling unstable equilibrium point method [16], potential energy boundary surface [17], the Boundary of stability region based Controlling Unstable equilibrium point (BCU) method [18], and extended equal area criteria method [19]. These methods have a relatively fast computational speed and can provide a quantitative index for stability assessment, and thus there are advantages for online application. A shortcoming of direct methods is that the assessment result may not incorporate sufficient details due to the simplification of system models.

8.1.2 PMU-based monitoring of power systems

The concept of phasor measurement units (PMUs) was proposed in the mid-1980s [20]. Algorithms for the application of PMUs in static-state estimation and adaptive protection were proposed [21–23]. The early prototype of PMU was built in 1992 at Virginia Tech [24]. In recent years, due to the increasing need for wide-area monitoring and control of the power grid, the number of PMUs installed on the grid has increased significantly. The new and accurate measurement device is an enabling technology for power grids to perform wide-area monitoring and control of system dynamics.

The PMU-based out-of-step relay is proposed and applied to the Florida–Georgia system [25–28]. The system is modeled as two interconnected equivalent generators, and they are measured by two PMU sets to monitor the angular difference between them. Based on the monitored angular difference, the equal-area criterion is utilized to assess transient stability of the power system after a power swing. In [29], synchronized phasor measurements are used for solving the states of a post-fault power system based on differential-algebraic equation model. The numerical results are used to evaluate system stability. Huang *et al.* [30] investigate the application of Extended Kalman Filter techniques for tracking dynamic states of power systems using phasor measurement data. In [31,32], PMU measurements at the substation level are applied to a distributed frequency domain optimization method for monitoring of oscillations. PMUs are also applied to the energy function based direct methods and decision-tree methods for online dynamic security assessment [33,34].

In this research, a new method is proposed for monitoring of wide-area rotor-angle stability based on PMU data in an online operational environment [35,36]. The maximal Lyapunov exponent (MLE), which is applied as a tool for prediction of out-of-step conditions [37], is connected with rotor-angle stability analysis. An efficient computational algorithm is proposed for calculation of MLE based on dynamic waveforms acquired from PMU data within a finite time window.

Furthermore, the observability properties of the system with the limited availability of PMUs are discussed. Consider a large-scale power system with only a small number of PMUs. With the available PMUs, is it possible to reconstruct all states through the available PMU measurements? This research provides the results to clarify different observability concepts for static and dynamic models. Nonlinear dynamic observability which is based on a rigorous theoretical foundation is applied to evaluate the level of observability for dynamic state calculation with partial PMU measurements. The proposed method is illustrated by a case study with a 3-bus system.

8.2 PMU-based monitoring and assessment of rotor-angle stability

8.2.1 Maximum Lyapunov exponent (MLE) method

In Ergodic theory [38], Lyapunov exponent is used to characterize whether a given system is "chaotic" and how chaotic it is. For two trajectories that start out very close to each other:

$$\begin{aligned} x(t) &= f^t(x_0) \\ x(t) + \Delta x(t) &= f^t(x_0 + \Delta x_0) \end{aligned} \tag{8.4}$$

all points in a neighborhood of $x(t)$ may converge toward it or separate exponentially as time goes on, which is called chaotic. Sensitivity of the system trajectory toward initial conditions can be quantified by

$$\|\Delta x(t)\| \approx e^{\lambda t}\|\Delta x_0\| \tag{8.5}$$

which is shown in Figure 8.1. The index λ is called the Lyapunov exponent which distinguished between exponential divergence or convergence of nearby trajectories.

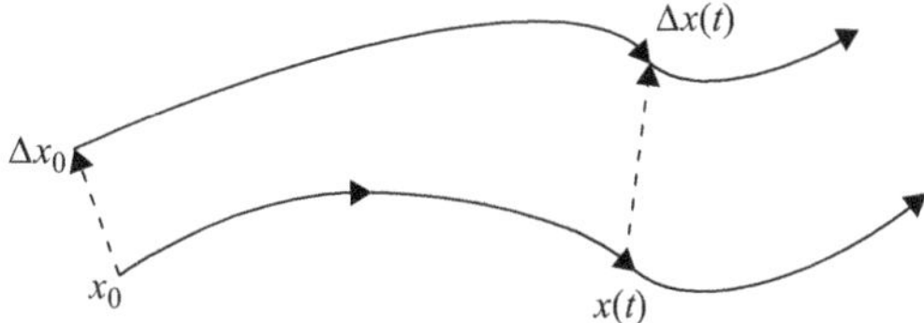

Figure 8.1 Perturbation of trajectories of a dynamic system

For a N-dimensional dynamical system represented by nonlinear differential equations:

$$\dot{x} = f(x) \tag{8.6}$$

Based on (8.5), MLE can be calculated as

$$\lambda = \lim_{t\to\infty} \frac{1}{t} \ln \frac{\|J^t(x_0)\Delta x_0\|}{\|\Delta x_0\|} \tag{8.7}$$

where $J^t(x_0)$ is the Jacobian matrix of the system, and Δx_0 is a randomly chosen vector in small scale, as shown in Figure 8.1. The relationship between MLE and the asymptotic behavior of the dynamical system is established in [39]. If MLE is found to be negative after the fault, the waveform following the disturbance is asymptotically stable. Otherwise, it is considered unstable.

8.2.2 MLE calculation algorithm

The flowchart of the proposed MLE calculation algorithm of a power system is shown in Figure 8.2.

In the flowchart, the procedure for calculation of MLE can be described in the following steps.

1. Spectral analysis

MLE is estimated over a specified, finite time window T with the length of seconds. It should be predetermined properly so that waveforms of the PMU measurements can reliably represent the features of system dynamics under a disturbance. In the algorithm, the spectral analysis is used to compute an appropriate time interval length. Credible disturbances can be simulated in the offline environment to obtain the power swing curves following each disturbance. The spectral analysis is performed for the power swing curves to determine a corresponding frequency range. Based on the result of the spectrum analysis, the window size T is selected as the inverse of the lower limit of the frequency range so that it covers at least one cycle of the power swing. The MLE estimated on the basis of T, $\lambda_{t=T}$, is an approximate value of λ, the theoretical MLE over an infinite time interval.

2. State calculator

The complete trajectories of all state variables over the time window T are required to calculate MLE. That is, rotor angles and angular speeds of all generators are to

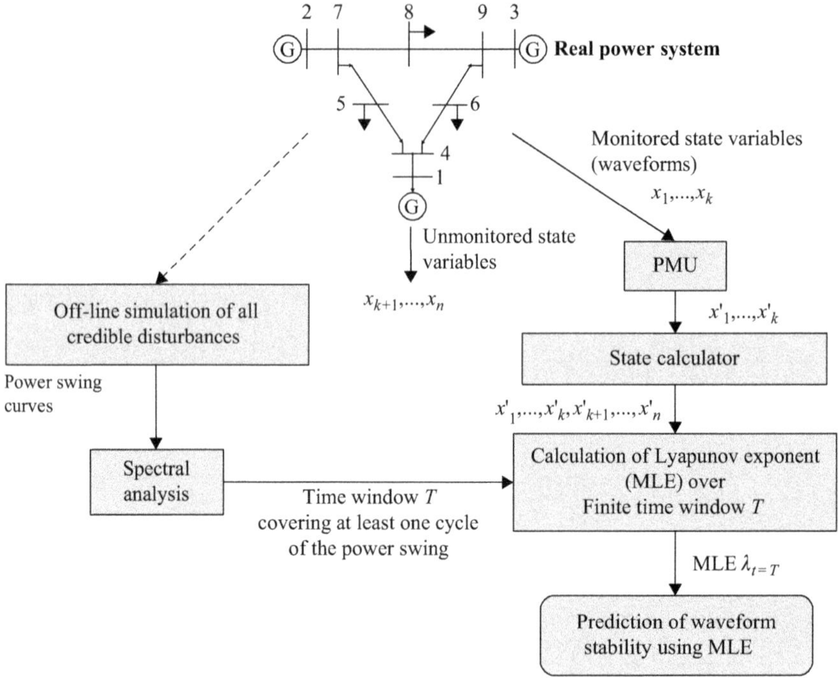

Figure 8.2 Flowchart of MLE calculation algorithm

be observed by PMU data. However, at present, there are only limited numbers of PMUs in the power system. To overcome the difficulty, the state calculator algorithm is developed to estimate the unmonitored state variables from the initial condition to the post-fault period.

The rotor-angle dynamics of power system is formulated as

$$\begin{cases} \dot{x}(t) = f(x(t)) \\ x(t_0) = x_0 \end{cases} \tag{8.8}$$

where $x(t) = [x_1(t), x_2(t), \ldots, x_n(t)]^T$ are state variables, x_0 is the initial state. Assuming that the waveforms of $x_1(t), \ldots, x_k(t)$ are monitored by PMUs, it is necessary to estimate the unmonitored state variables $x_{k+1}(t), \ldots, x_n(t)$ at a sequence of discrete points, $t = t_0, t_0 + \Delta t, t_0 + 2\Delta t, \ldots$. Numerical methods are needed for the solution of the above problem.

In [35], the implicit integration method with a trapezoidal rule is used to solve this problem. The state variables at instant $t = t_0 + \Delta t$ are estimated by integration of the dynamical system, as shown below:

$$x(\Delta t) = x(0) + \frac{\Delta t}{2}\{f(x(0)) + f(x(\Delta t))\} \tag{8.9}$$

The above equation can be formulated as $G[x(\Delta t)] = 0$. As this is a set of nonlinear simultaneous equations, an estimated value of $x(\Delta t)$ is obtained by the Newton–Raphson's method. Note that for the next time increments, $x(2\Delta t)$, $x(3\Delta t)$, ... can be approximated by the same procedure.

As $x_1, x_2, \ldots, x_k$ in $x(t)$ are monitored by PMUs at buses 1, 2, ..., k, $x_1(t)$, $x_2(t), \ldots, x_k(t)$ are known at each discrete points $t = t_0, t_0 + \Delta t, t_0 + 2\Delta t, \ldots$. The differential equations in (8.8) of $x_1, x_2, \ldots, x_k$ can, therefore, be eliminated and $x_1(t), x_2(t), \ldots, x_k(t)$ in other equations can be replaced by known PMU measurements $x_1'(t), x_2'(t), \ldots, x_k'(t)$:

$$\begin{cases} \dot{x}_i(t) = f(x_i(t), x_1'(t), x_2'(t), \ldots, x_k'(t)) & i = k+1, \ldots, n \\ x_i(t_0) = x_{i0} \end{cases} \tag{8.10}$$

Note that the order of the dynamic system is reduced from n to $(n-k)$ due to the direct measurements from the available PMUs at buses 1, 2, ..., k.

3. Computation of MLE

The standard method with Gram–Schmidt Orthonormalization (GSR) [40] is used to calculate MLE for the long-time average, after the estimation of the dynamic states. By this method, MLE can be represented as a mean growth rate of the perturbation between neighboring trajectories, as shown in Figure 8.3. By calculating the waveform changes for each time increment Δt, an average of the changes over a selected time window will be the approximate value of MLE:

$$\lambda = \lim_{t \to \infty} \frac{1}{t} \sum_i \Delta t_i \lambda_i \tag{8.11}$$

Based on a nonlinear dynamic model of the power system, the MLE can be calculated on the basis of the waveforms resulting from the dynamics of system states. The corresponding control actions can be determined based on the prediction. As PMUs can measure the phasors accurately and with synchronized time stamps, their installation on a power grid can considerably improve the performance of monitoring for system dynamics.

4. Prediction of system stability

The relationship between MLE and rotor-angle stability of power systems has been established and, therefore, MLE as an index of the waveform stability is founded on

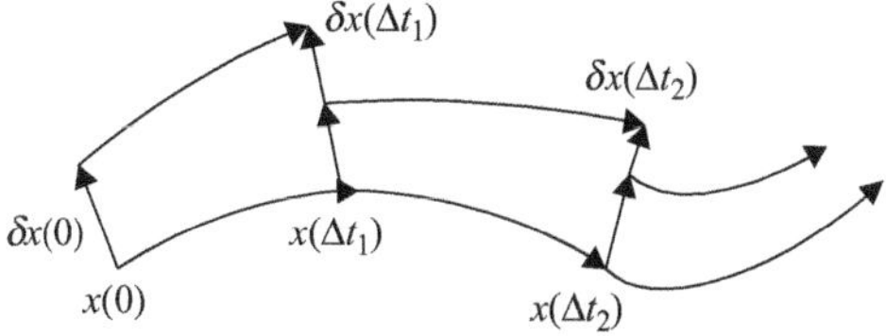

Figure 8.3 Standard method for GSR

a solid theoretical basis. Note that waveform stability is different from the usual stability around an equilibrium point of a nonlinear dynamic system. The former is to evaluate whether the waveform can stay close to a given waveform from the collected PMU data if a perturbation occurs to the initial operating condition. The latter, however, is to evaluate whether the system will approach a stable equilibrium point after a disturbance for a given post-fault system condition. The waveform stability does not assume the existence of a stable or unstable equilibrium point.

8.3 Observability for PMU-based monitoring of system dynamics

Observability analysis is a crucial task in model-based state monitoring and estimation. At this time, the total number of monitored buses is still modest compared with the large number of buses (tens of thousands) in a large interconnected grid such as the Western or Eastern Interconnection of North America. As MLE is calculated from available PMU measurements, it is questionable whether these PMU measurements contain sufficient information for estimation of the entire set of states and how reliable the results are. This is the issue of observability one needs to deal with.

The concepts of observability in power systems are defined for different contexts and in different ways. We need to clarify these concepts and apply appropriate concept for PMU-based monitoring of system dynamics.

8.3.1 Static observability

For static state estimation, a network is said to be observable if any flow in the network can be observed by some indication in the set of measurements [41]. For simplicity, consider the DC (direct current) power-flow model of the power system in Figure 8.4.

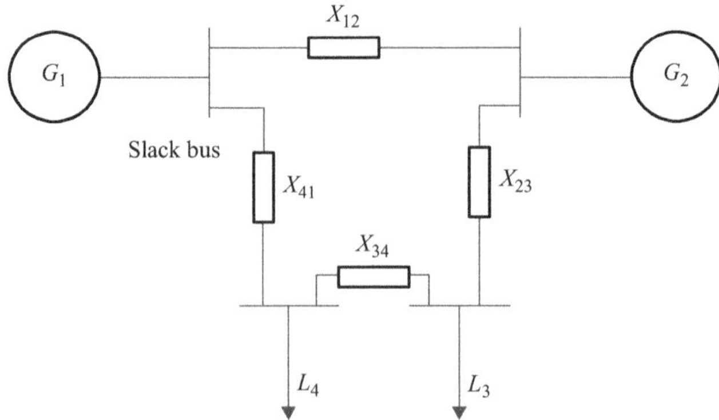

Figure 8.4 A simple power system

By applying DC power flow, the measurement in the system could be written as

$$z = H\theta \tag{8.12}$$

where z represents measurements, H represents the measurement matrix, and $\theta = \begin{bmatrix} \theta_2 & \theta_3 & \theta_4 \end{bmatrix}$ represents the state variables (bus angles) of the system. By the least-squared algorithm for static state estimation, θ can be estimated by

$$\theta^* = \left(H^T H\right)^{-1} H^T z \tag{8.13}$$

Therefore, if $H^T H$ is nonsingular, that is, the rank of H equals H's column number, the system is observable [42]. The observability for static state estimation is different depending on whether the number of measurements is greater than (or equal to) or less than the number of system states. Equation (8.13) applies to the case when the number of measurements is greater than or equal to the number of states. This solution also corresponds to the maximum likelihood estimate for jointly Gaussian noises of the measurement errors. Therefore, it is physically meaningful. If the number of measurements is lower than that of the state variables, it is possible to obtain a minimum norm estimate based on the available measurements. However, this minimum norm estimate is not physically meaningful.

8.3.2 Topology observability

For static state estimation with measurements provided by PMUs, however, it is possible to observe beyond the PMU buses. See Figure 8.5—there are 14 state variables, that is, 7 voltage magnitudes, and 7 angles, to be measured. Two PMUs are installed. However, a PMU installed at a specific bus is capable of measuring not only the bus voltage phasor, but also the current phasors along all lines incident with the bus. As a result, in addition to the voltage phasor at this bus, the voltage phasors of all neighboring buses can also be computed; hence, they also become observable.

If the complete set of bus voltage phasors can be determined by the number and location of PMUs, the system with this PMU placement scenario is called the system topological observability [43]. For a topologically observable network with

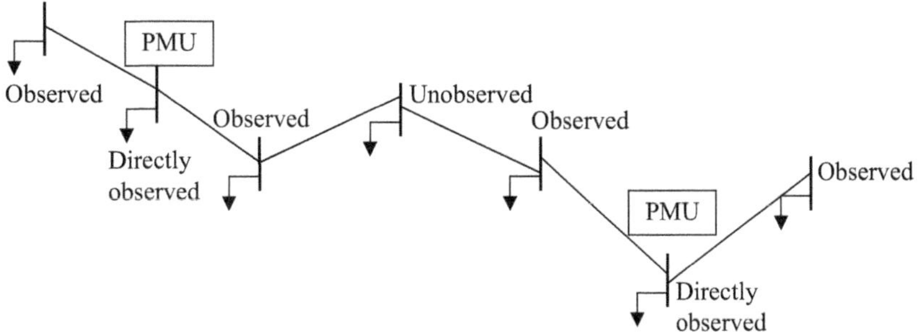

Figure 8.5 Observability for PMU-based static state estimation

only PMU measurements, the static state estimation becomes linear, as shown in the following equation:

$$\begin{pmatrix} z_V \\ z_C \end{pmatrix} = \begin{pmatrix} I & 0 \\ Y_{CL1} & Y_{CL2} \end{pmatrix} \begin{pmatrix} \dot{V}_{L1} \\ \dot{V}_{L2} \end{pmatrix} + \begin{pmatrix} \varepsilon_V \\ \varepsilon_C \end{pmatrix} \tag{8.14}$$

where $\dot{V}_{L1}$, $\dot{V}_{L2}$ are state vectors representing the bus voltage phasor at buses with and without PMU measurements; I is identical matrix. Y_{CL1} and Y_{CL2} are the measurement matrices with the network admittances; z_V and z_C are the measurement vectors with the bus voltages and line currents; ε_V and ε_C are the vector of measurement errors. According to Nuqui and Phadke [44], a minimum number of 1/5 to 1/4 of the system buses to be installed with PMUs should ensure the complete topological observability of the system.

8.3.3 Dynamic observability

A dynamic system can be represented by differential equations:

$$\begin{aligned} \dot{x}(t) &= f(x(t)) + u(x(t)) \\ z &= h(x(t)) \end{aligned} \tag{8.15}$$

where x and z represent state vector and measurement vector, f, u and h are state function vector, control function vector, and measurement function vector, respectively. The system is said to be observable if, for any possible sequence of state and control vectors, the current state can be determined in finite time using the outputs [45]. In other words, for an observable system, it is possible to use the system measurements to determine the states of the dynamic system. If a system is not observable, the current values of some of its states may not be determined through the output values.

To illustrate the concept of observability, the mathematical tool of Lie derivative is needed [46], which is summarized as follows:

Let f: $\mathbb{R}^n \to \mathbb{R}^n$ be a vector field in $\mathbb{R}^n$, that is

$$f = \begin{bmatrix} f_1(x) \\ \cdots \\ f_n(x) \end{bmatrix} \tag{8.16}$$

Let h: $\mathbb{R}^n \to \mathbb{R}$ be a scalar function. Then the Lie derivative of h with respect to f is

$$L_f h = \nabla h \cdot f(x) = \frac{\partial h}{\partial x} f = \sum_{i=1}^{n} \frac{\partial h}{\partial x_i} f_i \tag{8.17}$$

The higher order Lie derivatives can also be defined. By definition, the zero-order Lie derivative of h is

$$L_f^0 h = h \tag{8.18}$$

and the k-order derivative could be calculated by repeated derivative, that is,

$$L_f^k h = \frac{\partial\left(L_f^{k-1} h\right)}{\partial x} \cdot f \tag{8.19}$$

System (8.15) is locally observable at x_t, if the matrix:

$$O(x_t) = \left.\frac{\partial l(x)}{\partial x}\right|_{x=x_t} \tag{8.20}$$

has a full rank [46]. $O(x)$ is called the observability matrix, in which:

$$l(x_t) = \begin{bmatrix} L_f^0 h_1(x) \\ \cdots \\ L_f^0 h_p(x) \\ L_f^1 h_1(x) \\ \cdots \\ L_f^{n-1} h_p(x) \end{bmatrix} \tag{8.21}$$

The rank condition of (8.21) can be checked by the singular value decomposition method [47]. As the observability property is local, the smallest singular value of $O(x)$ will change along the trajectory of $x(t)$. Furthermore, the smallest singular value of $O(x)$ can be used to determine the level of observability of the system with different outputs.

If system (8.15) becomes linear, the observability matrix $O(x)$ will also reduce to linear case, which is well described in standard linear system textbooks, for example, [48].

8.3.4 PMU-based monitoring of power system dynamics

The static state estimation accounts only for the static system variables, that is, bus voltages and angles, without the dynamic system states. For control techniques that require information about these dynamic system states, the static state estimation methods and corresponding observability concepts are not sufficient.

It is important to monitor power system dynamic states such as the rotor speed to ensure a reliable operation and an efficient control of the power grid. As PMU data have a high sampling rate with high accuracy and synchronized time stamps, it can be acquired to approximate the continuous time waveforms and thus considerably improve monitoring of the system dynamics. The dynamics of the power system can be modeled as (8.15), and the PMU-measured waveform can be treated as the measurement function $h(x)$. With state calculation methods such as implicit integration method, the full states of the system are solvable, no matter how many measurements are available on the system. However, the accuracy of monitoring is questionable when the number of measurements is small compared to the number of states. As a result, the theory of dynamic observability can be applied to the

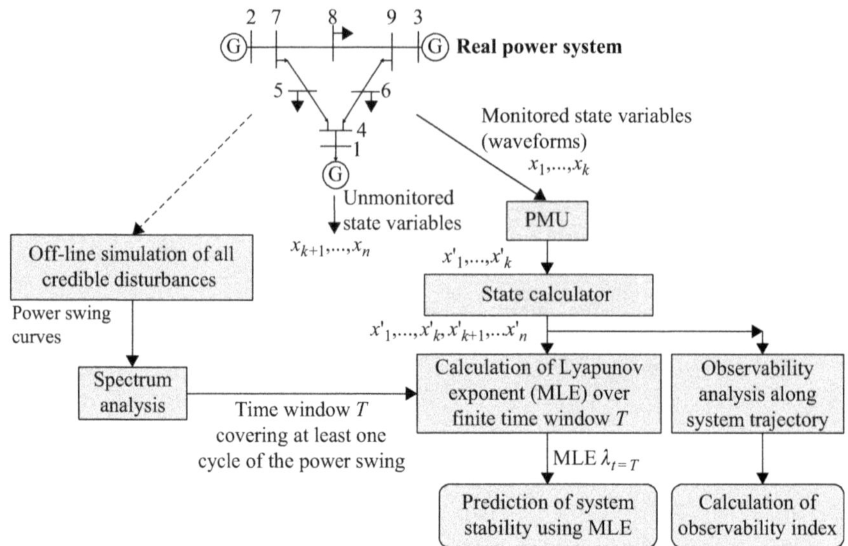

Figure 8.6 Flowchart of MLE calculation algorithm together with observability issue

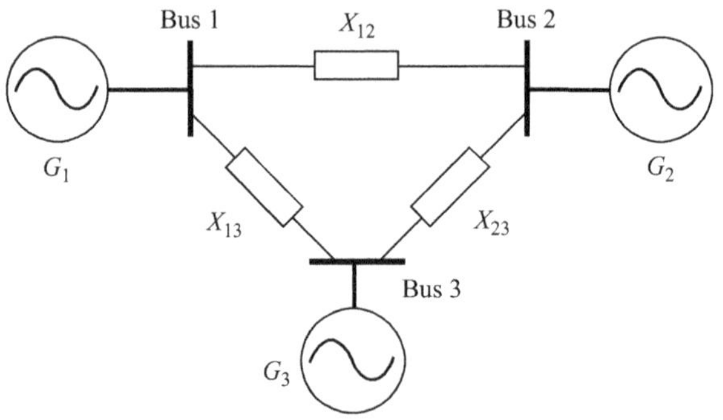

Figure 8.7 A 3-bus power system

PMU-based dynamic monitoring problem. The observability index serves as a metric for quality of the reconstructed dynamic states based on snapshots of PMU measurements.

Considering observability issue, the flowchart of the MLE calculation algorithm is revised as shown in Figure 8.6.

8.4 Example

The MLE calculation algorithm is illustrated by a 3-bus system, as shown in Figure 8.7.

The classical rotor dynamic equations are as follows:

$$P_{M1} = M_1\ddot{\delta}_1 + D_1\dot{\delta}_1 + \frac{E_1E_2}{X_{12}}\sin(\delta_1 - \delta_2) + \frac{E_1E_3}{X_{13}}\sin(\delta_1 - \delta_3)$$

$$P_{M2} = M_2\ddot{\delta}_2 + D_2\dot{\delta}_2 + \frac{E_1E_2}{X_{12}}\sin(\delta_2 - \delta_1) + \frac{E_2E_3}{X_{23}}\sin(\delta_2 - \delta_3)$$

$$P_{M3} = M_3\ddot{\delta}_2 + D_3\dot{\delta}_3 + \frac{E_1E_3}{X_{13}}\sin(\delta_3 - \delta_1) + \frac{E_2E_3}{X_{23}}\sin(\delta_3 - \delta_2)$$

The parameters of the 3-generator system are given by

$$\begin{aligned}
&D_1/M_1 = 0.5, D_2/M_2 = 0.3, D_3/M_3 = 0.2\\
&P_{M1}/M_1 = P_{M2}/M_2 = 0.85, P_{M3}/M_3 = -1.7\\
&E_1E_2/M_1 = E_1E_2/M_2 = 0.1, E_1E_3/M_1 = E_1E_3/M_3 = E_2E_3/M_2\\
&\qquad\quad = E_2E_3/M_3 = 1.0
\end{aligned} \tag{8.22}$$

By substitution of these parameters, a six-dimensional equation is obtained:

$$\begin{aligned}
\dot{\delta}_1 &= \omega_1\\
\dot{\omega}_1 &= 0.85 - 0.1 \cdot \frac{\sin(\delta_1 - \delta_2)}{X_{12}} - \frac{\sin(\delta_1 - \delta_3)}{X_{13}} - 0.5\omega_1\\
\dot{\delta}_2 &= \omega_2\\
\dot{\omega}_2 &= 0.85 - 0.1 \cdot \frac{\sin(\delta_2 - \delta_1)}{X_{12}} - \frac{\sin(\delta_2 - \delta_3)}{X_{23}} - 0.3\omega_2\\
\dot{\delta}_3 &= \omega_3\\
\dot{\omega}_3 &= -1.7 - \frac{\sin(\delta_3 - \delta_1)}{X_{13}} - \frac{\sin(\delta_3 - \delta_2)}{X_{23}} - 0.2\omega_3
\end{aligned} \tag{8.23}$$

Assume that $X_{12} = X_{13} = X_{23} = 1$. A three-phase fault followed by clearing of the fault is applied to transmission line 2-3. The fault is cleared by relay and breaker operations, which will last for a fraction of 1 s. Then it is assumed that the breaker recloses successfully at 1 s. The relative rotor angles calculated by simulation are shown in Figure 8.8.

Assuming that only one PMU is installed and available at generator 1, so that δ_1, ω_1 are monitored as $\delta_1{}', \omega_1{}'$, then the waveform of $\delta_1{}'$ is substituted into the last four equations, and the first two equations in (8.23) are eliminated. The three-phase fault is applied to each transmission line, respectively. MLE is calculated by the above algorithm together with the dynamic trajectories. The observability indices along the trajectory are also computed. The results are shown in Table 8.2 and Figure 8.9.

Similarly, assuming that two PMUs are installed at generator 1 and 2. The MLE and observability calculation results are shown in Table 8.3 and Figure 8.10.

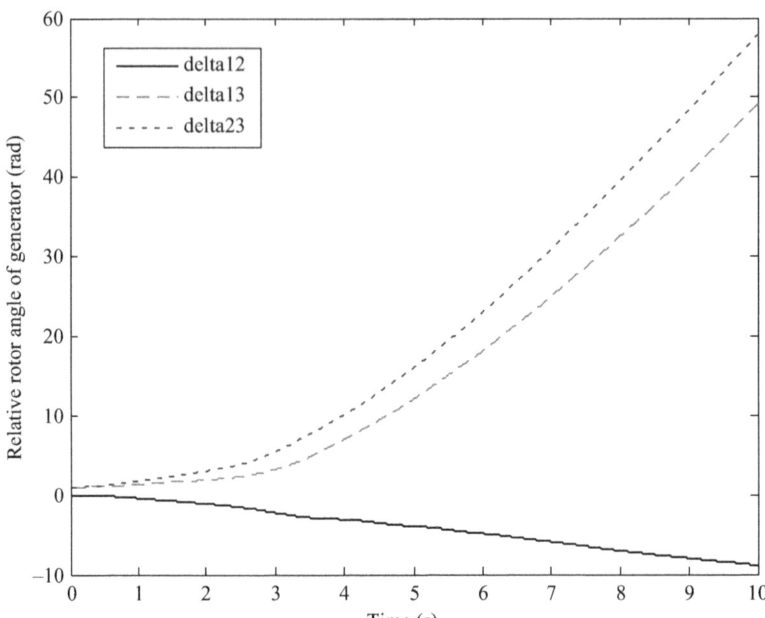

Figure 8.8 Simulation results of relative rotor angles

Table 8.2 MLE of different time window T with 1 PMU placement

3-Phase fault	Stability	MLE of $T = 1.25$	MLE of $T = 1.667$	MLE of $T = 2$
Line 1-3	Unstable	0.5848	0.7448	0.7798
Line 2-3	Unstable	0.0236	0.1647	0.3086
Line 1-2	Stable	−0.0531	−0.0788	−0.0870

The results indicate that

1. The MLE calculation results can predict rotor-angle transient stability correctly.
2. When there is only one available PMU in the system, the smallest singular values of the observability matrices are very close to zero, indicating that the system is poorly observable.
3. As the number of PMUs increases, the level of observability is also improved.
4. When only one PMU is available in the system, the system is hardly observable. However, the MLE calculation results are the same with the scenario with 2 PMUs. It indicates that for this simple 3-bus system, one PMU measurement is able to provide the necessary information for rotor-angle stability assessment. However, this observation is not expected to be true for large scale power systems.

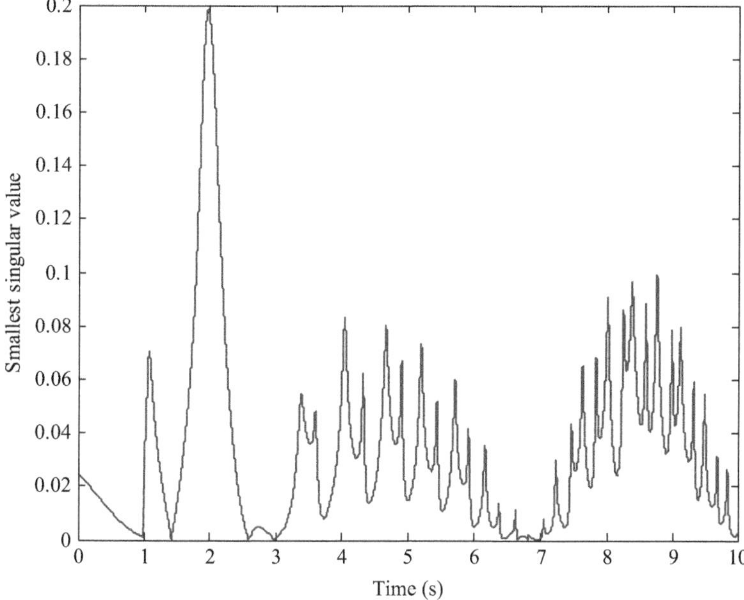

Figure 8.9 Smallest singular value of observability matrix with PMU placed at bus 1

Table 8.3 MLE of different time window T with two PMUs

3-Phase fault	Stability	MLE of $T = 1.25$	MLE of $T = 1.667$	MLE of $T = 2$
Line 1-3	Unstable	0.5848	0.7448	0.7798
Line 2-3	Unstable	0.0236	0.1647	0.3086
Line 1-2	Stable	−0.0531	−0.0788	−0.0870

8.5 Conclusion

The proposed MLE method is validated for the assessment of one of the vulnerable patterns of power systems: transient rotor-angle instability. The main idea is to calculate MLE along the system dynamic trajectory in order to predict a loss of synchronism of power systems. Observability of PMU-based power system analysis is studied with a purpose of developing a method to quantify how well the available PMU measurements can be used to monitor dynamics of a power system. The research results in an innovative application of PMU data. Once an unstable power swing is captured and rotor-angle instability is predicted by the above method, self-healing actions, such as power system partitioning control [49],

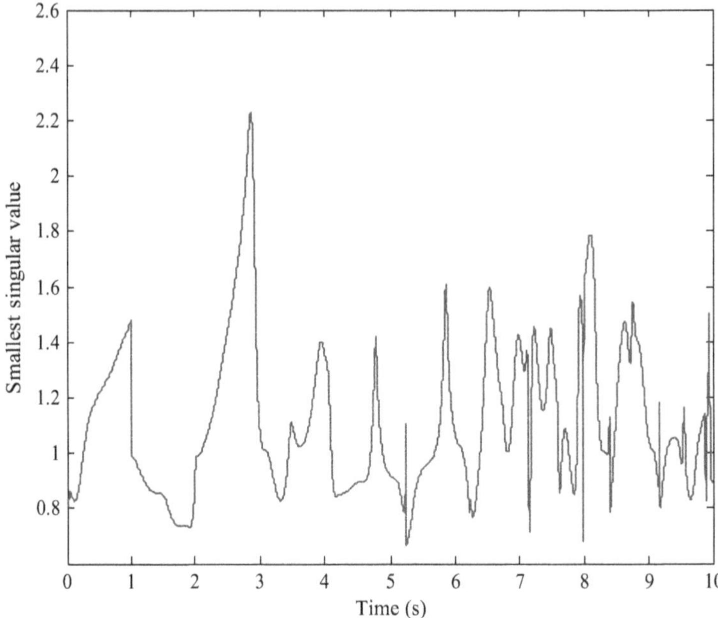

Figure 8.10 Smallest singular value of observability matrix with PMU placed at buses 1 and 2

can be initiated to reduce system vulnerability and avoid cascading failures that may result in a catastrophic blackout.

Acknowledgment

This research is sponsored by Electric Power Research Institute (EPRI), Palo Alto, CA, USA.

References

[1] Nye D. E. When the Lights Went Out: A History of Blackouts in America. MIT Press, Cambridge, MA, 2010.

[2] Parks B. Transforming the Grid to Revolutionize Electric Power in North America, U.S. Department of Energy, Edison Electric Institute: Fall 2003 Transmission, Distribution and Metering Conference, 2003.

[3] MoP. Report of the Enquiry Committee on Grid Disturbance in Northern Region on 30th July 2012 and in Northern, Eastern and North-Eastern Region on 31st July 2012. [Online] Available: http://www.powermin.nic.in/pdf/GRID_ENQ_REP_16_8_12.pdf, 2012.

[4] Romero J. J. Blackouts Illuminate India's Power Problems. IEEE Spectrum, 2012, 49(10): 11–12.
[5] FERC. Arizona-Southern California Outages on September 8, 2011: Causes and Recommendations. [Online] Available: http://www.ferc.gov/legal/staff-reports/04-27-2012-ferc-nerc-report.pdf, 2012.
[6] Hines P., Balasubramaniam K., Sanchez E. C. Cascading Failures in Power Grids. IEEE Potentials, 2009, 28(5): 24–30.
[7] Pahwa S., Hodges A., Scoglio C., Wood S. Topological Analysis of the Power Grid and Mitigation Strategies against Cascading Failures. IEEE the Fourth Annual Systems Conference, San Diego, USA, 2010.
[8] Yamashita K., Joo S. K., Li J., Zhang P., Liu C. C. Analysis, Control, and Economic Impact Assessment of Major Blackout Events. European Transactions on Electrical Power, 2008, 18: 854–871.
[9] Yamashita K., Li J., Zhang P., Liu C. C. Analysis and Control of Major Blackout Events. IEEE PES. Power Systems Conference and Exposition, Seattle, USA, 2009.
[10] Yamashita K., Li J., Liu C. C., Zhang P., Hofmann M. Learning to Recognize Vulnerable Patterns Due to Undesirable Zone-3 Relay Operations. IEEJ Transactions on Electrical and Electronic Engineering, 2009, 4: 322–333.
[11] Liu C. C., Jung J., Heydt G., Vittal V., Phadke A. G. The Strategic Power Infrastructure Defense (SPID) System: A Conceptual Design. IEEE Control Systems Magazine, 2000, 20(4): 40–52.
[12] Li H., Rosenwald G., Jung J., Liu C. C. Strategic Power Infrastructure Defense. Proceedings of the IEEE, 2005, 93(5): 918–933.
[13] Pai M. A., Kulkarni A. Y. A Simulation Tool for Transient Stability Analysis Suitable for Parallel Computation. IEEE the Fourth Conference on Control Applications, Albany, USA, 1995.
[14] Suyono H., Nor K. M., Yusof S. Component Based Development for Transient Stability Power System Simulation Software. Power Engineering Conference, Thessaloniki, Greece, 2003.
[15] Fouad A. A., Vittal V., Tae K. O. Critical Energy for Direct Transient Stability Assessment of a Multimachine Power System. IEEE Transactions on Power Apparatus and Systems, 1984, PAS-103(8): 2199–2206.
[16] Chiang H. D. A Theory-Based Controlling UEP Method for Direct Analysis of Power System Transient Stability. IEEE International Symposium on Circuits and Systems, Portland, OR, USA, 1989.
[17] Kakimoto N., Ohnogi Y., Matsuda H., Shibuya H. Transient Stability Analysis of Large-Scale Power System by Lyapunov's Direct Method. IEEE Transactions on Power Apparatus and Systems, 1984, PAS-103(1): 160–167.
[18] Chiang H. D., Wu F. F., Varaiya P. P. A BCU Method for Direct Analysis of Power System Transient Stability. IEEE Transactions on Power Systems, 1994, 9(3): 1194–1208.
[19] Xue Y., Van C. T., Ribbens-Pavella M. A Simple Direct Method for Fast Transient Stability Assessment of Large Power Systems. IEEE Transactions on Power Systems, 1988, 3(2): 400–412.

[20] Phadke A. G., Thorp J. S., Adamiak M. G. A New Measurement Technique for Tracking Voltage Phasors, Local System Frequency, and Rate of Change of Frequency. IEEE Transactions on Power Apparatus and Systems, 1983, PAS-102(5): 1025–1038.

[21] Thorp J. S., Phadke A. G., Karimi K. Real-Time Voltage Phasor Measurements for Static State Estimation. IEEE Transactions on Power Apparatus and Systems, 1985, PAS-104(11): 3098–3106.

[22] Phadke A. G., Thorp J. S., Karimi K. State Estimation with Phasor Measurements. IEEE Transactions on Power Systems, 1986, PER-6(2), 233–241.

[23] Thorp J. S., Phadke A. G., Horowitz S. H., Begovic M. M. Some Applications of Phasor Measurements to Adaptive Protection. IEEE Transactions on Power Systems, 1988, 3(2): 791–795.

[24] Phadke, A. G. Synchronized Phasor Measurements – A Historical Overview. IEEE/PES Transmission and Distribution Conference and Exhibition 2002: Asia Pacific, 1: 476–479.

[25] Centeno V., Phadke A. G., Edris A., Benton J., Gaudi M., Michel G. An Adaptive Out-of-Step Relay for Power System Protection. IEEE Transactions on Power Delivery, 1997, 12(1): 61–71.

[26] Centeno V., Phadke A. G., Edris A. Adaptive Out-of-Step Relay with Phasor Measurement. Sixth International Conference on Developments in Power System Protection, 1997, pub 1, no. 434: 210–213.

[27] Centeno V., Phadke A. G., Edris A., Benton J., Michel G. An Adaptive Out-of-Step Relay. IEEE Power Engineering Review, 1997, 17(1): 39–40.

[28] Centeno V., Ree J. de la, Phadke A. G. Michel G., Murphy R. J., Burnett R. Adaptive Out-of-Step Relaying Using Phasor Measurement Techniques. IEEE Computer Application in Power, 1993, 6(4): 12–17.

[29] Liu C. W., Thorp J. New Methods for Computing Power System Dynamic Response for Real-Time Transient Stability Prediction. IEEE Transactions on Circuits and Systems I: Fundamental Theory and Applications, 2000, 47 (3): 324–337.

[30] Huang Z., Schneider K., Nieplocha J., Zhou N. Estimating Power System Dynamic States Using Extended Kalman Filter. IEEE PES General Meeting, National Harbor, USA, 2014.

[31] Liu G., Ning J., Tashman Z., Venkatasubramanian V., Trachian P. Oscillation Monitoring System Using Synchrophasors. IEEE PES General Meeting, San Diego, USA, 2012.

[32] Ning J., Pan X., Venkatasubramanian V. Oscillation Modal Analysis from Ambient Synchrophasor Data Using Distributed Frequency Domain Optimization. IEEE Transactions on Power Systems, 2013, 28(2): 1960–1968.

[33] Chow J. H., Chakrabortty A., Arcak M., Bhargava B., Salazar A. Synchronized Phasor Data Based Energy Function Analysis of Dominant Power Transfer Paths in Large Power Systems. IEEE Transactions on Power Systems, 2007, 22(2): 727–734.

[34] Sun K., Likhate S., Vittal V., Kolluri V. S., Mandal S. An Online Dynamic Security Assessment Scheme Using Phasor Measurements and Decision Trees. IEEE Transactions on Power Systems, 2007, 22(4): 1935–1943.
[35] Yan J., Liu C. C., Vaidya U. PMU Based Monitoring of Rotor Angle Dynamics. IEEE Transactions on Power Systems, 2011, 26(4): 2125–2133.
[36] Yan J., Liu C. C., Vaidya U. A PMU-Based Monitoring Scheme for Rotor Angle Stability. IEEE PES General Meeting, San Diego, USA, 2012.
[37] Liu C., Thorp J., Lu J., Thomas R. J., Chiang H. D. Detection of Transiently Chaotic Swings in Power Systems Using Real-Time Phasor Measurements. IEEE Transactions on Power Systems, 1994, 9(3):1285–1292.
[38] Osedelec V. I. Multiplicative Ergodic Theorem: Lyapunov Characteristic Exponent for Dynamical Systems. Moscow Math, 1968, 19: 539–575.
[39] Eckman J. P., Ruelle D. Ergodic Theory of Chaos and Strange Attractors. Reviews of Modern Physics, 1985, 57: 617–656.
[40] Ramasubramanian K., Sriram M. S. A Comparative Study of Computation of Lyapunov Spectra with Different Algorithms. Physica D, 2000, 139(1–2): 72–86.
[41] Monticelli A., Wu F. F. Network Observability: Theory. IEEE Transactions on Power Apparatus and Systems, 1985, PAS-104(5): 1042–1048.
[42] Monticelli A. Electric Power System State Estimation. Proceedings of the IEEE, 2000: 262–282.
[43] Chakrabartiand S., Kyriakides E. Optimal Placement of Phasor Measurement Units for Power System Observability. IEEE Transactions on Power Systems, 2008, 23(3): 1433–1440.
[44] Nuqui R. F., Phadke A. G. Phasor Measurement Unit Placement Techniques for Complete and Incomplete Observability. IEEE Transactions on Power Delivery, 2005, 20(4): 2381–2388.
[45] Kalman R. E. On the Computation of the Reachable/Observable Canonical Form. SIAM Journal of Control and Optimization, 1982, 20: 258–260.
[46] Hermann R., Krener A. J. Nonlinear Controllability and Observability. IEEE Transactions on Automatic Control, 1977, AC-2(5): 728–740.
[47] Röbenack K., Reinschke K. J. An Efficient Method to Compute Lie Derivatives and the Observability Matrix for Nonlinear Systems. Proceedings of 2000 International Symposium on Nonlinear Theory and its Applications, Dresden, 2000, 2: 625–628.
[48] Chen C. T. Linear System Theory and Design (3rd edition). Oxford University Press, New York, NY, 1998.
[49] Li J., Liu C. C., Schneider K. Controlled Partitioning of a Power Network Considering Real and Reactive Power Balance. IEEE Transactions on Smart Grid, 2010, 1(3): 261–269.

Chapter 9

Synchrophasor applications for load estimation and stability analysis

Tushar, H. Lee[1], P. Banerjee[1] and A. K. Srivastava[1]

The use of advanced real-time monitoring and control technologies have made it possible to transform the existing power grid into smart grid. With the development of phasor measurement units (PMUs) and the availability of real-time measurements, it has become possible to develop data-driven power system applications to enhance the stability and reliability of the power grid. The application of synchrophasor measurements for online assessment of voltage stability, real-time load parameter estimation, and transient stability prediction is presented in this chapter. The different voltage stability monitoring tools used in the industry are reviewed and analyzed. PMU-based distributed voltage stability monitoring technique is also presented in this chapter. The load modeling tool used in the industry and comprehensive research effort in load modeling is discussed in this chapter. The adaptive search–based recursive least-square algorithm is presented for improved online estimation of load parameters. The existing techniques available in the literature for transient stability assessment are reviewed in this chapter. Application of maximum Lyapunov exponent (MLE) for PMU-based online transient stability detection is discussed. The effectiveness of MLE-based method is demonstrated on real-time simulation of New England 39 bus system.

9.1 Introduction

The electric power system is a highly complex man-made dynamical system. Monitoring the power system operating states is the first step toward stability and reliability of the power system. The power system is generally monitored by measuring key states, defined by the magnitude and phase angle of bus voltages. These states change continuously even during normal operation of the system and can be processed to track operating status of the grid given by (1) normal, (2) alert, and (3) emergency states. Set of preventive and emergency control actions are initiated by the system operator, under alert and emergency operating states of the

[1]School of Electrical Engineering and Computer Science, Smart Grid Demonstration and Research Investigation Lab (SGDRIL), Washington State University, Washington, DC, USA

system, to keep the system in stable and secure state. Supervisory control and data acquisition system–based energy management systems (SCADA/EMS) have been deployed in the power system networks, which utilizes remote terminal units (RTUs) in the field to acquire analog and digital signals from the network. The RTUs are located at substations and provide the data pertaining to the amplitude of bus voltages, branch currents, and line power flows, typically at a scan rate of 2–10 s [1,2]. These data, received at control center, are processed by a state estimator (SE). The results obtained from the SE are used for security analysis, optimal power dispatch, automatic generation control, and auxiliary control actions. The critical applications like dynamic voltage stability and transient stability assessment require much faster data refresh rate (100 ms–3 s) not supported by SCADA. Further, the data obtained from the SCADA measurements are not time stamped, and it is unable to directly measure the angle of the bus voltages. These limitations of the SCADA system have been addressed by the phasor measurement units (PMUs), which can provide the voltage and the current phasors with respect to a global reference sinusoid, synchronized to the global positioning system (GPS) clock at approximately 200 times faster than SCADA. The phasor data estimated by the PMUs are time stamped and archived by the phasor data concentrator (PDC), which is located at the substations or at the control center.

The archived phasor data in the control center are retrieved by different power system applications such as online voltage stability and transient stability assessment. A typical scenario of voltage collapse occurs in seconds, whereas transient instability can occur even faster [3]. The PMU data rate of 60 phasors in a second is particularly useful in these scenarios for initiating emergency control actions to rescue the system. The load parameter estimation on the other hand is a quasi-steady-state phenomenon, which needs to be updated in few minutes but can use high-resolution fast rate of data.

9.2 Voltage stability assessment using synchrophasor data

9.2.1 Review of existing PMU-based voltage stability

The voltage stability is one of the indicators that determine system strength to maintain required voltage range at all the nodes in all possible operating conditions and prevent voltage collapse leading to black/brown out. This application requires voltage information and/or system information to estimate voltage stability index (VSI) locally or widely. In recent years, many voltage stability algorithms are proposed using PMU data for higher accuracy and meeting real-time requirements [4–10]. Wang *et al.* [11] demonstrated the mechanism of voltage collapse phenomenon as the case of emergence of bifurcation followed by chaos in a dynamical system. The sign of Lyapunov exponents (LEs) of system states indicates chaos in the system. The bus voltage magnitudes are components of the states of the power system. The negative value of LEs represents convergence of the state trajectories, denoting stable system. The positive value designates divergence of state trajectories, leading to instability in the system [4] or chaos. The coupled single port circuit uses equivalent representation for multiple generators and loads [5]. The concept of couple single port circuit can reduce mismatches by modeling load as nonlinear and dynamic component. The security margins using multidimensional hyperplanes are defined in [6]. This work methodologically considers the effect of transmission and system constrains such as thermal limits, voltage stability, and oscillatory stability margins. A modified coupled single-port model–based voltage stability is proposed in [7]. The modified model is compensated using reactive power mismatching using mitigating factor. In [8], the Thevenin Equivalent parameters are updated by three consecutive local measurements of voltage and current phasors. The updated Thevenin Equivalent parameter is used to determine the voltage stability margin. The decision tree–based voltage stability using PMU data is proposed in [9]. The decision tree is composed of classification tree and regression tree. The classification tree is a database of the group of operating point (OP), which is created by offline study. The regression tree can detect voltage instability by referring to the classification tree. In [10], an artificial neural network (ANN)–based voltage stability is proposed. The multilayer perceptron networks are used to manage large input data in a nonlinear system.

9.2.2 Industrial voltage stability application

9.2.2.1 RTDMS by EPG

Real Time Dynamics Monitoring System (RTDMS) is a real-time software for synchrophasor visualization, monitoring critical metrics and analyze complex dynamical events that occur in the power system network. RTDMS is developed by Electric Power Group (EPG) and has been deployed at California Independent System Operator (CAISO), Midcontinent Independent System Operator (MISO), PJM, and Electric Reliability Council of Texas (ERCOT). The RTDMS interface displays the geographical distribution of the voltage magnitude and the angle difference of various interties along with frequency tracking. The modes and damping of power oscillation and voltage stability are some of the tools implemented in the RTDMS platform. The snapshot showing different metrics and margins related to voltage stability is shown in Figure 9.1. The voltage sensitivity with respect to change in power is estimated, which is compared to predefined voltage sensitivity to generate proximity indicators from the collapse point. The voltage sensitivity with respect to power is supposed to be equivalent to the slope of the real power-voltage (PV) curve. The safe operating limit of voltage sensitivity is typically around 4 kV/100 MW for 500 kV system. The deteriorating system condition would lead to increase in the voltage sensitivity as the OP of the system approaches the collapse point. This metric generate warnings and alarms sufficiently early for rescuing the system form voltage instability.

The voltage instability detector implemented in RTDMS is based on the source voltage model developed in [12]. To obtain the PV curve at the PMU location for a typical system in Figure 9.2, the system is modeled as a two-voltage source with different phase separation from the reference bus as shown in Figure 9.3. The Thevenin Equivalent parameters of the two ends of the network are estimated from multiple readings and solving the least-square problem.

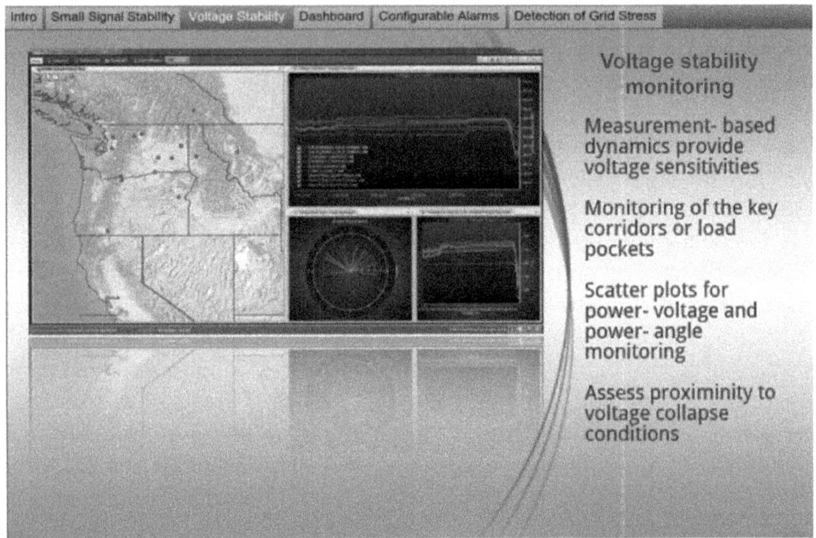

Figure 9.1 RTDMS implemented in CAISO, MISO, and ERCOT

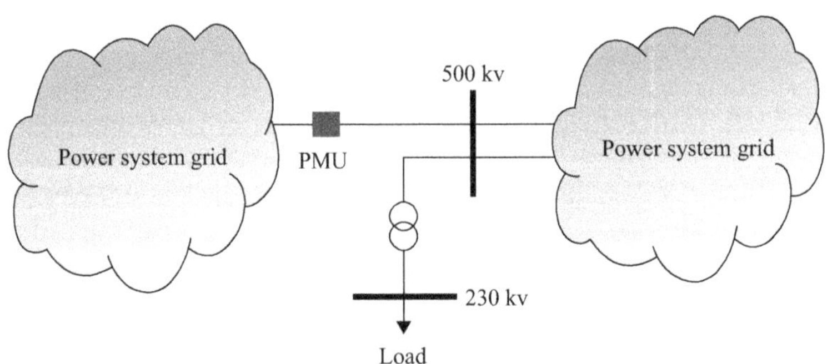

Figure 9.2 Two-source power system with synchrophasor measurement

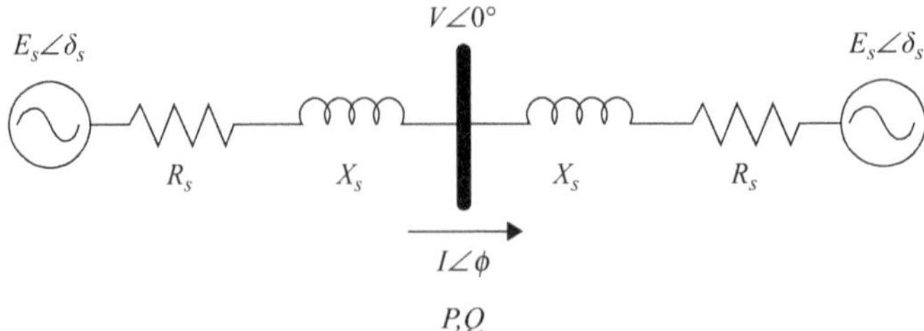

Figure 9.3 Double-voltage source model for two-source power system

The model parameters are calculated as follows:

$$P - jQ = VI^* = V\left(\frac{E_s\angle\delta_s - E_r\angle\delta_r}{Z\angle\theta}\right) \tag{9.1}$$

$$V = E_s\angle\delta_s - Z_s\angle\theta_s I^* \tag{9.2}$$

where

$$Z\angle\theta = (R_s + R_r) + j(X_s + X_r) \tag{9.3}$$

$$Z_s\angle\theta_s = R_s + jX_s \tag{9.4}$$

The main advantages of estimating voltage sensitivity using the double-voltage source are as follows:

- Topology independent
- Utilizes only PMU data
- Estimates power–voltage relationships from measured data
- Method uses system equivalency and is suited for local area

The main shortcoming of this technique is that it lacks the ability to perform contingency analysis.

9.2.2.2 Voltage stability using SDI

The voltage stability indicator commissioned by Terna as a pilot project of synchrophasor technology is based on S difference indicator (SDI). The measurement of apparent power at load end j and at time instant t of the transmission line, denoted by $\vec{S}_j^{(t)}$, is used for estimating SDI. $\vec{U}_j^{(t)}$ and $\vec{I}_{ji}^{(t)}$ are the voltage at bus j and the current flowing in line joining bus i and j, at the time instant t. The increase in the power losses at the proximity of the voltage instability causes the apparent power to increase at the sending end of the line, whereas the apparent power at the receiving end remains fairly unchanged. Consecutive measurements of the apparent power at the receiving end can be used for identifying the voltage collapse scenario. Two consecutive measurements of voltage and current phasors at time instant t and $t+1$ are subtracted, and the apparent power at $t+1$ is expressed as a function of the difference of the voltage and current phasors. The slow change in estimated difference of the apparent power denotes voltage collapse situation:

$$\vec{S}_j^{(t)} = \vec{U}_j^{(t)} \cdot \vec{I}_{ji}^{(t)^*} \tag{9.5}$$

$$\Delta\vec{U}_j^{(t+1)} = \vec{U}_j^{(t+1)} - \vec{U}_j^{(t)} \tag{9.6}$$

$$\Delta\vec{I}_{ji}^{(t+1)} = \vec{I}_{ji}^{(t+1)} - \vec{I}_{ji}^{(t)} \tag{9.7}$$

$$\vec{S}_j^{(t+1)} = \left(\vec{U}_j^{(t)} + \Delta\vec{U}_j^{(t+1)}\right) \cdot \left(\vec{I}_{ji}^{(t)} + \Delta\vec{I}_{ji}^{(t+1)}\right)^* \tag{9.8}$$

$$= \vec{U}_j^{(t)}\vec{I}_{ji}^{(t)^*} + \vec{U}_j^{(t)}\Delta\vec{I}_{ji}^{(t+1)^*} + \vec{I}_{ji}^{(t)^*}\Delta\vec{U}_j^{(t+1)} + \Delta\vec{U}_j^{(t+1)}\vec{I}_{ji}^{(t+1)^*} \tag{9.9}$$

$$\Delta \vec{S}_j^{(t+1)} \approx \vec{U}_j^{(t)} \Delta \vec{I}_{ji}^{(t+1)^*} + \vec{I}_{ji}^{(t)^*} \Delta \vec{U}_j^{(t+1)} \tag{9.10}$$

$$\mathrm{SDI} = 1 + \left| \frac{\vec{I}_{ji}^{(t)^*} \Delta \vec{U}_j^{(t+1)}}{\vec{U}_j^{(t)} \Delta \vec{I}_{ji}^{(t+1)^*}} \right| \cos\theta \geq 0 \tag{9.11}$$

Theoretically SDI would be 0 at the voltage collapse condition; however, margin of 0.2 is considered for practical operations. The proposed method, using SDI, works for transmission system that consumes reactive power but will not work for cables and lightly loaded overhead lines which consume very less reactive power.

9.2.2.3 WAVI in KEPCO system

The installation of PMUs in the metropolitan area of the KEPCO system is carried out in the first phase of the K-WAMS (Korean Wide Area Measurement and Monitoring System) project. The voltage stability prediction is identified as one of the applications of synchrophasor data to be developed for KEPCO system. The voltage stability index known as WAVI (wide-area voltage stability index) is developed for assessing proximity of the system to the voltage collapse point. WAVI requires at least two PMUs from different types of load areas, which are in this case metropolitan (Area 2) and non-metropolitan (Area 1). The bulk generation is installed in the non-metropolitan area in the south, and loads are concentrated on the metropolitan area in the north. The load centers in the metropolitan area are catered by six important transmission lines from the generation located in the south.

The bus observed by PMU in the metropolitan area is Bus 1 and one in the non-metropolitan area is Bus 2. The equivalent circuit representation of Figure 9.4 is presented in Figure 9.5.

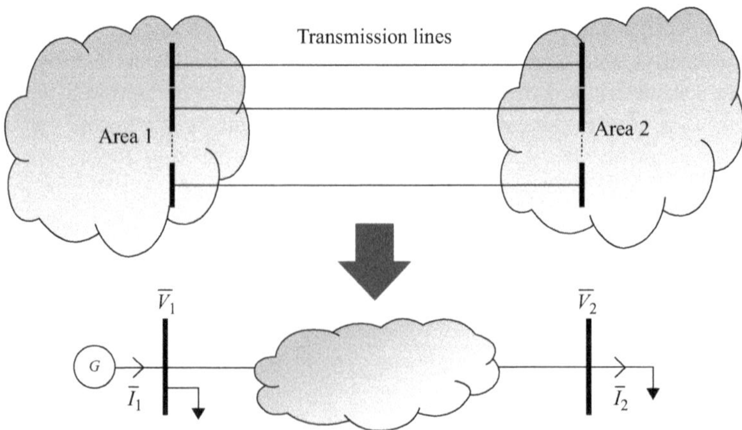

Figure 9.4 Typical equivalent circuit for two area system

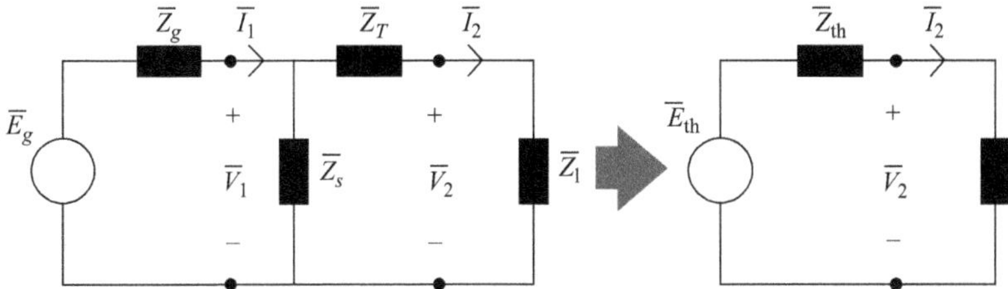

Figure 9.5 Reduced Thevenin Equivalent

The impedance Z_{app}, and Z_{th} is expressed as a function of v_1, i_1, v_2, i_2, and Z_g:

$$\overline{Z}_{th} = \overline{Z}_T + \frac{1}{(1/\overline{Z}_g) + (1/\overline{Z}_s)} \tag{9.12}$$

$$\overline{Z}_{app} = \overline{Z}_l \tag{9.13}$$

where

$$\overline{Z}_s = \frac{\overline{v}_1}{\overline{i}_1 - \overline{i}_2} \tag{9.14}$$

$$\overline{Z}_T = \frac{\overline{v}_1 - \overline{v}_2}{\overline{i}_1} \tag{9.15}$$

The estimation of Z_g is carried out once using the method presented in [13]. The WAVI is defined as the ratio of the impedance Z_{th} and Z_{app}:

$$\text{WAVI} = \frac{\overline{Z}_{th}}{\overline{Z}_{app}} \tag{9.16}$$

The normal system operation results in lesser value of Z_{th} as compared to Z_{app}, for which WAVI is in the range of 0.2. The voltage collapse scenario results in comparable values of Z_{th} and Z_{app}, for which WAVI is near to 1.0.

9.2.2.4 Voltage stability tool by quanta

The real-time assessment technique of voltage stability margin using local measurements proposed by Quanta Technology is called reactive power margin method. The power system network is modeled as a Thevenin Equivalent at the load end where measurements are acquired. The circuit diagram for the Thevenin Equivalent is shown in Figure 9.6, where E_{eq} and Z_{eq} are estimated using the VIP method presented in [14].

Based on the network theory, the following equation can be obtained:

$$|\overline{Z}_{Thev} + \overline{Z}_{app}| \times I = \overline{E}_{Thev} \tag{9.17}$$

where Z_{app} is measured at the load end using the phasor values of voltage and current. The unknown parameter Z_{th} (R_{th} and X_{th}) and E_{th} are obtained using

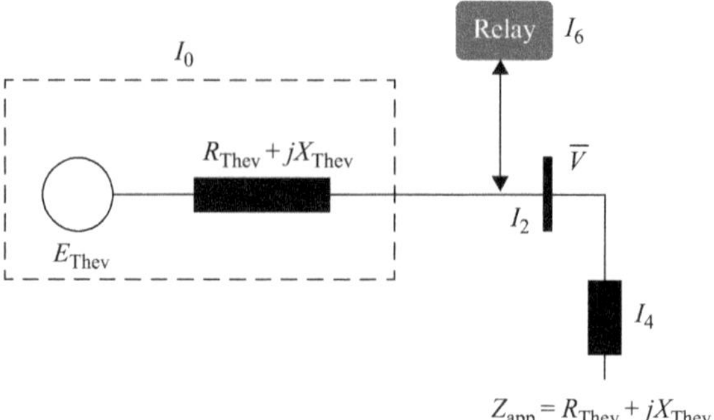

Figure 9.6 Thevenin equivalent with relay

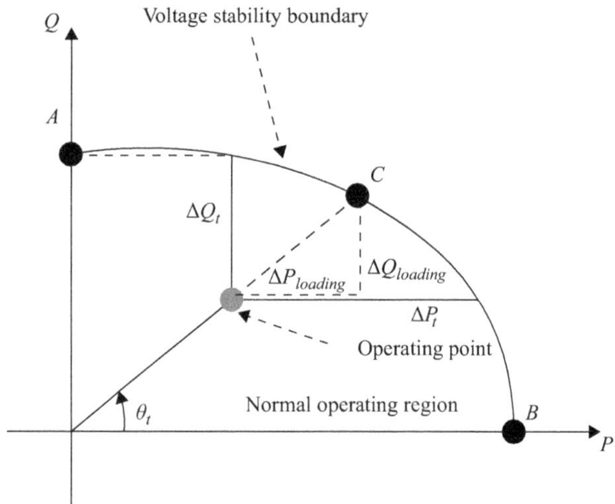

Figure 9.7 Voltage stability margin in PQ plane

multiple measurements of the Z_{app} and I, then applying a least-square estimation. where

A: $(0, Q^t_{max})$

B: $(P_{max,t}, 0)$

C: $(S_{max,t} \cos \theta, S_{max,t} \sin \theta)$

The voltage stability power margin on *PQ* plane is approximated by a parabola as shown in Figure 9.7, expressed as

$$Q_t = aP_t^2 + bP_t + c \tag{9.18}$$

The maximum mega volt ampere (MVA) corresponds to the current equivalent parameters can be expressed as

$$S_{\max} = \frac{\left|\overline{E}_{\text{eq}}\right|^2 \sqrt{1 + \tan^2\theta}}{2\left(R_{\text{eq}} + X_{\text{eq}} \tan\theta \pm \sqrt{\left(R_{\text{eq}}^2 + X_{\text{eq}}^2\right)(1 + \tan^2\theta)}\right)} \tag{9.19}$$

The margin of maximum power system is defined as

$$\Delta S = S_{\max} - S \tag{9.20}$$

The maximum values $P_{\max}$ and $Q_{\max}$ is obtained as

$$P_{\max,t} = \frac{\left|\overline{E}_{\text{eq},t}\right|^2}{2\left(R_{\text{eq},t} + \left|\overline{Z}_{\text{eq},t}\right|\right)} \tag{9.21}$$

$$Q_{\max,t} = \frac{\left|\overline{E}_{\text{eq},t}\right|^2}{2\left(X_{\text{eq},t} + \left|\overline{Z}_{\text{eq},t}\right|\right)}$$

where

$$\left|\overline{Z}_{\text{eq},t}\right| = \sqrt{R_{\text{eq},t}^2 + X_{\text{eq},t}^2} \tag{9.22}$$

For known values of theta, the parameters of the parabola *a*, *b*, and *c* are estimated for each time instant. The loading margin of real and reactive power is estimated by comparing the current OP and the envelope of the stability margin denoted by the parabola.

9.2.2.5 ROSE by VR energy

Region of stability existence (ROSE) [15] is a situational awareness tool developed by the VR energy that is implemented by Western Electricity Coordinating Council (WECC) [16] and Independent System Operator New England (ISONE) [17]. ROSE constructs the stability boundary using the SE data and updates it at the rate of SCADA measurements. The trajectory of the OP is updated at the PMU measurements rate. ROSE generates

1. Visualization of the current OP with respect to the boundary in the plane of two phase angles or power injections or power flows.
2. Generate warnings and alarms as the OP approaches the boundary.
3. Specify the range of secure operation in terms of the accepted *N–k* security criteria.
4. Details of the limit violations in a tabular format.

A typical visualization of the ROSE with boundary and operating trajectory is shown in Figure 9.8.

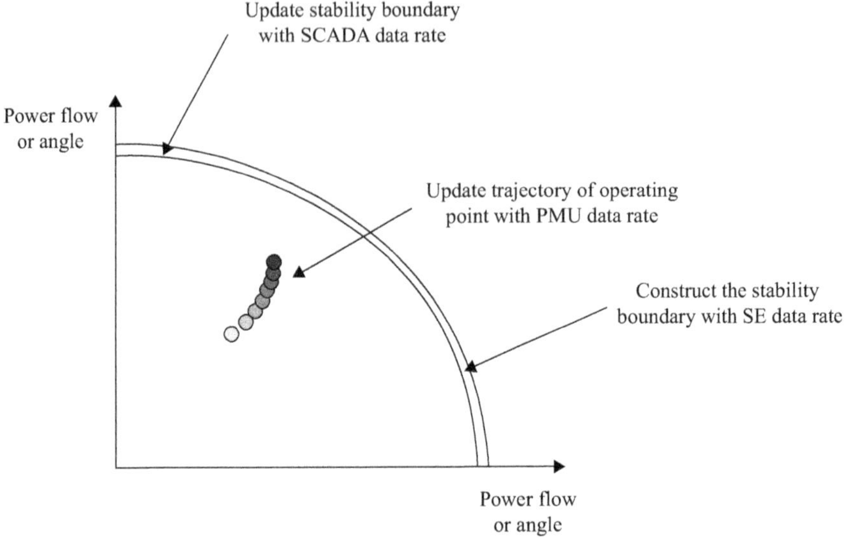

Figure 9.8 ROSE boundary and trajectory at the OP

9.2.3 Decentralized voltage stability assessment

The evolution of real-time application requires real-time devices such as PMU, PDC, and other time synchronized devices in power system. This mandates large data to be processed by the applications in a time-bound manner. Centralized real-time applications need to process a large number of measurements, which becomes computationally and economically ineffective. To solve these issues, the decentralized applications are explored for faster response and lower operating cost. The fuzzy logic–based decentralized voltage regulator using partitioning method and voltage sensitivity method are proposed in [18,19]. A pseudo-gradient evolutionary programming for controlling coordinated voltage is proposed in [20]. These methods mentioned in the literature mainly focus on the voltage control that indirectly can help with voltage stability. The main challenge in implementing decentralized application is efficient communication among the clusters and fault tolerant computation. To solve this issue, the decentralized application can be implemented using a platform like Distributed Coordination Computational Algorithms (DCBlocks) .

9.2.3.1 Distributed coordination platform

The DCBlocks platform is a middleware framework that enables clusters to communicate with each other for distributed power applications such as decentralized voltage stability (DVS). The DCBlocks is one of the decentralized platforms that provide expandability and faster decision-making process in small cluster instead of using centralized application for monitoring and control. The DCBlocks consist of a number of modules such as Group Management Block, ABCAST Block,

Leader Election Block, Consensus/Agreement Block, Mutual Exclusion Block, and Failure detection/tolerance [21].

Most of existing power system applications can be implemented in a decentralized framework. In this chapter, Thevenin-based voltage stability is implemented and tested as decentralized implementation similar to [22]. For implementing decentralized application, following steps are required: (1) initialization, (2) performing DVS monitoring, (3) control action to enhance voltage stability (optional).

9.2.3.2 Initialization of decentralized application

DVS assessment consists of a number of clusters (small or mid-size logical boundaries of power system network) based on electrical distance. Initial grouping method is performed for finding boundary of each clusters. Each cluster should contain at least one generator, transmission lines, loads, and reactive power source for short-/long-term voltage stability. These components in a cluster are close to each other electrically and geographically. After performing initialization, virtual buses are created between boundaries of two clusters. The virtual bus is connected with a virtual generator or virtual load as shown in Figure 9.9 to represent neighboring clusters. The virtual generator or virtual load is defined as follows: Power leaving from Bus i:

$$S_{ij} = V_i \times I_i^* \tag{9.23}$$

Power reaching to Bus j:

$$S_{ji} = V_j \times I_j^* \tag{9.24}$$

where V_i, I_i, V_j, and I_j are measured bus voltage and current at bus i and j, respectively. The negative (positive) of *real* (S_{ij}) value can represent generator (load). The clusters can be combined/separated on the basis of the specific criteria for efficient computation and control needs.

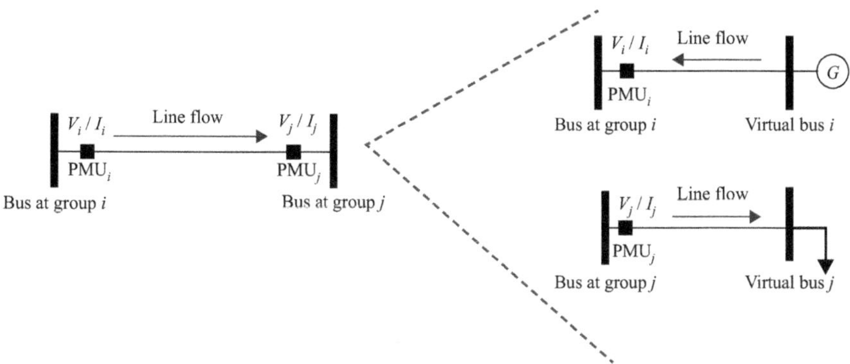

Figure 9.9 Illustration of conversion of tie line between two groups

9.2.3.3 Decentralized voltage stability monitoring method

The proposed DVS requires voltage and current phasor measurements and network topology. The Thevenin parameters (E_{th} and Z_{th}) are calculated using network topology and voltage and current phasors. The Thevenin parameters (E_{th} and Z_{th}) can be estimated as follows:

$$[Z_{th}] = [Z_{LL}] = \left[\left(Y_{LL} - Y_{LT}Y_{TT}^{-1}Y_{TL}\right)^{-1}\right] \tag{9.25}$$

where Y_{LL} indicate admittance element at load–load connection. Similarly, other rotations are LT (load–Tie connection), TT (Tie–Tie connection), GL (Generator–Load), GG (Generator–Generator), and GT (Generator–Tie):

$$v_{th_j} = \sum_{m=1}^{M} K_{LG_{jm}} v_{Gm} + \sum_{n=1,\, i\neq j}^{N} Z_{LL_{ji}} \left(\frac{-S_{Li}}{v_{Li}}\right)^*$$

where

$$[K_{LG}] = \frac{\left[Y_{LT}Y_{TT}^{-1}Y_{TG} - Y_{LG}\right]}{\left[Y_{LL} - Y_{LT}Y_{TT}^{-1}Y_{TL}\right]} \tag{9.26}$$

$$S_{Li} = P_{\text{measured at Load bus } i} + jQ_{\text{measured at Load bus } i}$$

The Thevenin-based voltage stability monitoring algorithm indicates VSI using maximum power-transfer theory. The maximum power transfer suggests that the maximum power is being transferred when the system impedance is complex conjugate of load impedance as shown in Figure 9.10.

The VSI can be calculated as follows:

$$\text{VSI}_S = 1 - \frac{S_{\text{max}_{\text{eachload}}} - S_{L_{\text{eachload}}}}{S_{\text{max}_{\text{eachload}}}} \tag{9.27}$$

$$\text{VSI}_P = 1 - \frac{P_{\text{max}_{\text{eachload}}} - P_{L_{\text{eachload}}}}{P_{\text{max}_{\text{eachload}}}} \tag{9.28}$$

$$\text{VSI}_Q = 1 - \frac{Q_{\text{max}_{\text{eachload}}} - Q_{L_{\text{eachload}}}}{Q_{\text{max}_{\text{eachload}}}} \tag{9.29}$$

$$\text{VSI} = \max\left(\text{VSI}_S, \text{VSI}_P, \text{VSI}_Q\right) \tag{9.30}$$

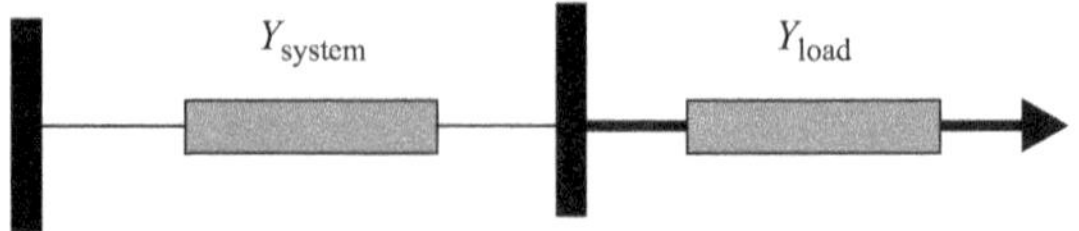

Maximum power transferred at $Y_{system} = Y_{load}$

Figure 9.10 Maximum power-transfer theory

where

$$S_{max} = E_{th} \times \left(\frac{E_{th}}{Z_{th}}\right)^{*} \tag{9.30a}$$

$$P_{max} = \text{real}(S_{max}) \quad Q_{max} = imag(S_{max})$$

When any of VSIs exceeds certain limit, the decentralized application starts communicating among cluster members and other cluster leader for deciding proper control action which can reduce voltage stability violation.

9.3 PMU-based load modeling

The representation of the load model has a significant impact on the system stability analysis [23,24]. One of the major reasons for inaccuracies in simulation studies for planning and operations are load models [25]. An accurate load model is necessary to correctly estimate the operational state of the system. High-resolution measurement from PMUs makes it possible to estimate load behavior models.

9.3.1 Definition of load modeling

Various definitions of load as recommended by IEEE Task force [24] are as follows:

- A device, connected to a power system, that consumes power. The term used for this definition is "load device."
- The total power (active and/or reactive) consumed by all devices connected to a power system. The term used for this definition is "system load."
- A portion of the system that is not explicitly represented in a system model but rather is treated as if it were a single power consuming device connected to a bus in the system model. The term used for this definition is "bus load."
- The power output of a generator or generating plant. The term used for this definition is "generator or plant load."

Following are some terms used to describe the composition of a load [24] are as follows:

Load component: A load component is the aggregate equivalent of all devices of a specific or similar type, for example, water heater, air conditioner, and fluorescent lighting.

Load class: A load class is a category of load, such as, residential, commercial, or industrial.

Load composition: A load composition is the fractional composition of the load-by-load components.

Load class mix: A load class mix is the fractional composition of the bus load-by-load classes.

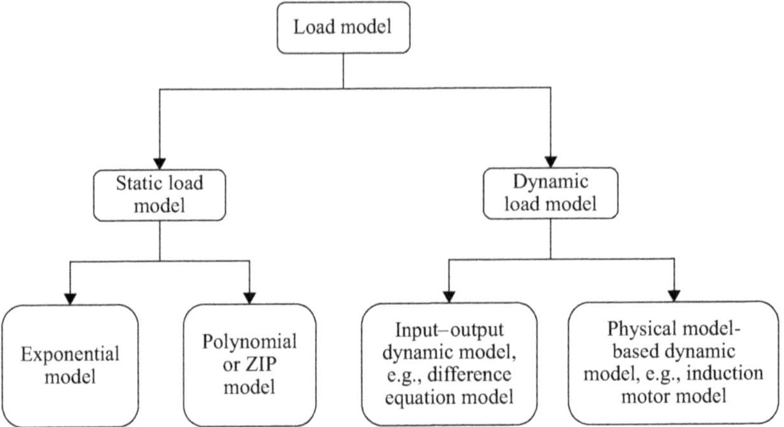

Figure 9.11 Classification of load model

Load characteristic: Load characteristic is determined by a set of parameters that characterize the behavior of a specified load, such as the variation of the active power with voltage. This term may be applied to a specified load device, a load component, a load class, or the total bus load.

9.3.2 Classification of load models

The load models are generally classified into two broad categories—static load models and dynamic load models [26] as shown in Figure 9.11.

9.3.2.1 Static load

A static load model describes the relation between active and/or reactive power at any time with voltage and/or frequency at the same instant of time. The static load model is time invariant. The most common forms of static load models for active and reactive power are polynomial load model and exponential load model. These models, if necessary, can also include a frequency dependent term [27].

Polynomial model or ZIP model

The static characteristics of a load can be classified as constant power (P), constant current (I), and constant impedance (Z) load, depending upon the relation of power to voltage [28]. Hence, a polynomial static load model is also known as ZIP model.

- Constant impedance load (Z): It is the part of load in which the active and reactive power of the load varies proportionally to the square of supply voltage magnitude—for example, heaters, incandescent lamps.
- Constant current load (I): It is the part of load in which the active and reactive power of the load varies in a direct proportion to supply voltage magnitude—for example, motors used in paper mill industry.
- Constant power load (P): It is the part of load in which the active and reactive power of the load remain constant irrespective of changes in supply voltage magnitude—for example, induction motors operating close to nominal voltage.

The polynomial form of static load model can be mathematically represented by the following equations:

$$P = P_0\left(Z_p\left(\frac{V}{V_0}\right)^2 + I_p\left(\frac{V}{V_0}\right) + P_p\right) \tag{9.31}$$

$$Q = Q_0\left(Z_q\left(\frac{V}{V_0}\right)^2 + I_q\left(\frac{V}{V_0}\right) + P_q\right) \tag{9.32}$$

where V_0 is the rated base voltage, V is the operating voltage, P_0 is the rated active base power at rated voltage, Q_0 is the rated reactive base power at rated voltage, P is the active power at operating voltage, Q is the reactive power at operating voltage.

Z_p, I_p, P_p are the ZIP coefficients for the active power with $Z_p + I_p + P_p = 1$.
Z_q, I_q, P_q are the ZIP coefficients for the reactive power with $Z_q + I_q + P_q = 1$.

Exponential model

The static load models can also be represented as exponential model. The exponential model is represented by the following equations:

$$P = P_0\left(\frac{V}{V_0}\right)^a \tag{9.33}$$

$$Q = Q_0\left(\frac{V}{V_0}\right)^b \tag{9.34}$$

where V_0, V, P_0, Q_0, P, and Q are the same as ZIP model. a and b are the non-integer exponents.

For special cases, in which a or b is equal to 0, 1, and 2, the load model will represent a constant power, constant current, or constant impedance load, respectively.

The frequency dependency of load characteristics can be represented in both polynomial and exponential model by multiplying with a frequency factor as shown in (9.35)–(9.38).

For active power:

$$P = P_0\left(Z_p\left(\frac{V}{V_0}\right)^2 + I_p\left(\frac{V}{V_0}\right) + P_p\right)(1 + k_{pf}\Delta f) \tag{9.35}$$

$$P = P_0\left(\frac{V}{V_0}\right)^a (1 + k_{pf}\Delta f) \tag{9.36}$$

For reactive power,

$$Q = Q_0\left(Z_q\left(\frac{V}{V_0}\right)^2 + I_q\left(\frac{V}{V_0}\right) + P_q\right)(1 + k_{qf}\Delta f) \tag{9.37}$$

$$Q = Q_0\left(\frac{V}{V_0}\right)^b (1 + k_{qf}\Delta f) \tag{9.38}$$

where Δf is the frequency deviation. Typically k_{pf} ranges from 0 to 3, and k_{qf} ranges from −2 to 0 [27].

9.3.2.2 Dynamic load

The dynamic loads, unlike static loads, are dependent on time. The response of power system following a disturbance spans for several time interval. The initial response that occurs immediately following the disturbance may be represented using a static load model. The dynamic loads, like induction motors, were approximated by static loads. This approximation is sufficient in some cases, but since load representation has critical effects in voltage stability studies, static loads need to be replaced by dynamic load models [26]. To predict a system response for longer time spans, like slow voltage recovery phenomena, etc., dynamic behavior of the loads has to be modeled using a dynamic load model.

A dynamic load is often expressed in the form of ordinary differential equations, or difference equations, or partial differential equations, or transfer functions. Dynamic load models are broadly classified as

- *Input–output dynamic models:* This class of models includes the difference equation–based dynamic models.

  The difference equation representation of the dynamic load is a numerical model that represents the time-varying nature of the load. It can be represented either as a first- or second-order model as follows:

  First-order model

$$\Delta P(k) = a_{p1}\Delta P(k-1) + c_{p0}\Delta V(k) + c_{p1}\Delta V(k-1) \tag{9.39}$$

$$\Delta Q(k) = a_{q1}\Delta Q(k-1) + c_{q0}\Delta V(k) + c_{q1}\Delta V(k-1) \tag{9.40}$$

  Second-order model

$$\begin{aligned}\Delta P(k) &= a_{p1}\Delta P(k-1) + a_{p2}\Delta P(k-2) + c_{p0}\Delta V(k)\\ &\quad + c_{p1}\Delta V(k-1) + c_{p2}\Delta V(k-2)\end{aligned} \tag{9.41}$$

$$\begin{aligned}\Delta Q(k) &= a_{q1}\Delta Q(k-1) + a_{q2}\Delta Q(k-2) + c_{q0}\Delta V(k)\\ &\quad + c_{q1}\Delta V(k-1) + c_{q2}\Delta V(k-2)\end{aligned} \tag{9.42}$$

- *Physical-model based dynamic models:* The induction motor load model is considered as a physical-model based load model.

  The mathematical formulation of induction motor model can be described by the following equations [29]:

$$\frac{dE'_d}{dt} = -\frac{1}{T'}\left[E'_d + (X - X')I_q\right] - (\omega - 1)E'_q \tag{9.43}$$

$$\frac{dE'_q}{dt} = -\frac{1}{T'}\left[E'_q + (X - X')I_d\right] - (\omega - 1)E'_d \tag{9.44}$$

$$\frac{d\omega}{dt} = -\frac{1}{2H}\left[T_0\left(A\omega^2 + B\omega + C + D\omega^E\right) - \left(E'_d I_d + E'_q Iq\right)\right] \tag{9.45}$$

$$I_d = \frac{1}{R_s^2 + X'^2}\left[R_s\left(U_d - E'_d\right) + X'\left(U_q - E'_q\right)\right] \tag{9.46}$$

$$I_q = \frac{1}{R_s^2 + X'^2}\left[R_s\left(U_q - E'_q\right) + X'\left(U_d - E'_d\right)\right] \tag{9.47}$$

$$T_L = T_0\left(A\omega^2 + B\omega + C + D\omega^E\right) \tag{9.48}$$

where H is the rotor inertia constant, T_L is the load torque equation, T_0 is the steady-state mechanical torque, ω is the rotor rotation speed, E'_d is the d-axis transient EMF of motor, E'_q is the q-axis transient EMF of motor, U_d is the d-axis bus voltage, U_q is the q-axis bus voltage, I_d is the d-axis stator current, I_q is the q-axis stator current.

The A, B, C, D, E coefficients in load torque equation need to satisfy $C = 1 - A\omega^2 - B\omega - D\omega^E$ equation. In these equations, X' is equivalent to $X_s + (X_m X_r/(X_m + X_r))$ and is the transient reactance, whereas X is equivalent to $X_s + X_m$ is the open circuit reactance.

9.3.2.3 Composite load (static load + dynamic load)

Generally, a composite load model (CLOD) approach is used by utilities. A CLOD is the combination of static load model and dynamic load model. Some of the composite models currently used [29] for research purposes are as follows:

- ZIP model augmented with induction motor (three state).
- ZIP model augmented with difference equation model (second order).
- Exponential model augmented with induction motor (three state).
- Exponential model augmented with difference equation model (second order).

As motors constitute about 60%–70% of the total load, most of the dynamic load is represented by an aggregate induction motor. Other load components that affect the stability studies are as follows [27]:

- Discharging the lamps and restart the lamp. The lamps that have mercury vapor, sodium vapor, and fluorescent lamps will affect the voltage recovery or delay the recovery.
- Protective relays especially thermal or overcurrent relays.
- Thermostatic control of loads (i.e., heaters and refrigerators).
- ULTCs on distribution transformers, voltage regulators, and voltage-controlled capacitor banks. These devices are used to mitigate a disturbance and to make the system return to the predisturbance levels.

Some of the load models used in industry are as follows.

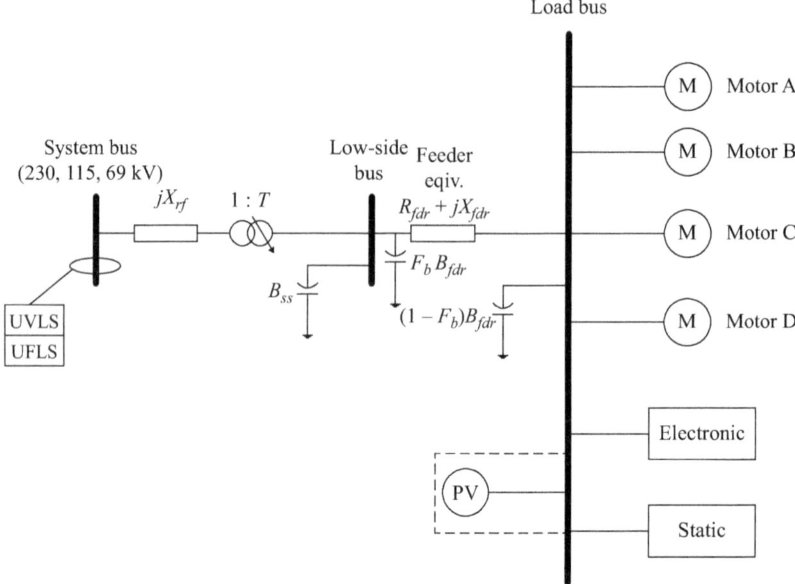

Figure 9.12 WECC CMPLDW load model

WECC composite load model

The CLOD (WECC Dynamic Composite Load Model (CMPLDW)) [30] developed by WECC and is implemented in production grade software like PSLF by GE, Siemens PTI PSS®E, Power World, PowerTech TSAT. The model is further refined and is presented in Figure 9.12.

This CLOD structure includes static loads, electronic loads, constant-torque three-phase motors (A), high inertia speed-squared load motors (B), low inertia speed-squared load motors (C), and constant-torque single-phase motors (D). A detailed modeling of each motor type is required in the CMPLDW, and the total input parameters required are more than 100. Model validation studies have been done to validate these load models. It was observed that these models can be tuned to reproduce past events like delayed voltage recovery, large underfrequency events, and interarea power oscillation with great accuracy.

Composite load model in PSSE

The CLOD used in PSS/E is an aggregation of the multiple models in the system. The structure of this model is shown in Figure 9.13.

This model consists of various submodels consisting of following components:

1. Induction motors: Motors are characterized by typical torque-speed, current-speed, and power factor-speed curves.
2. Discharge lighting: The percentage of discharge lightning is provided in this submodel. Above 0.75 pu of voltage, the real power is modeled as constant current, and the reactive power increases with exponential factor of 4.5. As voltage goes below 0.75 pu, both active and reactive power drop linearly

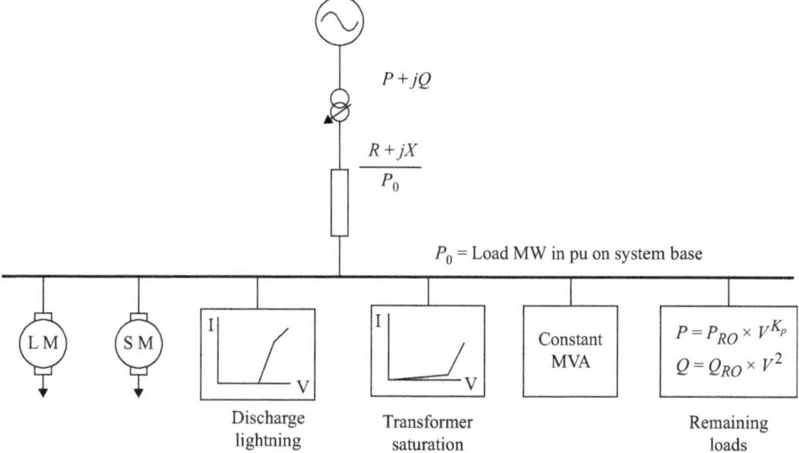

Figure 9.13 CLOD load model in PSS/E

with voltage until the value of voltage reaches 0.65 pu, below which the lighting gets extinguished.

3. Transformer losses: The percentage of the transformer losses is provided, which accounts for both core losses and copper losses.
4. Constant MVA: The percentage of the constant real and reactive power component of the load is provided.
5. Remaining loads: The real power is modeled as exponential function of voltage and the reactive power as constant impedance.

9.3.3 *Review of estimation of load model parameters*

Two basic approaches to obtain the load characteristics are physical-component based approach and measurement-data based approach [27]. These approaches are usually applied for aggregate load models, but these principles can also be applied for the formulation of individual loads.

9.3.3.1 Physical-component based approach

In this approach, the aggregate model is built from the models of individual components within the aggregated load [27,31,32]. This approach requires a survey of various components in power system. This makes it a very difficult task as it involves manual survey and uncertainty. Further, as the load data vary from bus to bus and are dependent on weather and time, it is necessary to often update the load data for each bus of the system.

9.3.3.2 Measurement-data based approach

In measurement-data based approach of load modeling, power meters, and PMUs are installed at substations to measure voltage and/or frequency variations. The main advantage of using a measurement-based model over the component-based approach is the availability of actual data from the system that makes it possible to track real-time variations in load behavior.

9.3.4 Advantages of using synchronized phasor measurements for load modeling

Various advantages of using PMUs for load modeling are as follows:

- PMUs can report data at very high rates. This can improve load model development and its validation process.
- The inputs required for load modeling like voltage, current, frequency, and rate of change of frequency can be directly obtained from PMUs.
- Any further signal processing is not required since PMUs directly provide all the necessary quantities required for estimation of load parameters.
- Since PMUs can provide real-time measurements with time-stamp, they can be used for centralized load monitoring and forming a database [33].
- The measurement noise of the PMU signal is very low [34]. There may still be some offset errors caused by current transformer and/or voltage transformers. But high-frequency measurement noise is not an issue with PMU data. Thus, PMUs can reliably process small variations in power systems.
- A certain level of voltage variation is required for obtaining better load responses. The normal variations in voltages are lower than 1% but since PMUs have a very low-noise level, accurate modeling of the load response to small voltage changes is possible [34].

9.3.5 Parameter identification techniques

Once a load model is chosen, the next step involves the estimation of its parameters, that is, to find a set of parameters for which the simulated results from the proposed load model best fits the measurements. There are two kinds of approaches for a nonlinear system identifications.

9.3.5.1 Analytical-based approach

In this approach, the unknown parameters of the load model are determined on the basis of experimental test results. Shackshaft *et al.* [35] performed a series of field tests and proposed a method for deriving the parameters of a simplified induction motor model. The load composition and characteristics are determined by conducting step response tests in which small voltage disturbances are imposed on a power station auxiliary system and consumer loads. These tests are extremely sensitive to measurement errors. The parameters obtained using small disturbances in such tests are not able to capture the parameters related to large disturbances.

9.3.5.2 Optimization-based approach

In this approach, the best fit parameters are searched that minimize the error function between the measured output variables and simulated ones. Various search algorithms have been applied in this type of approach. Search algorithms can be of two types—statistical techniques or heuristic techniques.

Some of the algorithms using statistical techniques are as follows:

- A least-square approach is used in [36] to estimate the parameters of the load model. The ordinary least squares method assumes that the measurement errors are independent and normally distributed with constant standard deviation. The predicted values are such that the sum of squared difference between the actual value and predicted value is small. The basic regression equation is given as

$$Y_k = H_k X + e_k \qquad k = 1, 2, \ldots, n \tag{9.49}$$

where $H_k = [H_{1k}, H_{2k}, \ldots, H_{nk}]^T$ is an input column vector, Y_k is an output column vector, X is a column vector for the parameters to be estimated, e_k is the error.
- In weighted least-square based parameter estimation, the problem is solved by considering unequal variances of errors. The weights are given to each measurement based on its variability. The main difficulty in this method is to find proper weights which also increase the complexity of the algorithm.
- An instrument variable (IV)–based technique was used in [37]. This method considers the co-relation between states to be estimated, measurements, and the residue error. A new variable "I" is defined such that
  1. It should be highly correlated with H
  2. It should not be correlated with error e

The two staged equation solving process is as follows:

$$\hat{H}_k = I_k Z \tag{9.50}$$

$$\hat{Y}_k = \hat{H}_k X + e \tag{9.51}$$

First Z is estimated from (9.50), and then X is obtained from (9.51). Hence, this is also called two-staged least squares approach.

Some of the algorithms using heuristic techniques are as follows:

- An adaptive simulated annealing (ASA) algorithm for parameter estimation is discussed in [38]. ASA performs better than the basic SA algorithm due to several substantial improvements that are discussed in [39].
- Uses a hybrid-algorithm for the estimation of load parameters. The hybrid algorithm including the combination of Levenberg–Marquardt and genetic algorithm was used to find the CLOD parameters in [29]. The advantage of using Levenberg–Marquardt approach is that it has a good local search ability, and the genetic algorithm has a better global search ability. The disadvantage of this hybrid approach is that since it involves genetic algorithm, it is slow and many parameters have to be chosen heuristically.
- ANNs are used for dynamic load modeling due to features like nonlinear mapping and generalization capabilities. ANN algorithms use the measurements to train and update the model structure adaptively, thereby eliminating the need to predefine the load model structure and to know the information on load composition and mix [40–42] use ANN-based technique for estimating the load parameters.

A method to reduce the number of parameters in a model based on a sensitivity method was proposed in [43]. In this study, trajectory sensitivity method is used to find the most important equivalent motor parameters for a composite model of load. This results in estimation of only 8 parameters instead of 13 parameters, thus reducing the identification time and enhancing the efficiency of the load modeling.

9.3.6 Estimation of static load parameters

In the following section, the identification and estimation techniques used for determining parameters of the static load for transmission level power system are analyzed. As the variations in the frequency during steady states are minimal in a power system, load changes due to frequency are not considered. Once the synchrophasor measurements are acquired, the identification process can start estimating the load parameters. In this study, the polynomial or the ZIP load model is chosen due to its significant physical meaning and also as it is one of the most widely accepted load models in the power system analysis software tools [44]. The dynamic loads can also be modeled as a constant ZIP model during steady state. For example, an active power load consisting of 30% motor and 70% of impedance can be modeled with a ZIP load model with $Z_p = 0.3$, $I_p = 0$, and $P_p = 0.7$ considering that the motor behaves as a constant power load during steady state.

The measured data should be examined for its suitability in parameter estimation. Following steps should be considered before including the measured data for estimation:

1. The data obtained during transients are not included for estimation purposes. Voltage magnitude can be processed to identify transient conditions and previous ZIP model could be used as an approximate ZIP model during transient conditions.
2. All measured values are taken around a steady-state condition. In some cases, frequency changes continuously after a sudden change in the system, whereas voltage and power may return to their original OP. These types of measurement data are discarded in the estimation procedure, and the previous value of ZIP could be used as an approximation.
3. All measured data are checked for bad data using data mining techniques. Bad data should be eliminated or given very less weighing factor during the estimation procedure.

9.3.6.1 Recursive least-square estimation

IEEE 14 bus system [45] is considered for analysis with load flow incorporating voltage-dependent load models. The simulations results are obtained using MATLAB®. To obtain the input measurements, the power flow program is modified by including a ZIP load model in the power flow equations. The ZIP load parameters are specified for each load bus in the IEEE 14 bus test system. The voltage magnitude, phase angle, real power, and reactive power are obtained as output of the power flow program that are treated as synchrophasor measurements.

These voltage magnitude and real power output data are used as measurement inputs to the load parameter estimation procedure. The voltage variations at the load bus may occur due to several reasons like tap change, contingency, load change, etc. The length of the window chosen is an important factor during the parameter estimation procedure. In this study, an initial window length of three data points is considered as there are three unknown parameters (Z, I, and P). During initialization of the estimation procedure, the window length is dynamically adjusted such that the system of equations become over determined.

The most commonly used estimation technique is the least-square method [36]. The objective function for the least-square optimization is to minimize the square of error between the model prediction and the observed value [46]:

$$J = \min |P_{\text{measured}} - P_{\text{estimated}}|^2 \tag{9.52}$$

The parameters to be estimated are Z, I, and P. The parameters are estimated using the following equation:

$$\hat{x} = (H^T R^{-1} H)^{-1} H^T R^{-1} b \tag{9.53}$$

where $\hat{x}$ is the estimated variables vector, R is the covariance matrix, b is the measurement vector, H is the Jacobian matrix of the state variable.

As there are more unknowns than the number of equations, to estimate the parameters online, a sliding window approach is used. Figure 9.14 shows how the

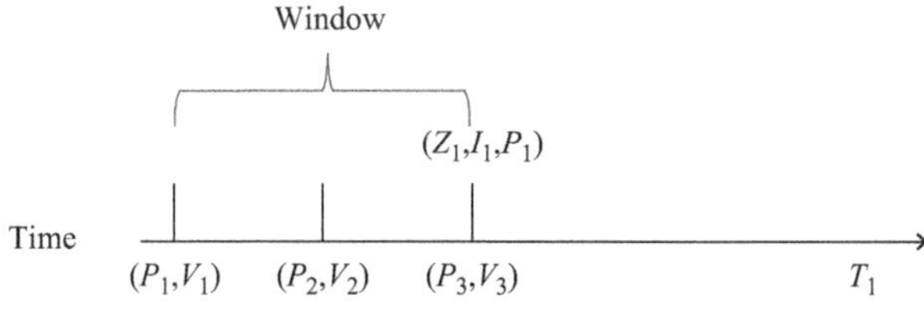

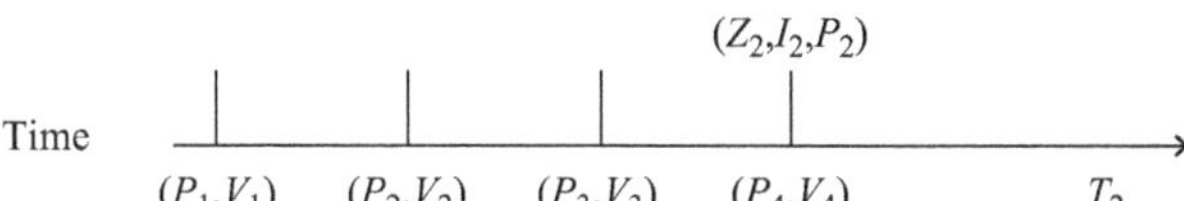

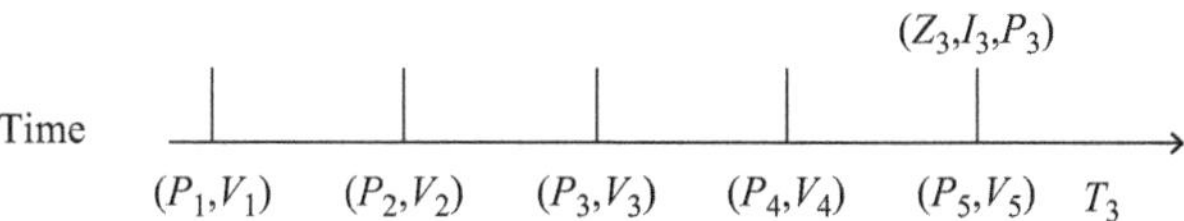

Figure 9.14 Sliding window approach

parameters are estimated at each time step. A minimum of three different voltage values are required to solve for three unknown variables (Z, I, and P).

In recursive least-square method, the solution from previous time step is also included in the computation of parameters for the present time step. This makes computation of parameters more accurate and efficient. The recursive least-square [46] algorithm can be written as follows:

$$P_k = \left[\lambda P_{k-1}^{-1} + \left(H_k^T R^{-1} H_k\right)\right]^{-1} \tag{9.54}$$

$$K_k = P_k H_k^T R^{-1} \tag{9.55}$$

$$\hat{x}_k = \hat{x}_{k-1} + K_k\left(b_k - H_k \hat{x}_{k-1}\right) \tag{9.56}$$

where P_k is the covariance matrix, K_k is known as the estimator gain matrix, λ is the forgetting factor.

P_0 and V_0 values are chosen such that these are continuously changing at every instant. The fact that P_0 is the measured output power when a voltage V_0 is applied to the system load, used in this approach. For instance, if a window size of "3" measurements is chosen, the first measurement instant of voltage and power are chosen as the rated base values respectively after bad data and transients detection. Mathematically, it can be shown as follows:

The basic regression equation is given by (9.57):

$$H_k X + e_k = Y_k \qquad k = 1, 2, \ldots, n \tag{9.57}$$

where $H_k = \left[H_{1k}, H_{2k}, \ldots, H_{nk}\right]^T$ is an input column vector, Y_k is an output column vector, X is a column vector for the parameters to be estimated, e_k is the error.

At time step = 3, the rated base voltage $V_0 = V_1$ and the rated base active power $P_0 = P_1$. Equation (9.57) can be written as

$$\begin{bmatrix} 1 & 1 & 1 \\ \left|\dfrac{V_2}{V_1}\right|^2 & \left|\dfrac{V_2}{V_1}\right| & 1 \\ \left|\dfrac{V_3}{V_1}\right|^2 & \left|\dfrac{V_3}{V_1}\right| & 1 \end{bmatrix} \begin{bmatrix} Z \\ I \\ P \end{bmatrix} = \begin{bmatrix} 1 \\ \dfrac{P_2}{P_1} \\ \dfrac{P_3}{P_1} \end{bmatrix} \tag{9.57a}$$

At time step = 4, the rated base voltage $V_0 = V_2$ and the rated base active power $P_0 = P_2$. Equation (9.57) becomes as follows:

$$\begin{bmatrix} 1 & 1 & 1 \\ \left|\dfrac{V_3}{V_2}\right|^2 & \left|\dfrac{V_3}{V_2}\right| & 1 \\ \left|\dfrac{V_4}{V_2}\right|^2 & \left|\dfrac{V_4}{V_2}\right| & 1 \end{bmatrix} \begin{bmatrix} Z \\ I \\ P \end{bmatrix} = \begin{bmatrix} 1 \\ \dfrac{P_3}{P_2} \\ \dfrac{P_4}{P_2} \end{bmatrix} \tag{9.57b}$$

Table 9.1 Summary of results using adaptive search–based algorithm vs recursive least-square approach—bus 14

Measurement instant	Voltage measurement (pu)	Active power measurement (pu)	Actual (*Z,I,P*) parameters	Estimated (*Z,I,P*) parameters using RLS method	Estimated (*Z,I,P*) parameters using adaptive search–based algorithm
1	1.0355	0.1490	(0,0,1)	–	–
2	1.0116	0.1490	(0,0,1)	–	–
3	0.9894	0.1490	(0,0,1)	(0,0,1)	(0,0,1)
4	1.0613	0.1490	(0,0,1)	(0,0,1)	(0,0,1)
5	1.0613	0.1490	(0,0,1)	(0,0,1)	(0,0,1)
6	1.0613	0.1490	(0,0,1)	(1,0,0)	(0,0,1)
7	1.0613	0.1490	(0,0,1)	(1,0,0)	(0,0,1)
8	1.0613	0.1490	(0,0,1)	(1,0,0)	(0,0,1)
9	1.0355	0.1490	(0,0,1)	(1,0,0)	(0,0,1)
10	1.0355	0.1490	(0,0,1)	(1,0,0)	(0,0,1)
11	1.0355	0.1490	(0,0,1)	(1,0,0)	(0,0,1)
12	1.0354	0.1501	(0,0.2,0.8)	(−38,0,39)	(0,0.21,0.79)
13	1.0354	0.1501	(0,0.2,0.8)	(−38,0,39)	(0,0.21,0.79)
14	1.0610	0.1508	(0,0.2,0.8)	(−38,0,39)	(0,0.2,0.8)
15	1.0610	0.1508	(0,0.2,0.8)	(−38,0,39)	(0,0.2,0.8)
16	1.0351	0.1517	(0.18,0.15,0.67)	(−18,0,19)	(0.18,0.15,0.67)
17	1.0351	0.1517	(0.18,0.15,0.67)	(−18,0,19)	(0.18,0.15,0.67)
18	1.0351	0.1517	(0.18,0.15,0.67)	(1,0,0)	(0.183,0.147,0.669)
19	1.0349	0.1527	(0.32,0.05,0.63)	(−17,0,18)	(0.33,0.04,0.63)
20	1.0349	0.1527	(0.32,0.05,0.63)	(−17,0,18)	(0.33,0.04,0.63)
21	1.0349	0.1527	(0.32,0.05,0.63)	(1,0,0)	(0.329,0.039,0.631)

The load is monitored at the 14th bus of the system. The estimation process is then initiated, and the results are summarized in Table 9.1. It is to be noted here that the current/power and voltage measurements used are obtained only for steady-state operating conditions of the system. But this algorithm does not give accurate estimates for those operating scenarios where ZIP parameters are changing.

Here, an "Operating scenario" can be defined as a single snap shot of the measurement window. The different operating scenarios that may occur during the process of parameter estimation are as follows:

1. Varying voltage, varying power, and constant ZIP parameters
2. Constant voltage, constant power, and constant ZIP parameters
3. Varying voltage, varying power, and varying ZIP parameters
4. Constant voltage, varying power, and varying ZIP parameters

It can be seen from Table 9.1 that the estimated parameters are not estimated accurately under all operating scenarios by using recursive least-square method.

9.3.6.2 Adaptive search–based algorithm

To overcome the above-mentioned limitations, another approach is proposed here. The basic flowchart of this proposed algorithm is shown in Figure 9.15. The estimation process is started once the voltage and power measurements are obtained. The process is initialized by having the voltage and power measurements at three

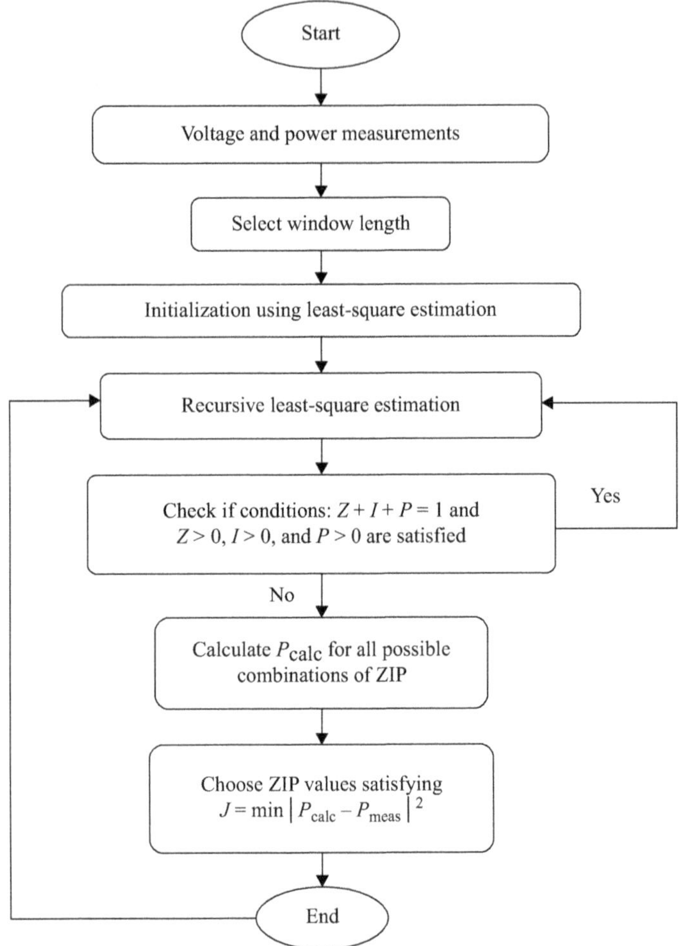

Figure 9.15 Flowchart for adaptive search–based algorithm

different voltage instances. After that, the estimation process is continued by using the recursive least-square technique. After the estimation of ZIP parameters by RLS method, the algorithm checks if the estimated ZIP parameters satisfy the following two conditions:

1. $Z+I+P=1$
2. $Z \geq 0, I \geq 0$, and $P \geq 0$

The RLS technique can be applicable if and only if the ZIP parameters remain constant. Hence if ZIP parameters have changed, the above-mentioned conditions will not hold true. If these conditions are violated, the algorithm shifts to the search-based estimation technique.

In this approach, the power is first calculated for all possible ZIP parameter values using the voltage measurements at that instant. The ZIP parameters are then selected that give the minimum square of error from the measured power values as given by (9.58):

$$J = \min|P_{\text{measured}} - P_{\text{calculated}}|^2 \tag{9.58}$$

If multiple parameter solutions exist, then the best solution is chosen by calculating the norm from the previous set of ZIP parameters. It can be seen that the ZIP parameters are estimated accurately under all operating conditions. Table 9.1 summarizes the results of this approach for different operating scenarios. The estimation of ZIP parameters is initialized after three measurement instances using the least-square approach. Until the 11th measurement instant, the recursive least-square method estimates the ZIP parameters accurately since there is no change in the ZIP parameters of the load. Once the ZIP change occurs at measurement instant = 12, the RLS method fails to accurately estimate the load parameters. This occurs during operating scenarios 3 or 4. These operating scenarios can be easily identified by looking into the necessary satisfactory conditions, that is, (1) $Z+I+P=1$ and (2) $Z \geq 0, I \geq 0$, and $P \geq 0$. If these conditions are not satisfied, the algorithm switches to a search-based estimation technique. It can be seen from Table 9.1 that the ZIP parameters are estimated accurately using the Adaptive Search–based algorithm. The estimation of load parameters is also done for bus 12 of the IEEE 14 bus system. The estimation results are summarized in Table 9.2 for all the operating scenarios.

From the above results, it can be concluded that the proposed algorithm can accurately estimate the ZIP parameters during all operating scenarios.

9.4 Transient stability assessment using synchrophasor data

The ability of a power system to maintain synchronism when subject to large disturbance like faults, tripping of generators, and large load changes is called large signal rotor angle stability or transient stability. Detection of transient stability in a power system network, under disturbances, is critical for secure operation of the system. The consequence of a disturbance in a large interconnected power system may be severe in case the post-disturbance OP is unstable, and if adequate preventive control action is not initiated within the short time interval.

The relative phase angle difference of the generator rotor angle monotonically increases during the first swing resulting due to insufficient availability of synchronizing torque. The system may become unstable depending on the initial state of the system and the magnitude of the disturbance. Typically, the bulk power transfer between interconnected power systems is limited on the basis of the transient stability limit. Faster assessment of transient stability for a predefined contingency is critical for preventing cascaded outages and large-scale blackouts. Fast rate of data provided by PMUs makes it feasible to analyze transient stability of the system.

Table 9.2 Summary of results using adaptive search–based algorithm vs recursive least-square approach—bus 12

Measurement instant	Voltage measurement (pu)	Active power measurement (pu)	Actual (*Z,I,P*) parameters	Estimated (*Z,I,P*) parameters using RLS method	Estimated (*Z,I,P*) parameters using adaptive search–based algorithm
1	1.0548	0.0641	(0.3,0.3,0.4)	–	–
2	1.0243	0.0623	(0.3,0.3,0.4)	–	–
3	1.0884	0.0660	(0.3,0.3,0.4)	(0.3,0.3,0.4)	(0.3,0.3,0.4)
4	1.0884	0.0660	(0.3,0.3,0.4)	(0.3,0.3,0.4)	(0.3,0.3,0.4)
5	1.0884	0.0660	(0.3,0.3,0.4)	(0.3,0.3,0.4)	(0.3,0.3,0.4)
6	1.1259	0.0682	(0.5,0.1,0.4)	(−0.477,0,1.477)	(0.5,0.1,0.4)
7	1.1259	0.0682	(0.5,0.1,0.4)	(−0.477,0,1.477)	(0.5,0.1,0.4)
8	1.1259	0.0682	(0.5,0.1,0.4)	(1,0,0)	(0.5,0.1,0.4)
9	1.1259	0.0682	(0.5,0.1,0.4)	(1,0,0)	(0.5,0.1,0.4)
10	1.1257	0.0699	(0.5,0.1,0.4)	(−70.9,0,71.9)	(0.5,0.1,0.4)
11	1.1257	0.0699	(0.5,0.1,0.4)	(−70.9,0,71.9)	(0.5,0.1,0.4)
12	1.1257	0.0699	(0.5,0.1,0.4)	(1,0,0)	(0.5,0.1,0.4)
13	1.0547	0.0647	(0.5,0.1,0.4)	(1,0,0)	(0.5,0.1,0.4)
14	1.0547	0.0647	(0.5,0.1,0.4)	(1,0,0)	(0.5,0.1,0.4)
15	1.0545	0.0657	(0.5,0.4,0.1)	(−60.1,0,61.1)	(0.5,0.4,0.1)
16	1.0545	0.0657	(0.5,0.4,0.1)	(−60.1,0,61.1)	(0.5,0.4,0.1)
17	1.0548	0.0641	(0.35,0.2,0.45)	(−65.5,0,66.56)	(0.35,0.2,0.45)
18	1.0548	0.0641	(0.35,0.2,0.45)	(−65.5,0,66.56)	(0.35,0.2,0.45)
19	1.0548	0.0641	(0.35,0.2,0.45)	(1,0,0)	(0.35,0.2,0.45)
20	1.0549	0.0637	(0.15,0.5,0.35)	(−197,263,−65)	(0.15,0.5,0.35)
21	1.0549	0.0637	(0.15,0.5,0.35)	(−197,263,−65)	(0.15,0.5,0.35)

9.4.1 *Review of existing techniques for transient stability*

Time domain approach, equal area criteria, and energy function are the traditional methods available for transient stability study. The availability of the synchrophasor measurements from different locations improved prediction of transient stability of large power system in real time. The development of several hybrid methods utilizing PMU data and machine learning are also carried out for online transient stability assessment.

9.4.1.1 Time domain approach

The power system network consists of generators, exciter, governors, stabilizers, transformers, transmission lines, cables, loads, compensators, etc. All these devices are mathematically represented as a set of differential and algebraic equations (DAEs), which is also known as the state space form of the system. The multi-dimensional state space representation of the dynamical behavior of the power system network can be represented as

$$\dot{x} = f(x, u, t) \tag{9.59}$$

$$y = g(x, u, t) \tag{9.60}$$

where the state variable x denotes the minimum set of vectors which can completely describe the system at any instant of time, and the future state can be determined without referring to other inputs, the input variable u denotes the vector of inputs to the system and y denotes the vector of output variables from the system. The function f and g are vectors of nonlinear function describing the behavior of the system based on state variables, input, and output vectors.

The set of DAEs represented by (9.60) are solved simultaneously using implicit integration techniques with a known initial condition. In implicit integration, like Backward Euler's Method, the state at the next time instant is a function of the state at that time and the state at next time instant. The computational requirement of the implicit method is higher; however, stability of the implicit method is guaranteed. The partitioned solution is another technique for solving the set of DAEs. The differential equation and algebraic equation are alternatively solved in each time step in this method. The variations of explicit integration methods like Explicit Euler Method, Open Multistep Formulas, Explicit Runge–Kutta Methods, and Predictor–Corrector Methods are used in this approach. The state at next time step is expressed as a function of state at the present time step, explicitly. The computational requirement for explicit integration is less; however, larger time step for integration may lead to instability of the solution. Simultaneous solution approach for solving DAEs requires the conversion of the differential equation into algebraic equation using implicit integration and combines with rest of the algebraic equation to solve for the unknown states. The resultant one set of nonlinear algebraic equation is solved using iterative Newton method. The computational burden involves Jacobian matrix formation and factorizing in each time step. The variable integration time step can be easily applied in this method that helps in reducing computational time for slow changing states. The overall speed of the time-domain method is slow, and high computational resource is required for large system.

9.4.1.2 Equal area criteria–based approach

The stability of one machine infinite bus (OMIB) system can be obtained without computing the swing curves using this method. The synchronous machine is modeled as a constant voltage source behind the synchronous reactance, and the mechanical power input is assumed to be constant. The assumptions arise by neglecting the impact of the exciter and the governor dynamics. After neglecting the damping and considering constant impedance model of the loads, the stable operation requires that the accelerating area is equal to the deceleration area. The expression of critical clearing time can be obtained from the equal area criteria. The application of this method is limited to OMIB system which is not applicable to generator having full controls and dynamics, and multi-swing stability analysis for multi-machine system.

To address these issues, Ribbens-Pavella *et al.* [47] proposed extended equal area criteria. The multi-machine power system is grouped into two sets—one consisting of critical machines and other consisting of noncritical machines. The two groups are reduced to two equivalent machines having aggregated parameters.

The acceleration criterion is used for identifying the cluster of critical machines. The machines having higher acceleration are grouped into one cluster. The two machine equivalent is analyzed as OMIB system that provides the critical clearing time and the stability margin of the multi-machine system. A fast prediction of transient stability using Taylor series approximation is proposed in [48]. Estimation of multi-swing transient stability limit using sensitivity-based iterative method for extended equal area criteria is presented in [49].

9.4.1.3 Energy function–based approach

The transient stability assessment using energy function does not require the solution of the system DAEs and is also known as direct method. The stability prediction using direct method is obtained much faster than the time-domain approach. The energy function proposed in [50] is based on scalar Lyapunov function which is presented as follows:

$$V = \frac{1}{2}\sum_{i=1}^{n} M_i\tilde{\omega}_i^2 - \sum_{i=1}^{n} P_i(\theta_i - \theta_i^s) - \sum_{i=1}^{n-1}\sum_{j=i+1}^{n}\left[c_{ij}\left(\cos\theta_{ij} - \cos\theta_{ij}^s\right) - \int_{\theta_i^s+\theta_j^s}^{\theta_i+\theta_j} D_{ij}\cos\theta_{ij}d(\theta_i+\theta_j)\right] \tag{9.61}$$

The expression of energy function in (9.61) consists of following terms:

1. $1/2\sum_{i=1}^{n} M_i\tilde{\omega}_i^2$ Total change in rotor kinetic energy relative to COA = total change in rotor kinetic energy minus change in COA kinetic energy.
2. $\sum_{i=1}^{n} P_i(\theta_i - \theta_i^s)$ Change in rotor potential energy relative to COA = change in rotor potential energy minus change in COA potential energy.
3. $c_{ij}(\cos\theta_{ij} - \cos\theta_{ij}^2)$ Change in magnetic stored energy of branch ij.
4. $\int_{\theta_i^s+\theta_j^s}^{\theta_i+\theta_j} D_{ij}\cos\theta_{ij}d(\theta_i+\theta_j)$ Change in dissipated energy of branch ij.

The criteria for system stability are that the energy function presented in (9.61) exhibits the property $V(x) > 0$; $\dot{V}(x) < 0$. For analyzing stability around the origin, (9.61) can be relaxed to $\dot{V}(x) \leq 0$. Under assumption of a classical generator model and neglecting transfer conductance, the energy function in (9.61) is reduced to

$$V = \frac{1}{2}\sum_{i=1}^{n} M_i\tilde{\omega}_i^2 - \sum_{i=1}^{n} P_i(\theta_i - \theta_i^s)\cdots - \sum_{i=1}^{n-1}\sum_{j=i+1}^{n}\left[c_{ij}\left(\cos\theta_{ij} - \cos\theta_{ij}^s\right) - \int_{\theta_i^s+\theta_j^s}^{\theta_i+\theta_j} D_{ij}\cos\theta_{ij}d(\theta_i+\theta_j)\right] \tag{9.62}$$

In [51], an improved Lyapunov function, considering speed governor, pole saliency and damping, for a single machine infinite bus system has been proposed. A similar Lyapunov method is extended for multi-machine system in [52] for predicting the transient stability. Pai *et al.* [53] have used the procedure given by Kalman, which utilizes the theorem of Popov on absolute stability of the nonlinear system to construct the Lurs type Lyapunov functions for a single machine system. The dynamics of the governor were also included in the construction of the Lyapunov function, which provides the region of the asymptotic stability of the post-fault system. Incorporation of the field flux linkage in the Lyapunov-based transient stability method is carried out in [54]. Lyapunov's direct method of stability analysis is also used in [55,56], which includes flux decay dynamics. Similarity of the Lyapunov's energy function with the transient energy function for classical machine model is shown in [57]. The fault location and transfer conductance are included in the transient energy function for stability analysis of classical machine model by Athaey *et al.* [50]. The effect of exciter within the first swing of the machines on the transient energy function is considered in [58]. The dynamics of the generator is represented by the two-axis model, and the exciter is represented with a single time constant. Chiang [59] demonstrated that general energy functions do not exist for multi-machine system when transfer conductance is considered in stability studies. However, under the assumption of low system loss, the existence of energy function over the compact set of state space can be derived.

The post-fault system, which may be a stable equilibrium point (SEP), is surrounded by many unstable equilibrium points (UEPs). The energy of the nearest UEP with respect to the post-fault SEP is the critical energy, V_{crit}. The controlling unstable equilibrium point (CUEP) technique proposed by Fouad and Stanton [60] determines the nearest UEP along the trajectory of the system disturbance. The angular speed at the UEP is zero, which results to the critical energy given by

$$V_{\text{crit}} = -\sum_{i=1}^{n} P_i\left(\theta_i^u - \theta_i^s\right) - \sum_{i=1}^{n-1}\sum_{j=i+1}^{n}\left[c_{ij}\left(\cos\theta_{ij}^u - \cos\theta_{ij}^s\right) + I_{ij}^u\right] \tag{9.63}$$

The CUEP is identified as the UEP with lowest normalized potential energy margin at the instant the disturbance is removed. The lowest normalized potential energy margin is expressed as the ratio of the potential energy difference of UEP and fault clearance with respect to the kinetic energy that contributes to the separation of critical generators. The estimation of transient stability region by analyzing the potential energy function, V_{p}, is presented in Kakimoto *et al.* [61] as well as Athay *et al.* [50] also known as potential energy boundary surface (PEBS). The stable operation of the system is possible till the system does not cross the boundary defined by orthogonal curves with respect to the equipotential surface V_p. The PEBS is defined by Chiang *et al.* [62] as the gradient of the stability boundary expressed as

$$\dot{\theta} = \frac{-\partial V_p}{\partial \theta} \tag{9.64}$$

193822 has shown that for the system with detailed models, the method of PEBS does not provide an accurate estimation of critical transient energy of the systems. Hence, they suggested that a modification in the generalized potential energy to properly includes the fast dynamics. The stability boundary of the original system and the reduced system are related and is used by thc BCU method [63]. The trajectory of the disturbance is analyzed to detect the crossing of the boundary of the reduced system. This boundary crossing point is used as the initial condition for integrating post-fault reduced system and initial guess for finding actual CUEP from (last) equating it to zero. The critical energy corresponds to CUEP is compared to value of the energy function at the time of disturbance and obtains the stability of the system. The following limitation of the direct method are known to exist:

1. The direct method may also converge to a wrong CUEP
2. Fails to predict the second swing rotor instability
3. Iterative calculation of CUEP fails to converge sometimes
4. BCU method assumes that the disturbance trajectory starts inside stability boundary and fails if such constraints are not met by the system.

9.4.1.4 Other methods

Advent of the synchrophasor technology opened a new paradigm of measurement-based online stability prediction. Stanton *et al.* [64] explained how the measurements, obtained from the PMU, could be utilized for the online calculation of the partial energies of the machines. They suggested to use the offline simulation for estimating the critical value of the transient energy and velocity, which is later utilized in predicting the transient stability. The LEs have been used for the identification of unstable swing by Liu *et al.* [65,66] using real-time PMU data. The application of PMU data for assessing transient stability of multi-machine system using equal area criteria is proposed in [67]. The application of PMU data and EMS model for online transient stability prediction is presented in [68]. A reduced power system model is built keeping intact all the generator buses for pre-fault and post-fault topology using EMS readings. The transient stability of the reduced system is estimated using time-domain simulation, and binary search is used for stability forecasting. The rotor angle referred to area center of inertia using support vector machine regression calculated on selected PMU data is presented in [69]. The transient stability indicator is proposed for each area based on the center of inertia angle for each area, computed for imminent out of step. The utilization of streaming PMU data for online transient stability prediction is proposed in [70]. Limited fault data are used in this work for predicting transient stability without solving the system DAEs. Critical generators pairs are selected in [71] on the basis of the phase angle difference using the PMU data. The margin of stability is calculated for these critical generators pairs using the correlation of the phase angle difference and the power transfer between these interfaces.

9.4.2 MLE-based transient stability assessment

Accurate estimation of the phase angle and the frequency provided by the PMUs can be utilized for online detection of the transient stability following a disturbance. LEs have been demonstrated, in some of the work, for stability assessment of the nonlinear dynamical systems. The stability predictions for the post-fault condition of the power system, using LEs, have been proposed in [72,73]. These works have used the Wolf's method for estimating LEs, which requires implicit integration of the variational equation. Determination of the voltage stability using maximum LE (MLE) has been proposed in [74] for real-time application. This chapter proposes a MLE-based technique for online assessment of the transient stability. The sign and the magnitude of the MLE are used as the criteria for the divergence of the state space trajectories and, thereby, used for online transient stability detection. The estimation of the MLE, following a disturbance has been done using the PMU data, without forming the variational equation of the state space representation. The effectiveness of the proposed method for assessing the transient stability is demonstrated through simulation of faults of different impedance and clearing times at various locations in New England (NE) 39 bus systems using real-time digital simulator (RTDS).

9.4.2.1 Estimation of MLE from PMU data

The time series data of the state variables, of finite length N, measured by the PMUs can be stored in a matrix, which is denoted by $\mathbf{Y} = \left[\mathbf{X}^1 \cdots \mathbf{X}^N\right]^T$. The LEs are defined as the average exponential rate of divergence of the nearby trajectories, which are separated initially by infinitesimally small distance in the phase plane. The method proposed by Rosenstein *et al.* [75] approximates the separate nearby trajectories, by searching the nearest neighbor in the phase plane. The nearest neighbor for a kth point is denoted by $\mathbf{X}^{k'}$ such that:

$$d^k(0) = \min_{k=1:N} \|\mathbf{X}^k - \mathbf{X}^{k'}\| \tag{9.65}$$

where $d^k(0)$ is the initial distance between the kth point and its nearest neighbor, and $\|\cdot\|$ denotes the Euclidean norm. The nearest neighbors in the phase plane are also constrained by the mean period, which is considered 0.0167 s. This constraint is considered so that the nearest neighbors occur on the separate orbits of the attractor. The dynamics of the system is tracked by monitoring the distance of the nearest-neighbor points evolved after j time steps. The distance between the nearest neighbor after jth time step is denoted by

$$d^k(j) = \|\mathbf{X}^{k+j} - \mathbf{X}^{k'+j}\| \tag{9.66}$$

According to [75], the MLE of the system is calculated from the change in the distance of the nearest neighbor along the locus of the state space trajectory. The logarithm of the distance between the nearest-neighboring points are averaged over each point in time series of the data of a finite window length. The logarithm of the average distance between the nearest neighbor at jth time instant is related to the MLE, as given below:

$$\langle \ln d^k(j) \rangle \approx \ln d^k(0) + \lambda_1 (j\Delta t) \tag{9.67}$$

where λ_1 is the MLE, and Δt is the sampling time of the time series data.

A large power system consists of several generators interconnected by long transmission lines. The steady-state operation of the power system is characterized by the synchronous operation of all the rotating machines in the network. The state space representation of an ith synchronous machine in a power system network, having m generators at kth time instant, is expressed as follows:

$$\dot{\delta}_i^k = \omega_0 \Delta\omega_i^k, \quad \Delta\dot{\omega}_i^k = h_i\left(\delta^k, \delta\omega_i^k\right) \tag{9.68}$$

where δ_i^k is the ith generator internal voltage in radian, ω_0 is the synchronous speed in rad/s, $\Delta\omega_i^k$ is the ith generator per unit speed deviation from the synchronous speed, $\delta^k = \left[\delta_1^k \ldots \delta_m^k\right]^T$, and h_i is a nonlinear function describing the dynamics of the ith generator. The state vector of the system at kth time instant is denoted by $\mathbf{X}^k = \left[\delta_1^k \ldots \delta_m^k, \Delta\omega_1^k \ldots \Delta\omega_m^k\right]^T$

During the steady state operation, the right side of (9.68) is zero. The system deviates from the steady-state value on the occurrence of a fault, following which the state vector converges to a SEP provided that the system is transiently stable under the post-fault scenario. The state vector diverges and leads to instability if the system is transiently unstable under post-fault condition.

The sign of the MLE serves as an indicator to the stability of the nonlinear systems. The Wolfs method [76] has been, generally, used for the estimation of the MLE in the power system applications. The estimation of the LEs is performed by forming the state space equations of the power system, from which the variational equation is derived. The Jacobian matrix from the state space model of the multi-machine power system is formed to estimate the evolution of the state vector trajectories. The LEs estimate the exponential rate of convergence and divergence of the nearby trajectories in the state space. The method of integrating the variational equation for the post-fault power system conditions depends on the size of the Jacobian matrix of the state space equation. The estimation of the MLE using Rosenstein's method [75] has the following advantages.

1. It does not require to construct the variational equation using Jacobian of the state space model.
2. It does not require to perform Gram–Schmidt Orthogonalization of the state vectors at each time step.
3. It is completely based on the measurement of the state variables.
4. It directly measures the MLE from the measured state variables.

The typical data frame transmission rate per second for a PMU installed in a 60-Hz power system is 60, which is also $1/\Delta t$. The MLE at the time $j\Delta t$ is given by rearranging the expression given in (9.67).

9.4.2.2 Test study

The number of initial nearest neighbors used for the MLE estimation affect the performance of the stability assessment. More initial neighbors results in accurate MLE estimation with increased latency, and few initial neighbors results in inaccurate stability indication. The optimum initial neighbors considered in [74] for voltage stability application is around 120 resulting in latency of 2 s, which is quite high for transient stability application. Rigorous simulations of MLE estimation for various disturbances lead to the appropriate nearest neighbors count to 65, which is used in this work. The one axis flux decay dynamic model of the synchronous generator has been used from the library component of RTDS. The data for exciter and governor available in Appendix D of [77] are used in this work. The generator rotor angles are usually considered for assessing the transient stability [73]. But, with the consideration of 1 s latency, the voltage phase angle of the generator buses are considered in this work to obtain approximate results. The magnitude of the speed deviation vectors in a power system network, following a contingency, is observed to be insignificant as compared to the phase angle of the generators. Hence, only the voltage phase angles at the generator buses, under various fault scenarios obtained through RTDS simulations, are considered in this work as the state vectors. These are stored for consecutive 65 phasors, which is used as initial nearest neighbors. The MLE is estimated and reported at the time instant of the next phasors arriving after the initial 65 phasors. The accumulation of the initial 65 phasors starts soon after the fault is cleared in the system.

The RTDS is a real-time power system and power electronics digital simulation framework providing input output interface to a hardware controller. The architecture for estimating MLE, using RTDS in software in the loop architecture, is shown in Figure 9.16. This shows a synchrophasor technology–based wide-area monitoring system (WAMS) set up in the lab. The RTDS, shown in Figure 9.16,

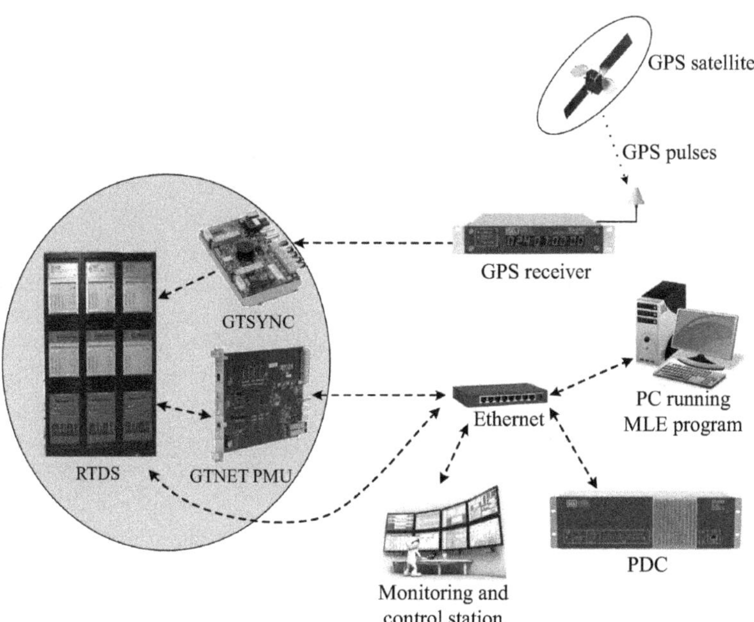

Figure 9.16 Block diagram of the WAMS setup using RTDS for estimating MLE

is the multi-processor simulator consisting of a large number of PB5 processor cards for parallel computing. The RTDS simulates the multi-machine power system test networks and is capable of exchanging analog and digital data to the physical devices in real time. The RTDS also supports the network protocol interface for exchanging the control signal and data to the network interfaced physical devices. The MLE estimation in the power system network requires time stamped phasor data of the voltage from all the generator buses at a central computer. The GPS satellite transmits the Universal Time Coordinate (UTC) time information to the GPS clock receiver, having accuracy in the order of 1 μs. The GPS clock sends the timing information to the GTSYNC card in the RTDS. The GTSYNC card controls the clock pulses of the RTDS inside the simulation, and, thereby, synchronizes the clock ticks of the RTDS simulation with the UTC clock of the satellite.

The GTNET card is a network interface card in the RTDS, which is loaded with the PMU protocol to transmit PMU data of the power system network to the PDC connected to the Ethernet network by a network switch. The voltage and current phasors at a bus in the power system network is estimated by using GTNET PMU card of the RTDS from the sampled values of the instantaneous voltage and current signals. The algorithm applied for estimating the voltage and current phasors inside the GTNET PMU card is proprietary, but compliant to the synchrophasor standard [78]. The phasor data of the voltage and current are transmitted to the PDC through the Ethernet network, in which the phasor data are time aligned, and transmitted to a computer for the MLE estimation. The computer receives the phase angles of the generator busses, which are buffered to form a vector of data length 65. The MLE is estimated following a disturbance, and the result is transmitted to the monitoring and control station. The system operator sitting at the control station is assisted with the estimated MLE to initiate emergency control action, if instability in the system is observed. The control action may be transmitted back to the system using analog, digital, and network interface for rescuing the system from collapse. The visualization of the MLE is only carried out in this chapter, and the control action transmitted back to the system is not implemented in this work. The results of the RTDS simulations are obtained on NE 39 bus systems are given below.

The first case considered in the NE 39 bus system, using real-time simulator, is a three phase to ground bus fault having impedance 0.01 ohm for the fault duration of 0.14 s and 0.15 s at bus 18, causing the system to remain stable in the first and becoming unstable in the second case. The phase angles of the generator buses for the fault duration of 0.14 s are shown in Figure 9.17. The unstable case corresponding to the fault duration of 0.15 s results in phase angle of bus 38 to diverge from the group of buses 33, 34, 39 and another group of buses 31, 32 from the remaining generator buses, as shown in Figure 9.18. The positive MLE, shown in Figure 9.19, denotes unstable case for the corresponding fault duration.

The fault at bus 4 is also considered for estimation of the MLE by creating a fault of duration 0.18 s and 0.19 s, which correspond to the stable and the unstable cases in the NE 39 bus system. The phase angles of the generator buses are shown in Figure 9.20 for the stable case, and those for the unstable case are shown

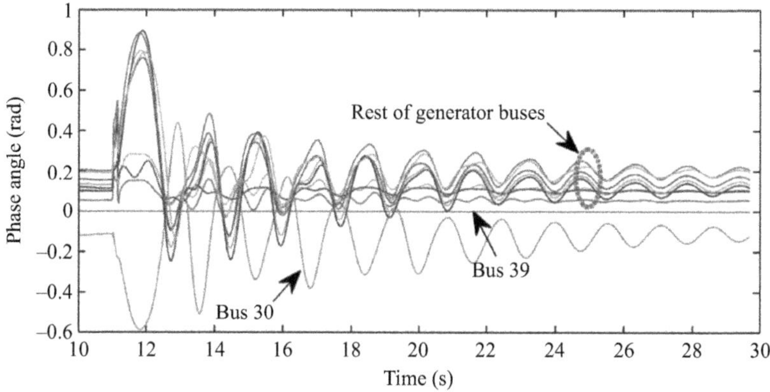

Figure 9.17 Phase angles at generator buses for a fault of 0.14 s at bus 18

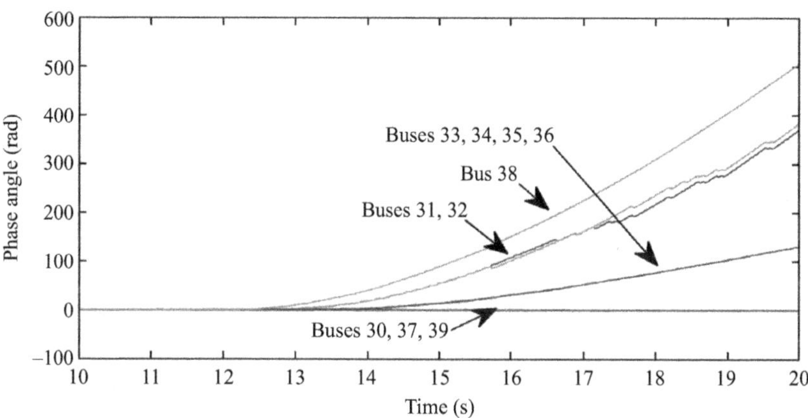

Figure 9.18 Phase angles at generator buses for a fault of 0.15 s at bus 18

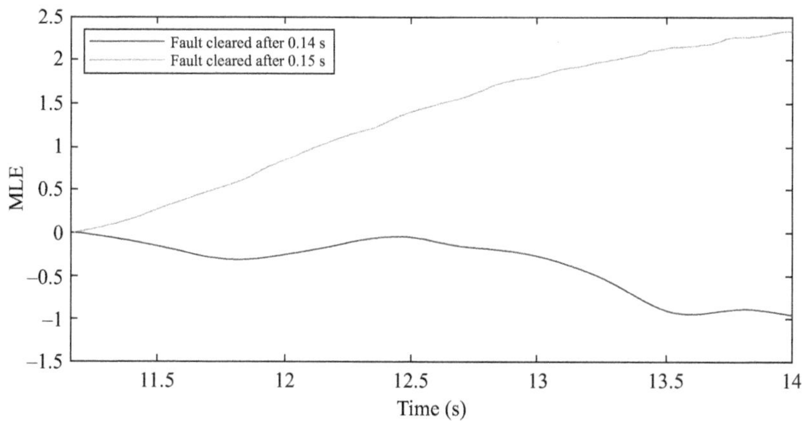

Figure 9.19 MLE for a fault of 0.14 s and 0.15 s at bus 18

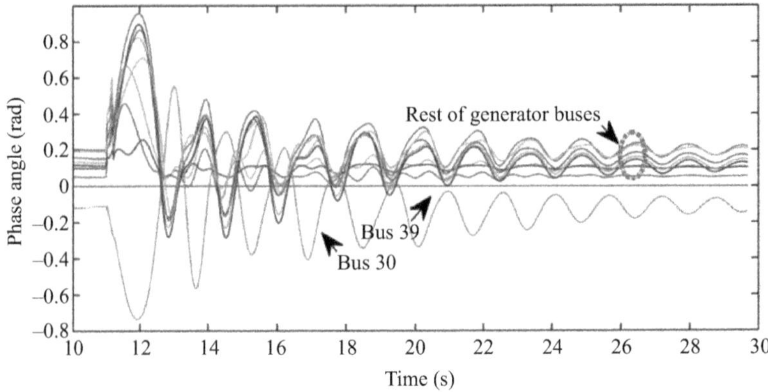

Figure 9.20 Phase angles at generator buses for a fault of 0.18 s at bus 4

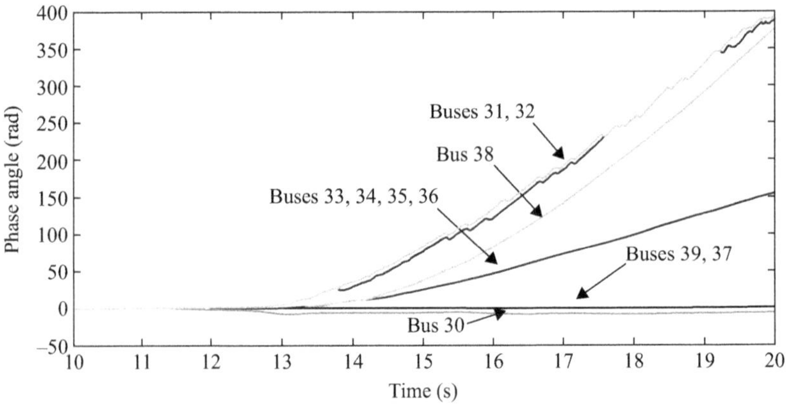

Figure 9.21 Phase angles at generator buses for a fault of 0.19 s at bus 4

in Figure 9.21. The MLE values for both the stable and the unstable cases are shown in Figure 9.22, which becomes positive for the unstable case.

The above analysis has demonstrated the application of the MLE for detection of the impending transient stability utilizing the generator voltage phasor measurements and its time variation data. The MLE assumes negative value under system stable condition and positive value under the unstable condition. The online estimation of the MLE for NE 39 bus systems have been demonstrated, through simulations using software in the loop configuration on the RTDS. The faults are created at different locations, and subsequently cleared at different time duration, making the system stable for some cases and unstable for other cases. The MLE is estimated from the phase angle of all the generator buses, assumed to be measured by the PMUs at the generator terminals after the fault is cleared. The transient stability of the system is determined within one second using the sign of the estimated MLE.

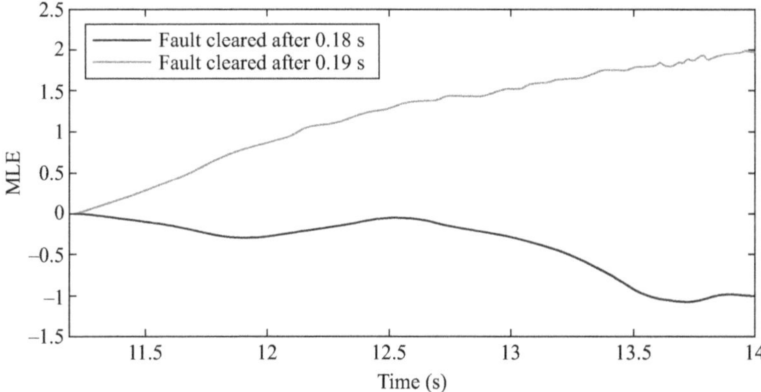

Figure 9.22 MLE for a fault of 0.18 s and 0.19 s at bus 4

The unstable cases in NE 39 bus systems using RTDS platforms demonstrate that the MLE becomes strictly positive following the disturbance in the power system network. The present length of data vector of 65 samples for computing the MLE could be effectively used for detecting the slow to moderate transition to the instability. However, the proposed technique will require further modification if the system quickly becomes unstable in less than 1 s, following a disturbance. The early indication of instability can be used by the system operator to initiate emergency controls or to trigger automatic sequence of protections to save the grid from possible blackout condition.

The consideration of the generator bus voltage phase angle, instead of generator rotor angle, does not degrade the performance of the algorithm. The indication of the stability provided by the MLE is correctly captured for all the test cases considered in this work by using the generator bus voltage phase angle measurements.

9.5 Conclusion

The application of synchrophasor measurements for online voltage stability, load parameter estimation, and online transient stability is presented in this chapter. The state-of-the-art techniques for voltage stability deployed in the ISOs and utility are reviewed with the underlying principal. The extension of the centralized voltage stability to decentralized scheme is also presented in this chapter. The distributed voltage stability architecture discussed in this chapter is highly scalable and fault tolerant.

The importance of load modeling is described in detail. There are various approaches of modeling a load, among which the measurement-based load modeling approach is discussed in detail. Various load models and parameter estimation techniques have also been discussed. The polynomial/ZIP load model, which is one of the widely used load model is discussed with interface to the existing power system simulation software. Some of the techniques like least-squares estimation and adaptive search–based algorithm have been presented in detail with simulation results. The difficulties faced during the estimation process of the load parameters have also been discussed.

The existing techniques available in the literature for transient stability assessment are presented in this chapter. The deployment of online transient stability application in the utility and industry is still being explored because of practical limitations like data availability, algorithm accuracy, and criticality of communication network. The online transient stability assessment using MLE, presented in this chapter, uses only bus angle from the PMU data. This technique can be easily extended for large systems as the computational burden is lower than the energy function–based method.

References

[1] Y. Kokai, F. Masuda, S. Horiike, and Y. Sekine, "Recent development in open systems for EMS/SCADA," *International Journal of Electrical Power and Energy Systems*, vol. 20, no. 2, pp. 111–123, 1998.

[2] A. Bose, "Smart transmission grid applications and their supporting infrastructure," *IEEE Transactions on Smart Grid*, vol. 1, no. 1 (Jun), pp. 11–19, 2010.

[3] P. Kundur, J. Paserba, V. Ajjarapu, *et al.*, "Definition and classification of power system stability IEEE/CIGRE joint task force on stability terms and definitions," *IEEE Transactions on Power Systems*, vol. 19, no. 3 (Aug), pp. 1387–1401, 2004.

[4] Y. Wang, I. Pordanjani,W. Li, *et al.*, "Voltage stability monitoring based on the concept of coupled single-port circuit," *IEEE Transactions on Power Systems*, vol. 26, no. 4 (Nov), pp. 2154–2163, 2011.

[5] Y. Makarov, P. Du, S. Lu, *et al.*, "PMU-based wide-area security assessment: concept, method, and implementation," *IEEE Transactions on Smart Grid*, vol. 3, no. 3 (Sep), pp. 1325–1332, 2012.

[6] S. Dasgupta, M. Paramasivam, U. Vaidya, and V. Ajjarapu, "Real-time monitoring of short-term voltage stability using PMU data," *IEEE Transactions on Power Systems*, vol. 28, no. 4 (Nov), pp. 3702–3711, 2013.

[7] J.-H. Liu, and C.-C. Chu, "Wide-area measurement-based voltage stability indicators by modified coupled single-port models," *IEEE Transactions on Power Systems*, vol. 29, no. 2 (Mar), pp. 756–764, 2014.

[8] S. Abdelkader, and D. Morrow, "Online tracking of Thevenin equivalent parameters using PMU measurements," *IEEE Transactions on Power Systems*, vol. 27, no. 2 (May), pp. 975–983, 2012.

[9] C. Zheng, V. Malbasa, and M. Kezunovic, "Regression tree for stability margin prediction using synchrophasor measurements," *IEEE Transactions on Power Systems*, vol. 28, no. 2 (May), pp. 1978–1987, 2013.

[10] D. Zhou, U. Annakkage, and A. Rajapakse, "Online monitoring of voltage stability margin using an artificial neural network," *IEEE Transactions on Power Systems*, vol. 25, no. 3 (Aug), pp. 1566–1574, 2010.

[11] H. O. Wang, E. H. Abed, and A. M. A. Hamdan, "Bifurcations, chaos, and crises in voltage collapse of a model power system," *IEEE Transactions on Circuits and Systems I: Fundamental Theory and Applications*, vol. 41, no. 4, pp. 294–302, 1994.

[12] M. Parniani, J. Chow, L. Vanfretti, B. Bhargava, and A. Salazar, "Voltage stability analysis of a multiple-infeed load center using phasor measurement data," in *Power Systems Conference and Exposition, 2006. PSCE'06. 2006 IEEE PES*, Oct 2006, pp. 1299–1305.

[13] D. Julian, R. Schulz, K. Vu, W. Quaintance, N. Bhatt, and D. Novosel, "Quantifying proximity to voltage collapse using the voltage instability predictor (VIP)," in *Power Engineering Society Summer Meeting, 2000. IEEE*, vol. 2, 2000, pp. 931–936.

[14] K. T. Vu, Patent US 6,219,591 B1.

[15] M. Vaiman, M. Vaiman, S. Maslennikov, E. Litvinov, and X. Luo, "Calculation and visualization of power system stability margin based on PMU measurements," in *Smart Grid Communications (SmartGridComm), 2010 First IEEE International Conference on*, Oct 2010, pp. 31–36.

[16] S. Malik, M. Vaiman, and M. Vaiman, "Implementation of rose for real-time voltage stability analysis at WECC RC," in *T D Conference and Exposition, 2014 IEEE PES*, April 2014, pp. 1–5.

[17] S. Maslennikov, E. Litvinov, M. Vaiman, and M. Vaiman, "Implementation of rose for on-line voltage stability analysis at ISO new England," in *PES General Meeting—Conference Exposition, 2014 IEEE*, July 2014, pp. 1–5.

[18] H. Mehrjerdi, S. Lefebvre, M. Saad, and D. Asber, "A decentralized control of partitioned power networks for voltage regulation and prevention against disturbance propagation," *IEEE Transactions on Power Systems*, vol. 28, no. 2 (May), pp. 1461–1469, 2013.

[19] H. Mehrjerdi, S. Lefebvre, M. Saad, and D. Asber, "Coordinated control strategy considering effect of neighborhood compensation for voltage improvement in transmission systems," *IEEE Transactions on Power Systems*, vol. 28, no. 4 (Nov), pp. 4507–4515, 2013.

[20] J. Wen, Q. Wu, D. Turner, S. Cheng, and J. Fitch, "Optimal coordinated voltage control for power system voltage stability," *IEEE Transactions on Power Systems*, vol. 19, no. 2 (May), pp. 1115–1122, 2004.

[21] T. K. G. Coulouris, J. Dollimore, and G. Blair, *Distributed Systems: Concepts and Design*. 5th ed. Boston, MA: Addison-Wesley, 2011.

[22] A.S. Hyojong Lee, S. Niddodi, and D. Bakken, "Decentralized voltage stability control in the smart grid using distributed computing architecture," in *IEEE Industry Applications Society Annula Meeting (IAS),* October 2016.

[23] C. Taylor, *Power System Voltage Stability*. New York, NY: McGraw-Hill, 1994.

[24] "Load representation for dynamic performance analysis [of power systems]," *IEEE Transactions on Power Systems*, vol. 8, no. 2 (May), pp. 472–482, 1993.

[25] "Using PMU data to increase situational awareness," PSERC Final Project Report, Tech. Rep., September 2010.

[26] "Measurement-Based Load Modeling," EPRI, Tech. Rep., September 2006.

[27] P. Kundur, *Power System Stability and Control*. New York, NY: McGraw-Hill, 1994.

[28] S. Kalinowsky, and M. Forte, "Steady state load-voltage characteristic field tests at area substations and fluorescent lighting component characteristics," *IEEE Transactions on Power Apparatus and Systems*, vol. PAS-100, no. 6, pp. 3087–3094, 1981.

[29] H. Bai, P. Zhang, and V. Ajjarapu, "A novel parameter identification approach via hybrid learning for aggregate load modeling," *IEEE Transactions on Power Systems*, vol. 24, no. 3 (Aug), pp. 1145–1154, 2009.

[30] "Composite Load Model for Dynamic Simulations," Tech. Rep., June 2012.

[31] T. Ohyama, A. Watanabe, K. Nishimura, and S. Tsuruta, "Voltage dependence of composite loads in power systems," *IEEE Transactions on Power Apparatus and Systems*, vol. PAS-104, no. 11, pp. 3064–3073, 1985.

[32] D. Kosterev, A. Meklin, J. Undrill, *et al.*, "Load modeling in power system studies: WECC progress update," in *Power and Energy Society General Meeting—Conversion and Delivery of Electrical Energy in the 21st Century, 2008 IEEE*, July 2008, pp. 1–8.

[33] P. Regulski, P. Wall, Z. Rusidovic, and V. Terzija, "Estimation of load model parameters from PMU measurements," in *Innovative Smart Grid Technologies Conference Europe (ISGT-Europe), 2014 IEEE PES*, Oct 2014, pp. 1–6.

[34] G. Ledwich and C. F. Moyano, "Synchrophasors for load modelling in Australia," in *Power and Energy Society General Meeting, 2011 IEEE*, July 2011, pp. 1–6.

[35] G. Shackshaft, O. Symons, and J. Hadwick, "General-purpose model of power-system loads," *Proceedings of the Institution of Electrical Engineers*, vol. 124, no. 8, pp. 715–723, 1977.

[36] Q. Liu, Y. Chen, and D. Duan, "The load modeling and parameters identification for voltage stability analysis," in *International Conference on Power System Technology, 2002. Proceedings. PowerCon 2002*, vol. 4, 2002, pp. 2030–2033.

[37] S. Zhu, J. Zheng, and G. Luo, "Effect of load modeling on voltage stability," in *IEEE Power Engineering Society Summer Meeting, 2000*, vol. 1, 2000, pp. 395–400.

[38] V. Knyazkin, C. Canizares, and L. Soder, "On the parameter estimation and modeling of aggregate power system loads," *IEEE Transactions on Power Systems*, vol. 19, no. 2 (May), pp. 1023–1031, 2004.

[39] L. Ingber, "Adaptive simulated annealing (ASA): lessons learned," *Control and Cybernetics*, vol. 25, pp. 33–54, 1996.

[40] A. Alves da Silva, C. Ferreira, and G. Torres, "Dynamic load modeling based on a nonparametric ANN," in *International Conference on Intelligent Systems Applications to Power Systems, 1996. Proceedings, ISAP'96*, Jan 1996, pp. 55–59.

[41] D. Chen and R. Mohler, "Load modelling and voltage stability analysis by neural networks," in *American Control Conference, 1997. Proceedings of the 1997*, vol. 2, June 1997, pp. 1086–1090.

[42] T. Hiyama, M. Tokieda, W. Hubbi, and H. Andou, "Artificial neural network based dynamic load modeling," *IEEE Transactions on Power Systems*, vol. 12, no. 4 (Nov), pp. 1576–1583, 1997.

[43] J. Ma, D. Han, R. M. He, Z. Y. Dong, and D. J. Hill, "Reducing identified parameters of measurement-based composite load model," *IEEE Transactions on Power Systems*, vol. 23, no. 1 (Feb), pp. 76–83, 2008.

[44] *Powertech Labs Inc., DSA Tools*, Surrey, British Columbia, Canada, April 2012.

[45] R. Christie, *IEEE Power System Test Cases Archive* [Online]. Available from: www.ee.washington.edu/research/pstca, Aug 1999.

[46] D. Simon, *Optimal State Estimation: Kalman, H Infinity, and Nonlinear Approaches*. Hoboken, NJ: Wiley-Interscience, 2006.

[47] Y. Xue, T. V. Custem, and M. Ribbens-Pavella, "Extended equal area criterion justifications, generalizations, applications," *IEEE Transactions on Power Systems*, vol. 4, no. 1 (Feb), pp. 44–52, 1989.

[48] F. Da-Zhong, T. S. Chung, and A. K. David, "Fast transient stability estimation using a novel dynamic equivalent reduction technique," *IEEE Transactions on Power Systems*, vol. 9, no. 2 (May), pp. 995–1001, 1994.

[49] M. Yin, C. Y. Chung, K. P. Wong, Y. Xue, and Y. Zou, "An improved iterative method for assessment of multi-swing transient stability limit," *IEEE Transactions on Power Systems*, vol. 26, no. 4 (Nov), pp. 2023–2030, 2011.

[50] T. Athay, R. Podmore, and S. Virmani, "A practical method for the direct analysis of transient stability," *IEEE Transactions on Power Apparatus and Systems*, vol. PAS-98, no. 2 (Mar), pp. 573–584, 1979.

[51] J. L. Willems, "Improved Lyapunov function for transient power-system stability," *Proceedings of the Institution of Electrical Engineers*, vol. 115, no. 9, pp. 1315–1317, 1968.

[52] J. L. Willems and J. C. Willems, "The application of Lyapunov methods to the computation of transient stability regions for multimachine power systems," *IEEE Transactions on Power Apparatus and Systems*, vol. PAS-89, no. 5, pp. 795–801, 1970.

[53] M. A. Pai, M. A. Mohan, and J. G. Rao, "Power system transient stability: regions using Popov's method," *IEEE Transactions on Power Apparatus and Systems*, vol. PAS-89, no. 5, pp. 788–794, 1970.

[54] G. S. Hope, S. T. Nichols, and J. Carr, "Measurement of transfer functions of power system components under operating conditions," *IEEE Transactions on Power Apparatus and Systems*, vol. PAS-96, no. 6 (Nov), pp. 1798–1808, 1977.

[55] F. S. Prabhakara and A. H. El-Abiad, "A simplified determination of transient stability regions for Lyapunov methods," *IEEE Transactions on Power Apparatus and Systems*, vol. PAS-94, no. 2 (Mar), pp. 672–689, 1975.

[56] N. Krishnamoorthy and V. G. Rau, "Transient stability analysis of a multimachine power system with flux decay and feedback type voltage regulator," *Electric Machines & Power Systems*, vol. 13, no. 5, pp. 315–327, 1987 [Online]. Available from: http://dx.doi.org/10.1080/07313568708909250.

[57] D. C. Hu and V. M. Venkatasubramanian, "New wide-area algorithms for detection and mitigation of angle instability using synchrophasors," in *IEEE Power Engineering Society General Meeting, 2007*, June 2007, pp. 1–8.

[58] A. A. Fouad, V. Vittal, Y.-X. Ni, *et al.*, "Direct transient stability assessment with excitation control," *IEEE Transactions on Power Systems*, vol. 4, no. 1 (Feb), pp. 75–82, 1989.

[59] H. D. Chiang, "Study of the existence of energy functions for power systems with losses," *IEEE Transactions on Circuits and Systems*, vol. 36, no. 11, pp. 1423–1429, 1989.

[60] A. A. Fouad and S. E. Stanton, "Transient stability of a multi-machine power system. Part I: investigation of system trajectories," *IEEE Transactions on Power Apparatus and Systems*, vol. PAS-100, no. 7, pp. 3408–3416, 1981.

[61] N. Kakimoto, Y. Ohsawa, and M. Hayashi, "Transient stability analysis of multimachine power system with field flux decays via Lyapunov's direct method," *IEEE Transactions on Power Apparatus and Systems*, vol. PAS-99, no. 5 (Sep), pp. 1819–1827, 1980.

[62] H. D. Chiang, F. F. Wu, and P. P. Varaiya, "Foundations of the potential energy boundary surface method for power system transient stability analysis," *IEEE Transactions on Circuits and Systems*, vol. 35, no. 6, pp. 712–728, 1988.

[63] H. -D. Chiang, F. F. Wu, and P. P. Varaiya, "A BCU method for direct analysis of power system transient stability," *IEEE Transactions on Power Systems*, vol. 9, no. 3 (Aug), pp. 1194–1208, 1994.

[64] S. E. Stanton, C. Slivinsky, K. Martin, and J. Nordstrom, "Application of phasor measurements and partial energy analysis in stabilizing large disturbances," *IEEE Transactions on Power Systems*, vol. 10, no. 1 (Feb), pp. 297–306, 1995.

[65] C. W. Liu, J. S. Thorp, J. Lu, R. J. Thomas, and H. D. Chiang, "Detection of transiently chaotic swings in power systems using real-time phasor measurements," *IEEE Transactions on Power Systems*, vol. 9, no. 3 (Aug), pp. 1285–1292, 1994.

[66] C. W. Liu, and J. Thorp, "Application of synchronised phasor measurements to real-time transient stability prediction," *IEE Proceedings on Generation, Transmission and Distribution*, vol. 142, no. 4 (Jul), pp. 355–360, 1995.

[67] B. B. Monchusi, Y. Mitani, L. Changsong, and S. Dechanupaprittha, "PMU based power system stability analysis," in *TENCON 2008—2008 IEEE Region 10 Conference*, Nov 2008, pp. 1–5.

[68] M. Liu, H. Sun, B. Zhang, L. Yao, M. Han, and W. Wu, "PMU measurements and EMS models based transient stability on-line forecasting," in *Power Energy Society General Meeting, 2009. PES'09. IEEE*, July 2009, pp. 1–8.

[69] J. C. Cepeda, J. L. Rueda, D. G. Colome, and D. E. Echeverria, "Real-time transient stability assessment based on centre-of-inertia estimation from phasor measurement unit records," *IET Generation, Transmission Distribution*, vol. 8, no. 8, pp. 1363–1376, 2014.

[70] J. Hazra, R. K. Reddi, K. Das, D. P. Seetharam, and A. K. Sinha, "Power grid transient stability prediction using wide area synchrophasor measurements," in *Innovative Smart Grid Technologies (ISGT Europe), 2012 Third IEEE PES International Conference and Exhibition on*, Oct 2012, pp. 1–8.

[71] Y. Wu, M. Musavi, and P. Lerley, "Synchrophasor-based monitoring of critical generator buses for transient stability," *IEEE Transactions on Power Systems*, vol. 31, no. 1, pp. 287–295, 2016.

[72] J. Yan, C. C. Liu, and U. Vaidya, "PMU-based monitoring of rotor angle dynamics," *IEEE Transactions on Power Systems*, vol. 26, no. 4 (Nov), pp. 2125–2133, 2011.

[73] D. P. Wadduwage, C. Q. Wu, and U. D. Annakkage, "Power system transient stability analysis via the concept of Lyapunov exponents," *Electric Power Systems Research*, vol. 104, pp. 183–192, 2013.

[74] S. Dasgupta, M. Paramasivam, U. Vaidya, and V. Ajjarapu, "Real-time monitoring of short-term voltage stability using PMU data," *IEEE Transactions on Power Systems*, vol. 28, no. 4 (Nov), pp. 3702–3711, 2013.

[75] M. T. Rosenstein, J. J. Collins, and C. J. D. Luca, "A practical method for calculating largest Lyapunov exponents from small data sets," *Physica D: Nonlinear Phenomena*, vol. 65, no. 12, pp. 117–134, 1993.

[76] A. Wolf, J. B. Swift, H. L. Swinney, and J. A. Vastano, "Determining Lyapunov exponents from a time series," *Physica D: Nonlinear Phenomena*, vol. 16, no. 3, pp. 285–317, 1985.

[77] P. M. Anderson and A. A. Fouad, *Power System Control and Stability*. Iowa City, IA: Iowa State University Press, 1977.

[78] "IEEE standard for synchrophasor measurements for power systems," *IEEE Std. C37.118.1-2011 (Revision of IEEE Std. C37.118-2005)*, Dec 2011.

Chapter 10

State estimation in the presence of synchronized measurement

Sanjeev Kumar Mallik[1], *Saikat Chakrabarti*[1] *and Sri Niwas Singh*[1]

State estimation plays a critical role for the wide-area monitoring system at the control center. The power system network, geographically spread over a large area, is monitored and controlled using the measurements collected at regular intervals. A number of application functions are performed based on the estimated states of the system to ensure secure and reliable operation of the power system. The amplitude and angle of the voltage phasor at each bus constitute the states of the system. The measurements are collected from the field and telemetered to the control center, where they are processed, typically using standard-weighted least squares static state estimator, to estimate the states of the system.

10.1 State estimation

Since the first introduction of static state estimation as a data processing algorithm [1–3] in 1970s, for converting meter readings and other available information into estimates of the states, the state estimator (SE) algorithms have traversed a long path of development. This chapter discusses the inclusion of synchronized measurements from phasor measurement units (PMUs) into the static SE.

The states of a power system are usually estimated using equivalent single-phase asynchronous measurements, assuming the system to be balanced. In the SE, typically the positive sequence model is considered. The measurements collected from the field are telemetered to the control center and then processed through the SE to provide the estimate of the states. The measurements from the metering devices constitute the core inputs for the estimation of the power system states, and hence, there is a need to pay attention to the measuring device technology. Power system measurements may be classified into two categories based on the measuring device technologies, namely, supervisory control and data acquisition (SCADA) system measurements and the measurements from PMUs.

[1]Department of Electrical Engineering, Indian Institute of Technology Kanpur, Kanpur 208016, India

10.1.1 SCADA measurements

A SCADA system involves collection of data through monitoring devices and transferring them to the control center via a communication system. The first SCADA systems used panel of meters, lights, and strip chart recorders for data acquisition. With the technology advancement, remote terminal units (RTUs) and intelligent electronic devices (IEDs) have become integral parts of the modern SCADA system. IEDs are the sensors designed with the intelligence of programmable logic controllers (PLCs) and distributed control systems. An RTU is a microprocessor-based data acquisition and control unit, used to control and acquire data from the measuring devices [4]. This acquired data is telemetered to the control center.

SCADA-based EMS has been widely deployed in power system network. It utilizes RTUs and IEDs in the field to acquire measurement data through the measuring devices installed at the substations. The acquired measurements are transmitted to the control center through the communication network. The conventional SCADA measurements consist of real and reactive power injection measurements at the bus, real, and reactive power flow measurements through the branch, magnitude of the current flowing through the branch, as well as the voltage magnitude at the bus. These measurements are scanned typically at a rate of 2–10 s. The breaker and switch statuses are also acquired and transmitted to the control center in order to construct the network topology.

10.1.2 Phasor measurement unit

The synchrophasor technology provides time-stamped voltage and current phasors from different locations in a power system network. The time stamps are derived from the global positioning system (GPS) clock. The PMU is a measuring device based on synchronized measurement technology [5,6]. The concept of the PMU came from the symmetrical component distance relay and the synchronization of sampling clock [7]. The first prototype of the PMU was developed at Virginia Tech, USA. On the basis of this prototype, Macrodyne Co. had started the production of the PMU commercially. Now, there are a large number of PMU manufacturers in the market.

The analog voltage and the current from the field through the instrument transformers are the input to the PMU. The PMU then samples the waveforms and computes phasors using discrete Fourier transform type of algorithms. One-pulse-per-second timing signal obtained from the GPS is used to time-stamp these computed phasors. These phasors may be stored locally or transmitted to remote locations using a communication network. Phasor data concentrators collect these measurements for further applications or transmission to other locations. The PMU data standard is specified in IEEE Std. C37.118 [8,9].

Measurement accuracy of a PMU is expressed in terms of total vector error (TVE), defined as [8,10]:

$$\%\text{TVE} = \frac{\left|\overline{\mathbf{X}}_{\text{est}} - \overline{\mathbf{X}}_{\text{ideal}}\right|}{\left|\overline{\mathbf{X}}_{\text{ideal}}\right|} \times 100 \tag{10.1}$$

where $\overline{\mathbf{X}}_{est}$ and $\overline{\mathbf{X}}_{ideal}$ are the estimated phasor and the true phasor, respectively, for the measurements at a given instant. According to IEEE C37.118 Std., all the PMUs installed in the power system network should have less than 1% TVE under steady state. The TVE in the PMU measurements may be due to the error in the phasor magnitude, angle, and inaccuracy in time synchronization.

Due to the ability to provide time-synchronized phasor measurements at high refresh rates, PMUs are being increasingly deployed in power systems. It is envisaged that there will be increasing use of the PMUs in state estimation and its corresponding functions [6].

10.1.3 State estimation functions

State estimation is an integral part of the energy management system (EMS). Based on the estimated states, subsequent actions are taken at the control center. There are a number of functions performed by the SE.

- *Topology processing:* It collects the status data from the circuit breakers and switches placed in the whole power system network to configure one-line diagram of the system.
- *Observability analysis:* It ensures whether the unique SE solution can be achieved or not, with the available set of measurements. If the system is not observable, it identifies the unobservable branches and the observable islands in the systems. A number of techniques have been evolved by the researchers for the observability analysis [11–20].
- *SE solution:* It provides an estimate for the system states by processing the given set of measurements.
- *Bad data processing:* It detects the bad data in the measurements, identifies the bad data, and processes or eliminates the bad data.
- *Parameter estimation:* It identifies the most likely parameters in the network.

These functions are integrated in the SE software of the EMS at the control center. This chapter focuses mainly on the state estimation solution techniques in presence of the PMU measurements.

10.1.4 Measurement model for static state estimation

The measurements for the power system network can be modeled as follows:

$$\mathbf{z} = \mathbf{h}(\mathbf{x}) + \mathbf{e}_m \tag{10.2}$$

where $\mathbf{z}$ is the vector of m measurements consisting of SCADA measurements as well as PMU measurements; $\mathbf{h}(\mathbf{x})$ is the relationship between the measurement vector $\mathbf{z}$, and the state vector $\mathbf{x}$ consists of voltage magnitudes and phase angles at the buses; $\mathbf{e}_m$ is the error vector for the measurements.

The state estimation problem in the presence of equality constraints can be formulated by minimizing the following objective function:

$$J(\mathbf{x}) = \frac{1}{2}[\mathbf{z} - \mathbf{h}(\mathbf{x})]^T \mathbf{W}[\mathbf{z} - \mathbf{h}(\mathbf{x})] \tag{10.3}$$

Subject to

$$\mathbf{c}(\mathbf{x}) = 0 \tag{10.4}$$

where $\mathbf{W}$ is the diagonal weight matrix (assuming the measurements to be independent), and $\mathbf{c}(\mathbf{x})$ represents the zero injection power measurements (both real and reactive power injection).

The state estimation solution is achieved by iteratively solving the following set of equations:

$$\begin{bmatrix} \mathbf{G} & \mathbf{C}^T \\ \mathbf{C} & \mathbf{0} \end{bmatrix} \begin{bmatrix} \Delta \mathrm{x} \\ \lambda_L \end{bmatrix} = \begin{bmatrix} \mathbf{H}^T \mathbf{W}(\mathbf{z} - \mathbf{h}(\mathbf{x})) \\ -\mathbf{c}(\mathbf{x}) \end{bmatrix} \tag{10.5}$$

where at kth iteration, $\Delta\mathbf{x} = \mathbf{x}^{k+1} - \mathbf{x}^k$ is the change in the state vector; λ_L is the vector of Lagrangian multiplier corresponding to zero injection measurements; $\mathbf{H} = (\partial \mathbf{h}(\mathbf{x})/\partial \mathbf{x})$ and $\mathbf{C} = (\partial \mathbf{c}(\mathbf{x})/\partial \mathbf{x})$ are the measurement Jacobian matrices; $\mathbf{G} = \mathbf{H}^T\mathbf{W}\mathbf{H}$ is the gain matrix corresponding to weighted least squares (WLS)-SE. Above equation can be written as follows:

$$\mathbf{E}\mathbf{y} = \mathbf{b}$$

where $\mathbf{E}$ is the so-called information matrix, $\mathbf{y}$ is the augmented state vector, and the vector $\mathbf{b}$ is defined as follows:

$$\mathbf{b} = \left[\left[\mathbf{H}^T \mathbf{W}(\mathbf{z} - \mathbf{h}(\mathbf{x}))\right]^T ; -\mathbf{c}(\mathbf{x})^T \right]^T. \tag{10.6}$$

10.1.5 Assigning weights to the measurements

Weights are usually assigned to the measurements in proportion to their accuracy. Quantitatively speaking, a widely used approach is to assign weights inversely proportional to the variation in the measurements. The manufacturer usually specifies the maximum error in the measurements obtained from the measurement devices. In the absence of any measurement error distribution specified by the manufacturer, the guide to the uncertainty in measurements (GUM) prescribes adoption of uniform measurement error distribution [21]. An alternative error distribution, namely Gaussian distribution, is also widely used for the measurements. Following subsections briefly discuss the assignment of measurement weights for these two types of error distributions.

10.1.5.1 Gaussian error distribution

Gaussian error distribution is specified by the mean and the standard deviation (SD) in the measurements. The mean value of any error in the measurements is assumed to be zero. Typical values of the SD that are widely used in the literature for different types of measurements are listed in Table 10.1 [11,22]. However, these values may differ for different metering devices and may be specified by the manufacturer.

Table 10.1 Typical standard deviation in various measurements

Conventional measurements		PMU measurements	
Power injection	**Power flow**	**|V| or |I|**	**Phase angle**
0.01 pu	0.008 pu	0.002 pu	0.0017 rad

Table 10.2 Typical values of maximum measurement uncertainties

Conventional measurements		PMU measurements		
Power injection	**Power flow**	**|V|**	**|I|**	**Phase angle**
3%	3%	0.02%	0.03%	0.01 degree

The variance in the *i*th measurement, R_i, for the Gaussian error distribution can be expressed as follows:

$$R_i = \mathrm{SD}_i^{\,2} \tag{10.7}$$

where SD_i is the SD in the *i*th measurement.

10.1.5.2 Uniform error distribution

The maximum measurement uncertainty of a measuring device is usually specified by the manufacturer as a percentage of the meter reading, typical value of which is listed in Table 10.2. However, these values may differ for different metering devices, depending on the manufacturer specifications.

In absence of any specified probability distribution for the measurement uncertainty, the standard uncertainty in the direct measurement can be expressed in terms of maximum specified uncertainty, by assuming a uniform probability distribution for the measurement uncertainty over the entire range, as prescribed in [21,23]. For the *i*th measurement, for example, the standard uncertainty or SD in measurement can be expressed as follows:

$$u_i = \frac{\Delta u_i}{\sqrt{3}} \tag{10.8}$$

where Δu_i is the specified maximum uncertainty in the *i*th measurement.

Example 10.1: The voltage magnitude measured by a PMU is 1.04 pu. Using Table 10.2, determine the standard uncertainty in this magnitude measurement.

Solution: The standard uncertainty for this measurement is as follows:

$$u_{V_i} = \frac{1.04 \times 0.02/100}{\sqrt{3}} = 1.2 \times 10^{-4}$$

Similarly, the standard uncertainty in the phase angle measurement will be

$$u_{\theta_i} = \frac{0.01 \times \pi}{180 \times \sqrt{3}} = 1.0077 \times 10^{-4}$$

The variance in the *i*th measurement, R_i, can then be expressed as follows:

$$R_i = u_i^2 \tag{10.9}$$

The weight matrix, **W**, in (10.3) is the inverse of the error covariance matrix, **R**, which is a diagonal square matrix, with R_i as (i,i)th element. During state estimation process, direct measurements may need to be converted into different form. For example, the voltage phasor computed at a bus by using voltage phasor, directly measured at the other end and current phasor through the line, is a derived or pseudo-measurement. The combined standard uncertainty in the derived measurements may be evaluated using the classical theory of propagation of measurement uncertainty [21].

10.1.5.3 Calculation of combined uncertainty

Let k independent measurement variables be $v_1, v_2, \ldots, v_k$ in the system. A measurement y is derived from the available k independent measurement variables. The combined standard uncertainty, u_y, for the measurement y can be obtained by combining the standard uncertainties of the independent variables, as shown below:

$$u_y = \sqrt{\sum_{j=1}^{k} \left(\frac{\partial y}{\partial v_j}\right)^2 \left(u_{v_j}\right)^2} \tag{10.10}$$

where u_{v_j} is the standard uncertainty in the measurement variable v_j.

10.2 Conventional state estimation

After collecting all the measurements, observability analysis is carried out to determine whether the system is observable with the given set of measurements or not. After ensuring the observability, the measurements are processed through an SE algorithm to provide an estimate of the power system states. Depending on the type of measurements, the SE can be classified into three categories, namely conventional SE, linear SE, and hybrid SE. The SE may include both conventional and synchronized measurements and may be classified as preprocessed hybrid SE and post-processed sequential SE. Conventional SE is discussed in this section.

Conventional SE processes the conventional SCADA measurements to estimate the voltage phasor angle and magnitude at all the buses. Bus power injection and branch power flow (both real and reactive), as well as the magnitude of the bus voltage and the branch current, form the conventional SCADA measurement set. The measurement vector and its corresponding Jacobian (assuming state vector in polar form, voltage phasor angle, and magnitude vector, δ and $\mathbf{v}$) for the SE process can be written as follows:

$$\mathbf{z}_{\text{conventional}} = \begin{bmatrix} \mathbf{z}\mathbf{P}_{\text{inj}} \\ \mathbf{z}\mathbf{Q}_{\text{inj}} \\ \mathbf{z}\mathbf{P}_{\text{flow}} \\ \mathbf{z}\mathbf{Q}_{\text{flow}} \\ \mathbf{z}_{|\mathbf{V}|} \\ \mathbf{z}_{|\mathbf{I}|} \end{bmatrix} \tag{10.11}$$

$$\mathbf{H}_{\text{conventional}} = \begin{bmatrix} \frac{\partial \mathbf{z}\mathbf{P}_{\text{inj}}}{\partial \boldsymbol{\delta}} & \frac{\partial \mathbf{z}\mathbf{P}_{\text{inj}}}{\partial \mathbf{V}} \\ \frac{\partial \mathbf{z}\mathbf{Q}_{\text{inj}}}{\partial \boldsymbol{\delta}} & \frac{\partial \mathbf{z}\mathbf{Q}_{\text{inj}}}{\partial \mathbf{V}} \\ \frac{\partial \mathbf{z}\mathbf{P}_{\text{flow}}}{\partial \boldsymbol{\delta}} & \frac{\partial \mathbf{z}\mathbf{P}_{\text{flow}}}{\partial \mathbf{V}} \\ \frac{\partial \mathbf{z}\mathbf{Q}_{\text{flow}}}{\partial \boldsymbol{\delta}} & \frac{\partial \mathbf{z}\mathbf{Q}_{\text{flow}}}{\partial \mathbf{V}} \\ \frac{\partial \mathbf{z}_{|\mathbf{V}|}}{\partial \boldsymbol{\delta}} & \frac{\partial \mathbf{z}_{|\mathbf{V}|}}{\partial \mathbf{V}} \\ \frac{\partial \mathbf{z}_{|\mathbf{I}|}}{\partial \boldsymbol{\delta}} & \frac{\partial \mathbf{z}_{|\mathbf{I}|}}{\partial \mathbf{V}} \end{bmatrix} \tag{10.12}$$

Once the measurement vector and its corresponding Jacobian is formed, an iterative method is used to estimate the states for the conventional SE by solving (10.5) in presence of equality constraints, since the relationship between the measurement vector and the state vector is nonlinear. The solution of the conventional SE involves matrix inversion of **E**. The convergence and the accuracy must be ensured while solving the iterative equation of the WLS-SE. The convergence of the SE is mathematically represented in term of the condition number of the gain matrix. The high condition number of the gain matrix may deteriorate the accuracy and convergence. Therefore, it is essential to address the numerical conditioning issues associated with the SE problems, which is indicated by the condition number. The condition number of the matrix is defined as follows:

$$\text{cond}(\mathbf{E}) = \|\mathbf{E}\|\|\mathbf{E}^{-1}\| \tag{10.13}$$

where $\|\mathbf{E}\|$ represents the matrix norm of **E**. The second-order norm is typically used for the SE problems. Higher order norm are not considered mainly due to large computational effort. However, in some of the literature, first-order norm is used to simplify the analysis of the condition number [24,25]. A high value of the condition number usually signifies numerical ill-conditioning. Some of the commonly encountered factors causing numerical ill-conditioning of the state estimation problem, as identified by the researchers, are existence of negative line reactance, existence of very small series impedance branch, presence of numerous power injection measurements, low measurement redundancy, size of the system, position of the slack bus, differential weight assigned for the measurements, inclusion of zero bus power injection as an equality constraints, and many more. To address these numerical conditioning issues, a number of techniques [24,26–29] have been developed to find the solution of (10.5) by the researchers. The majority of the existing SE methods are based on the inversion and decomposition of the gain matrix, **G** or information matrix, **E**. A new regularization method to solve the numerical instability of an equality constrained SE using *L*-curve method is discussed later [29]. The complete flowchart for the conventional SE is shown in Figure 10.1.

10.3 Linear state estimation

The synchrophasor technology has brought a revolution in the SE due to its capability to measure time stamped and synchronized phasor data at a high refreshing rate (as high as one phasor per cycle). The PMUs also provide direct measurements of the voltage and current phase angle with respect to a global reference obtained from the GPS clock. In short, the time stamped synchronized measurement of voltage and current phasors with high accuracy and high reporting rate make the PMUs preferable over the conventional SCADA measurements. In the presence of sufficient number of PMUs, the WLS-SE becomes linear and noniterative [30].

In linear estimator, the power system network is observed through PMU measurements only. In this case, the measurement vector, **z**, and its corresponding Jacobian can be expressed as follows:

$$\mathbf{z}_{\text{PMU}} = \begin{bmatrix} \mathbf{z}_{\mathbf{V}_{\text{real}}} \\ \mathbf{z}_{\mathbf{V}_{\text{imaginary}}} \\ \mathbf{z}_{\mathbf{I}_{\text{real}}} \\ \mathbf{z}_{\mathbf{I}_{\text{imaginary}}} \end{bmatrix}_{\text{PMU}} \tag{10.14}$$

$$\mathbf{H}_{\text{PMU}} = \begin{bmatrix} \mathbf{H}_{11} & \mathbf{0} \\ \mathbf{0} & \mathbf{H}_{22} \\ \mathbf{H}_{31} & \mathbf{H}_{32} \\ \mathbf{H}_{41} & \mathbf{H}_{42} \end{bmatrix} \tag{10.15}$$

In general, the voltage and current phasor measurements from the PMU are available in polar form, which needs to be converted into rectangular form (as expressed in (10.14)) for the linear estimator process, and the corresponding weight of the converted measurement is evaluated using classical theory of propagation of uncertainty, as discussed earlier. In case of linear estimator, the real and reactive parts of the voltage at all buses constitute the state vector, **x** ([**e**;**f**]). With the given measurement vector, $\mathbf{z}_{\text{PMU}}$, the measurement Jacobian with respect to the state vector (represented by (10.15)) is independent of the state vector, and hence, the state vector can be directly evaluated in a noniterative manner:

$$\mathbf{x}_{\text{PMU}}^{\text{est}} = \begin{bmatrix} \mathbf{e} \\ \mathbf{f} \end{bmatrix} = \left(\mathbf{H}_{\text{PMU}}^{T} \mathbf{W}_{\text{PMU}} \mathbf{H}_{\text{PMU}}\right)^{-1} \mathbf{H}_{\text{PMU}}^{T} \mathbf{W}_{\text{PMU}} \mathbf{z}_{\text{PMU}} \tag{10.16}$$

where $\mathbf{W}_{\text{PMU}}$ is the weight matrix corresponding to the PMU measurements.

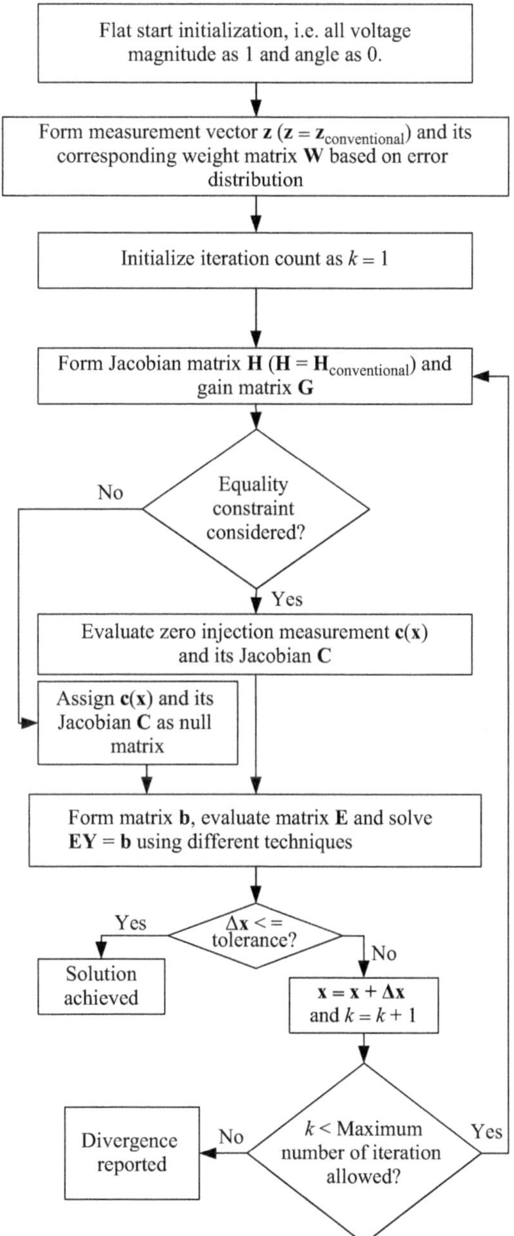

Figure 10.1 Conventional state estimation flowchart

For the linear estimator, sufficient number of PMUs should be employed in the power system network. Due to the communication as well as financial constraints, limited numbers of the PMUs are being installed in the network that may not make the system completely observable. Therefore, the measurements in term of voltage and current phasors data from the PMUs should be embedded with the conventional SCADA measurements in the SE to exploit the advantages of the PMUs. In the fusion of the PMU current and voltage phasor data with the conventional SCADA measurements, two approaches are reported in the literature, namely postprocessing sequential estimator and preprocessing hybrid estimator.

10.4 Postprocessing sequential state estimation

The postprocessing algorithm for the SE combines the conventional and the linear SE in a two-stage process in a sequential manner and hence may be called as sequential SE. The first stage is exactly similar to the conventional SE processing the conventional SCADA measurements to estimate the voltage magnitude and angle at all buses of the system, as described in Section 10.3. The estimated states, $\mathbf{x}_{\text{stageI,polar}}$ from the conventional SE, are in polar form. The error covariance matrix ($n \times n$ matrix) for the estimated states can be expressed as [30]:

$$\text{covariance}\left(\mathbf{x}_{\text{stageI,polar}}\right) = \left(\mathbf{H}_{\text{conventional}}^{T}\mathbf{W}_{\text{conventional}}\mathbf{H}_{\text{conventional}}\right)^{-1} \tag{10.17}$$

where $\mathbf{W}_{\text{conventional}}$ is the weight matrix corresponding to the conventional SCADA measurements. The measurement Jacobian for the conventional SCADA measurements, $\mathbf{H}_{\text{conventional}}$, is calculated at the estimated states from stage I, $\mathbf{x}_{\text{stageI,polar}}$. The variance matrix, $\mathbf{R}_{\text{stageI,polar}}$, of the estimated voltage phasors from the conventional SE can be obtained by the corresponding diagonal element of the matrix expressed in (10.17), by ignoring the off-diagonal elements.

In a sequential SE, the estimates from the existing SCADA software are combined with PMU measurements in stage II, to obtain improved estimates of the states. The complete measurement vector and its Jacobian for stage II estimator may be expressed as follows:

$$\mathbf{z}_{\text{II}} = \begin{bmatrix} \begin{pmatrix} \mathbf{V}_{\text{real}} \\ \mathbf{V}_{\text{imaginary}} \end{pmatrix}_{\text{stageI}} \\ \begin{pmatrix} \mathbf{V}_{\text{real}} \\ \mathbf{V}_{\text{imaginary}} \end{pmatrix}_{\text{PMU}} \\ \begin{pmatrix} \mathbf{I}_{\text{real}} \\ \mathbf{I}_{\text{imaginary}} \end{pmatrix}_{\text{PMU}} \end{bmatrix} \tag{10.18}$$

$$\mathbf{H}_{\text{II}} = \begin{bmatrix} \mathbf{H}_{11} & \mathbf{0} \\ \mathbf{0} & \mathbf{H}_{22} \\ \mathbf{H}_{31} & \mathbf{0} \\ \mathbf{0} & \mathbf{H}_{42} \\ \mathbf{H}_{51} & \mathbf{H}_{52} \\ \mathbf{H}_{61} & \mathbf{H}_{62} \end{bmatrix} \tag{10.19}$$

For the linear SE in stage II, all the voltage phasors are either estimated from stage I or collected from the PMUs. The current phasors from the PMU must be in rectangular form. As the states are estimated from conventional SE (stage I) in polar form, they need to be converted into rectangular form for the inclusion in the measurement vector of stage II. The variance, $\mathbf{R}_{\text{stageI,rect}}$, corresponding to the converted states from stage I in rectangular form can be evaluated using classical theory of propagation of uncertainty.

The measurement vector from stage II includes the real and reactive part of the voltage phasor obtained from stage I, and the real and reactive part of the voltage and the current phasor from the PMU, as expressed in (10.18). The complete variance matrix for the stage II can be written as follows:

$$\mathbf{R}_{\text{II}} = \begin{bmatrix} \mathbf{R}_{\text{stageI,rect}} & \mathbf{0} & \mathbf{0} \\ \mathbf{0} & \mathbf{R}_{\text{V,PMU}} & \mathbf{0} \\ \mathbf{0} & \mathbf{0} & \mathbf{R}_{\text{I,PMU}} \end{bmatrix} \tag{10.20}$$

where $\mathbf{R}_{\text{V,PMU}}$ and $\mathbf{R}_{\text{I,PMU}}$ are the variance matrix corresponding to the voltage and the current phasor measurements, respectively, obtained from the PMUs placed in the network. The weight matrix for the stage II is the inverse of $\mathbf{R}_{\text{II}}$. The measurement Jacobian expressed by (10.19) is independent of the states of the system. Therefore, the final states of the power system can be estimated directly without involving any iteration:

$$\mathbf{x}_{\text{II}} = \begin{bmatrix} \mathbf{e} \\ \mathbf{f} \end{bmatrix} = \left(\mathbf{H}_{\text{II}}\mathbf{R}_{\text{II}}^{-1}\mathbf{H}_{\text{II}}\right)^{-1}\mathbf{H}_{\text{II}}\mathbf{R}_{\text{II}}^{-1}\mathbf{z}_{\text{II}} \tag{10.21}$$

The complete algorithm for the sequential SE is shown in Figure 10.2. In the sequential SE, the states obtained from the conventional SE are in polar form, which are converted into rectangular form along with their variance or weight. The converted estimated states are combined with the PMU measurements (voltage and current phasor in rectangular form) and processed through linear SE to estimate the final states of the system, i.e., the real and reactive part of the voltage phasor at all buses. In this sequential SE, the stage I estimator is a conventional, iterative, and nonlinear process, whereas the stage II estimator is a noniterative and linear process. The sequential approach is preferred due to the reusability of the existing SE software [30]. However, this method is relevant when the number of PMUs is small in the system. As the number of PMUs will increase, the natural approach will be to combine the conventional and PMU measurements in a single stage. This type of hybrid estimators are referred here as the preprocessing estimator.

10.5 Preprocessing hybrid state estimation

In Sections 10.3 and 10.4, the voltage and current phasor from the PMUs were considered for the estimator in rectangular form. In hybrid SE, the voltage and current phasor from the PMU may be considered either in polar or rectangular

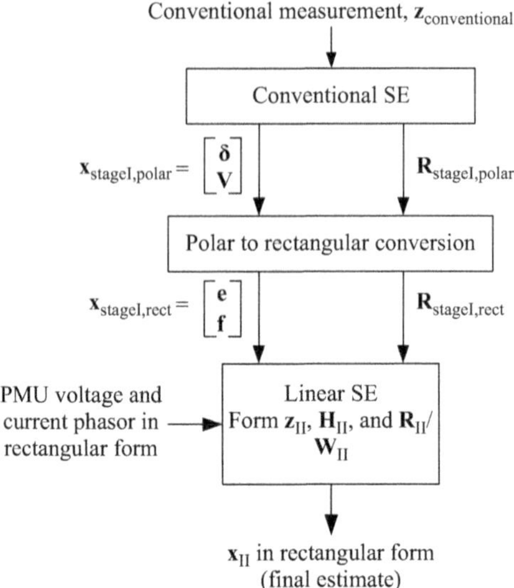

Figure 10.2 Sequential state estimation flowchart

form. In the hybrid SE, the PMU measurements are combined with the conventional SCADA measurements in a preprocessing algorithm, where the estimation process is similar to the conventional SE. The hybrid SE differs with the conventional SE in the formation of measurement vector and its corresponding Jacobian. The hybrid SE is iterative and nonlinear with respect to the states of the system. The states are estimated by solving the set of linear equations (10.5) or (10.6) in each iteration. For estimating the states, the measurement vector and its corresponding Jacobian must be defined.

For the hybrid SE, the voltage magnitude and angle at all buses constitute the states of the system and can be expressed as follows:

$$\mathbf{x} = \begin{bmatrix} \boldsymbol{\delta} \\ \mathbf{V} \end{bmatrix} \tag{10.22}$$

The conventional measurement vector consists of the real and reactive power injection at the buses as well as the real and reactive power flow through the line, and expressed as follows:

$$\mathbf{z}_{conventional} = \begin{bmatrix} \mathbf{zP}_{inj} \\ \mathbf{zQ}_{inj} \\ \mathbf{zP}_{flow} \\ \mathbf{zQ}_{flow} \end{bmatrix} \tag{10.23}$$

The voltage and the current phasor are the measurements available from the PMU. Due to the iterative nature of the solution of hybrid SE, the inclusion of PMU in the hybrid SE has become important for ensuring the convergence and the accuracy of the estimated states. As the states of the system are in polar form, the voltage phasor measurements from the PMU are included directly in polar form in the measurement vector. The inclusion of the PMU currents in the hybrid SE has been a major challenge for the researchers due to ill-conditioned gain matrix, which may result in divergence problems. There are three possible ways to include the PMU current measurements in the state estimation problem [31]:

- *Current phasor magnitude and angle:* In this case, certain elements (partial derivatives) of the Jacobian matrix may become undefined (divided by zero) or extremely large at flat start or specific loading conditions. Hence, the inclusion of the current phasor in polar form is not preferred for the hybrid SE.
- *Real and imaginary part of the current phasor:* The problem associated with the polar form of current phasor is alleviated if the current phasor is included in rectangular form. In general, the rectangular form of PMU current phasor is considered for the hybrid SE.
- *Pseudo-voltage form derived from the PMU current and voltage phasor:* For a line between bus i and j, if the PMU is located at bus i, then the voltage phasor at bus j can be calculated using the voltage phasor at bus i, current phasor flowing from bus i to bus j, and the line parameters.

As the PMU current is a major concern for power system engineers in terms of the SE convergence, the researchers have attempted to use the PMU current in pseudo-voltage form. The pseudo-voltage calculation involves the voltage and current measurements along with the network parameter information. Due to the uncertainty in the network parameter, which is very difficult to evaluate, there is a need to convert the PMU current in other pseudo forms depending only on the measurements. Pseudo-flow measurement $(\mathbf{V}_i\mathbf{I}_{ij}^*)$ derived from the PMU current and voltage phasor is one such form that is to be considered by the researchers. For the pseudo-flow form, the accuracy of the pseudo-measurement depends on the PMU measurement only. With the direct inclusion of the PMU current in rectangular form, the gain matrix may become ill-conditioned due to high weight associated with the PMU current measurements. The convergence issues associated with the PMU currents can be resolved by inclusion of PMU current as branch flow pseudo-measurements, which are reflected by less number of iteration [32].

For the hybrid measurement configurations, the PMU current measurement is being used in the SE either in polar or in rectangular form. The rectangular form of PMU current is frequently used for the state estimation. By considering the voltage phasor magnitude and current, as well as real and imaginary part of the current phasor collected from PMUs, the measurement vector corresponding to the PMU measurements can be expressed as follows:

$$\mathbf{z}_{\text{PMU}} = \begin{bmatrix} \mathbf{z}_{\delta} \\ \mathbf{z}_{\text{V}} \\ \mathbf{z}_{\mathbf{I}_{\text{real}}} \\ \mathbf{z}_{\mathbf{I}_{\text{imaginary}}} \end{bmatrix} \tag{10.24}$$

The complete measurement vector consisting of conventional measurements expressed by (10.23) and the PMU measurements expressed by (10.24), can be written as follows:

$$\mathbf{z}_{\text{Hybrid}} = \begin{bmatrix} \mathbf{z}_{\text{conventional}} \\ \mathbf{z}_{\text{PMU}} \end{bmatrix} \tag{10.25}$$

The measurement Jacobian and the weight matrix corresponding to the measurement vector expressed by (10.25) can be written as follows:

$$\mathbf{H}_{\text{Hybrid}} = \begin{bmatrix} \mathbf{H}_{\text{conventional}} \\ \mathbf{H}_{\text{PMU}} \end{bmatrix} \tag{10.26}$$

$$\mathbf{W}_{\text{Hybrid}} = \begin{bmatrix} \mathbf{W}_{\text{conventional}} & 0 \\ 0 & \mathbf{W}_{\text{PMU}} \end{bmatrix} \tag{10.27}$$

where $\mathbf{H}_{\text{conventional}}$ and $\mathbf{H}_{\text{PMU}}$ are the measurement Jacobian corresponding to the conventional measurements and the PMU measurements, respectively; $\mathbf{W}_{\text{conventional}}$ and $\mathbf{W}_{\text{PMU}}$ are the measurement weights corresponding to the conventional measurements and the PMU measurements, respectively.

The algorithm for the hybrid SE is similar to the algorithm shown in Figure 10.1. The difference lies in the formation of the measurement vector ($\mathbf{z} = \mathbf{z}_{\text{Hybrid}}$), the corresponding Jacobian ($\mathbf{H} = \mathbf{H}_{\text{Hybrid}}$), and the weight matrix ($\mathbf{W} = \mathbf{W}_{\text{Hybrid}}$). Due to the poor conditioning of the gain matrix, a number of techniques have been developed for the SE to process the conventional measurements using inversion/decomposition/factorization methods, and these methods are also extended to the hybrid measurement set. In the next section, a regularization based method for solving the linear set of equations, $\mathbf{E}\mathbf{y} = \mathbf{b}$ expressed in (10.6) during each iteration, is discussed. This method is better equipped in handling numerical ill-conditioning problems in the SE execution.

10.6 Regularization method

The inversion of a matrix having large condition number is prone to numerical problems. A numerically robust technique is, therefore, needed to handle the ill-conditioning issues. In this section, a methodology based on Tikhonov regularization is presented to handle the ill-conditioning of the state estimation problem.

A substantial amount of research work has been done in the field of regularization and optimization of inverse ill-posed problems. The least squares method is one of the methods to solve the problem of minimization of $\|\mathbf{E}\mathbf{y} - \mathbf{b}\|^2$ (refer (10.6)). The least square solution for (10.6) is also known as pseudo inverse solution. The output of least squares optimization is affected by the data error and rounding error [33]. Regularization methods can be used to reduce these errors in solving the inverse problems. Tikhonov regularization is a widely used method for solving ill-conditioned problems and is described below in the context of solving (10.6) [34]:

$$\min_{\mathbf{y}} \left\{ \|\mathbf{Ey} - \mathbf{b}\|^2 + \lambda^2 \|\mathbf{L}(\mathbf{y} - \mathbf{y}_0)\|^2 \right\} \tag{10.28}$$

where $\mathbf{L}$ is referred to as a regularization operator and λ is a regularization parameter. The common choice of $\mathbf{L}$ is the identity matrix $\mathbf{I}$. The vector $\mathbf{y}_0$ is an a priori estimate of $\mathbf{y}$. In the absence of a-priori information, $\mathbf{y}_0$ is assumed to be zero. With these assumptions applied, (10.28) is simplified as follows:

$$\min_{\mathbf{y}} \left\{ \|\mathbf{Ey} - \mathbf{b}\|^2 + \lambda^2 \|\mathbf{y}\|^2 \right\} \tag{10.29}$$

The first term $\|\mathbf{Ey} - \mathbf{b}\|^2$ is known as the least square or residual norm, and $\|\mathbf{y}\|^2$ is the regularized or solution norm. From (10.29), it is clear that the estimate of $\mathbf{y}$ depends on the regularization parameter. The unregularized solution ($\lambda = 0$) of (10.29) corresponds to the pseudo-inverse solution. Pseudo-inverse solution only considers the idea of the curve fitting, i.e., minimization of the least square norm. It does not consider the solution norm, i.e., the accuracy of the estimate of $\mathbf{y}$. Tikhonov regularization provides the solution by minimizing the least square norm as well as the solution norm, and the solution is controlled by the regularization parameter. The next step is to analyze the effect of regularization on the solution. If too much regularization is imposed on the solution, then it will not fit the given data $\mathbf{b}$. On the other hand, if very small regularization is imposed, the fit will be good, but the solution will be dominated by the contribution from the data errors. Hence, there is a trade-off between the size of the regularized solution and the quantity of the fit. This trade-off is controlled by selection of proper regularization parameter. Some of the well-known methods for selecting the regularization parameter are listed below:

- Discrepancy principle
- Generalized cross validation (GCV) method
- *L*-curve method

According to the discrepancy principle, the regularization parameter λ is chosen so that the discrepancy is satisfied assuming accurate estimation in ε [35]:

$$\|\mathbf{b} - \mathbf{Ey}\| = \varepsilon \tag{10.30}$$

where ε is the norm of the error in $\mathbf{b}$. The discrepancy principle faces difficulty in the absence of the information on ε. GCV method is based on cross validation of each data point as reported in [36]. The GCV method is not satisfactory when the errors are highly correlated. The *L*-curve method is robust and has the ability to handle perturbations caused by correlation noise. Hence, the *L*-curve method for selection of regularization parameter has gained attention in recent years and is considered for estimating the regularization parameter.

10.6.1 Selection of regularization parameter by L-curve method

L-curve is a log–log plot between the squared norm of the regularized solution and the squared norm of the residual for a range of values of the regularization parameter. The name is derived from the fact that this plot resembles letter "*L*" closely. Taking the following notations:

$$\eta = \|\mathbf{y}\|^2, \quad \rho = \|\mathbf{E}\ \mathbf{y} - \mathbf{b}\|^2 \tag{10.31}$$

$$\hat{\eta} = \log\eta, \quad \hat{\rho} = \log\rho \tag{10.32}$$

L-curve is the plot of $\hat{\eta}/2$ versus $\hat{\rho}/2$. With the above notations expressed in (10.31) and (10.32), the solution norm as well as least square norm depends on **y**, which is the solution of (10.29), expressed as follows:

$$\mathbf{y} = \left(\mathbf{E}^T\mathbf{E} + \lambda^2\mathbf{I}\right)^{-1}\mathbf{E}^T\mathbf{b} \tag{10.33}$$

The sparsity of the matrix $(\mathbf{E}^T\mathbf{E} + \lambda^2\mathbf{I})$ used in (10.33) is reduced, and the evaluation of **y** is computationally expensive. Hence, a computationally efficient tool is required to simplify the calculation of two norms associated with **y**, and singular value decomposition (SVD) is one such tool. The SVD component of the matrix **E** can be expressed as follows:

$$\mathbf{E} = \sum_{i=1}^{\mathrm{ns}} u_i\sigma_i{v_i}^{\mathrm{T}} \tag{10.34}$$

$$\sigma_1 \geq \sigma_2 \geq \sigma_3 \geq \cdots \geq \sigma_{\mathrm{ns}} \geq 0 \tag{10.35}$$

where u_i and v_i are the left and right singular vectors for the corresponding singular value σ_i; ns is the number of nonzero singular values of **E**. After inserting the value of SVD component of **E** into (10.33), Tikhonov solution is given as follows:

$$\mathbf{y} = \sum_{i=1}^{\mathrm{ns}} f_i \frac{u_i^T b}{\sigma_i} v_i \tag{10.36}$$

$$f_i = \frac{\sigma_i^2}{\sigma_i^2 + \lambda^2} \tag{10.37}$$

where $f_1, f_2, \ldots, f_{\mathrm{ns}}$ are the Tikhonov filter factors. With this SVD formulation, the solution and residual norm can be expressed as follows:

$$\eta = \|\mathbf{y}\|^2 = \sum_{i=1}^{\mathrm{ns}} \left(f_i \frac{u_i^T b}{\sigma_i}\right)^2 \tag{10.38}$$

$$\rho = \|\mathbf{E}\ \mathbf{y} - \mathbf{b}\|^2 = \sum_{i=1}^{\mathrm{ns}} \left((1 - f_i)u_i^T b\right)^2 \tag{10.39}$$

For different values of λ, the solution norm as well as least square norm based on (10.38) and (10.39) may be plotted on a logarithmic scale to depict L-curve. L-curve for the unperturbed data is mostly flat at the value $\|\mathbf{y}\|$, except for larger values of the residual norm, where the curve approaches the axis corresponding to the residual norm. L-curve corresponding to the error in **b** is overall very steep, located slightly left of ε except for small value of λ, where it approaches the axis corresponding to the regularized norm [34]. Combining both the component of the data, it is concluded that L-curve is dominated by pure noise in case of small λ and by the unperturbed data for large value of λ. The above conclusion is valid only when L-curve is plotted on a logarithmic scale. In linear scale, L-curve is always convex irrespective of **b**.

A graphical strategy to select the regularization parameter uses the location of the corner of this curve. The point corresponding to the suitable λ lies on the corner of the L-curve. The logic behind this choice is that the corner separates the flat and vertical parts of the curve, where the solution is dominated by the regularization errors and perturbation errors, respectively [34]. Graphically, it is very difficult to locate the corner point of the L-curve with sufficient accuracy. Hence, there is a need to convert the graphical approach into a numerical technique for locating the corner point. Hansen [34] suggested a numerical technique for finding the corner of the L-curve by tracing the point of maximum curvature. The curvature $k(\lambda)$ can be found by,

$$k(\lambda) = \frac{\hat{\rho}'\hat{\eta}'' - \hat{\rho}''\hat{\eta}'}{\left((\hat{\rho}')^2 + (\hat{\eta}')^2\right)^{3/2}} \tag{10.40}$$

where the prime denotes the differentiation with respect to λ. The point of maximum curvature corresponds to the corner of the L-curve, where the values of both the regularized norm and residual norm are low simultaneously. The formulation of the curvature involves the evaluation of first- and second-order derivative of $\hat{\eta}$ and $\hat{\rho}$ with respect to λ [29].

With this formulation, the curvature is evaluated for different values of λ, depending upon the singular values of matrix **E**, to make sure that the maximum curvature point is included. Out of these values of the curvature, the value of λ corresponding to the maximum curvature is evaluated. Let this value be λ_c, where c is the index for λ. Further, the maximum curvature is evaluated by using Golden section search between two points λ_{c-1} and λ_{c+1}.

10.6.2 Regularized state estimation problem

The iterative equation for the state estimation problem (10.5) or (10.6) is solved through Tikhonov regularization, and the regularization parameter is estimated using L-curve criterion. A corrective step is taken at the postestimation stage to include zero injection. Without this corrective step, the SE results for the bus having zero injection measurement will become erroneous.

The flowchart for finding the solution for $\mathbf{Ey} = \mathbf{b}$ is shown below.

[STEP 1] Formulate SVD component of $\mathbf{E}$, as shown in (10.34).

[STEP 2] Generate a number of values of λ. Let this number be 200. Then:

$$\lambda_{200} = \sigma_{\min} \tag{10.41}$$

$$\frac{\lambda_i}{\lambda_{i+1}} = \left(\frac{\sigma_{\max}}{\sigma_{\min}}\right)^{1/200-1} \quad \forall\, i = 199:\, -1 : 1 \tag{10.42}$$

[STEP 3] Evaluate the least square norm as well as the solution norm for all 200 values of λ.

[STEP 4] Plot L-curve for all λ.

[STEP 5] For each λ, the value of the curvature is evaluated.

[STEP 6] Find the maximum curvature and its corresponding λ (say, λ_c).

[STEP 7] Search λ_{opt} between λ_{c-1} and λ_{c+1} for finding out the absolute maximum curvature point using Golden section search method.

[STEP 8] Evaluate the solution $\mathbf{y}$ for the corresponding λ_{opt}. From $\mathbf{y}$, separate out $\Delta\mathbf{x}$ required for SE algorithm.

From Step 6, the regularization parameter estimated may not correspond to the maximum curvature. But, it is certain to have maximum curvature between λ_{c-1} and λ_{c+1}. Hence, there is a need to refine the search for regularization parameter between λ_{c-1} and λ_{c+1}. For this purpose, Golden section search method is attempted while considering the parabolic nature of the curvature.

After completing the regularization based algorithm for the conventional or hybrid state estimation problem, the estimated value of injection at zero injection bus may not satisfy the equality constraint. To satisfy the equality constraints, the zero injection bus voltage is modified after completing the regularization algorithm. The voltage phasor at *iz*th bus having zero injection can be expressed in terms of nonzero injection bus voltage phasors as follows:

$$\overline{\mathbf{V}}_{iz} = -\frac{1}{\overline{\mathrm{Y}}_{iz,iz}} \sum_{\substack{j=1 \\ j \neq iz}}^{n} \overline{\mathrm{Y}}_{iz,j} * \overline{\mathbf{V}}_j \tag{10.43}$$

where $\overline{\mathrm{Y}}_{iz,j}$ is the (iz, j)th element of the complex bus admittance matrix. $\overline{\mathrm{V}}_j$ is the complex voltage at bus j.

10.7 Case study

In this section, the regularized state estimation is implemented and tested on IEEE 14-bus test system [37]. The system is used for investigating the performance of the regularized method with some of the other existing methods. In this chapter, the regularization based SE is compared with LU factorization based SE (referred as

LU-SE) [38]. In LU factorization based SE, iterative equation (10.6) is solved by using LU factorization of the information matrix, **E**.

PMU measurements include voltage and current phasors in polar form at the buses where it is deployed. The PMU current phasor measurements, thus collected, are converted into rectangular form, and the corresponding weights are evaluated using the principle of combined measurement uncertainty [31]. The PMU inputs for the SE methods are PMU current phasor in rectangular form and PMU voltage phasor in polar form. Zero injection measurements are treated as equality constraints in this chapter. One thousand Monte Carlo simulations are carried out for testing the accuracy of the SE for IEEE 14-bus test system. A flat start is assumed and the convergence tolerance is set to 10^{-6}. All simulations are carried out on an Intel Core processor i7-2600 CPU at 3.4 GHz and 4 GB of RAM.

The true value of the state vector is acquired from load flow solutions. The true measurements, such as real and reactive power injections as well as power flows, and line currents in polar form, are evaluated from the true state vector as well as line parameter information. The measurements are generated by adding Gaussian errors to the true value of the measurements. One thousand Monte-Carlo trials are performed by randomly selecting the measurements following a Gaussian distribution from a range defined by the mean and the SD of the distribution. The SD of the error in the conventional measurements (real and reactive power injection as well as power flow) and the PMU measurements (voltage as well as current phasor magnitude and angle) is considered as listed in Table 10.1. All the measurement errors are assumed to have zero mean.

The performance of the regularized SE is evaluated by its absolute error (average and maximum) in voltage phasor (magnitude in pu as well as angle in radian). The behavior of the SE is evaluated by the performance indices, namely the filter effect and the total statistical error variance (TSEV) [22]. The filter effect (FE) of the SE algorithm is defined as follows:

$$\mathrm{FE} = \frac{\sum_{i=1}^{m} \left(\mathrm{S}_i^E - \mathrm{S}_i^T\right)^2}{\sum_{i=1}^{m} \left(\mathrm{S}_i^M - \mathrm{S}_i^T\right)^2} \tag{10.44}$$

where S_i^E, S_i^T and S_i^M are the final estimated measurements, true value of measurements, and the measurements at each Monte Carlo run, respectively. In this chapter, the average FE over the entire Monte Carlo run is reported for the performance analysis.

TSEV of the estimated state variables of size n averaging over the T Monte Carlo run is expressed as follows:

$$\mathrm{TSEV} = \sum_{i=1}^{n} \left(\frac{1}{T} \sum_{\mathrm{mc}=1}^{T} \left(\mathbf{x}(i)_{\mathrm{mc}} - \mathbf{x}(i)_t\right)^2 \right) \tag{10.45}$$

where $\mathbf{x}(i)_{mc}$ is the vector of the estimates of the ith state variable in the (mc)th Monte Carlo trial. $\mathbf{x}(i)_t$ is the true value of ith state.

The regularized SE is implemented and tested on the IEEE 14-bus system for two cases. In Case I, a set of measurements consists of the injection meters, the flow meters, and some PMUs. In Case II, the network is covered by few injection and flow meters without any PMU.

10.7.1 Case I for IEEE 14-bus test system

The real and reactive power injection measurements at all buses except the zero injection buses and flow measurements (real as well as reactive) at all lines are part of the conventional measurements for this case. Two PMUs at buses 1 and 6 are placed for the validation of the regularized SE.

The regularized SE is demonstrated for the second iteration of 550th Monte Carlo sample (randomly selected). The singular value of **E** obtained for this particular iteration and sample are found to vary from 2.5569×10^{-5} (σ_{min}) to 1.1355×10^{8} (σ_{max}). The least square norm as well as the solution norm are evaluated for 200 points of λ such that $\lambda_1 \geq \lambda_2 \geq \lambda_3 \geq \cdots \geq \lambda_{200}$. The ratio of two consecutive value of λ is evaluated by (10.42).

The values of both the norms are plotted on logarithmic scale for the second iteration of a 550th Monte Carlo trial and shown in Figure 10.3 (only few values of λ are indicated in this figure). The indicated values in the plot correspond to different values of the regularization parameter. Its corresponding curvature

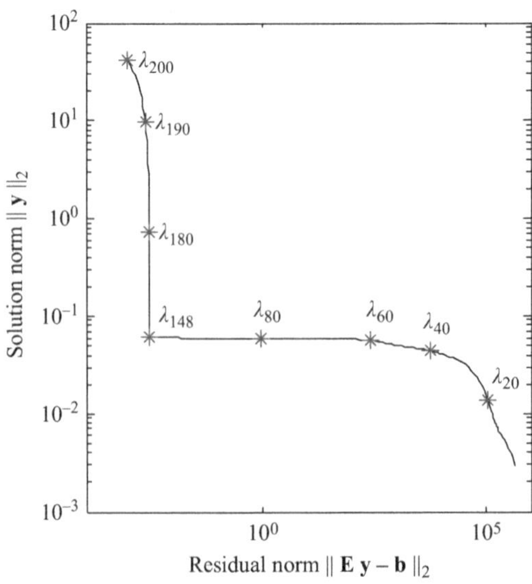

Figure 10.3 L-curve during second iteration of 550th Monte Carlo sample of IEEE 14-bus system for different value of λ

represented by (10.40) is plotted against different values of λ, as shown in Figure 10.4. From Figure 10.4, the maximum point of $k(\lambda)$ is captured to be 1.3173×10^5 at $\lambda_{148} = 0.0516$. This maximum value of $k(\lambda)$ may not be the exact maximum point and hence further processing is required to evaluate the exact maximum point within the narrow search between $\lambda_{147} = 0.0597$ and $\lambda_{149} = 0.0446$. The golden-section-search method is employed to find the maximum value of $k(\lambda)$ in the interval [0.0446 0.0597] of λ. The regularization parameter selected corresponding to Figure 10.4 is taken as 0.0485, with the maximum curvature 1.3542×10^5. For this particular 550th Monte Carlo trial, the values of regularization parameter selected in all iterations are listed in Table 10.3.

With two PMUs at buses 1 and 6, the performance of the regularized estimator is compared with the LU-SE and listed in Table 10.4. The accuracy of the estimator is

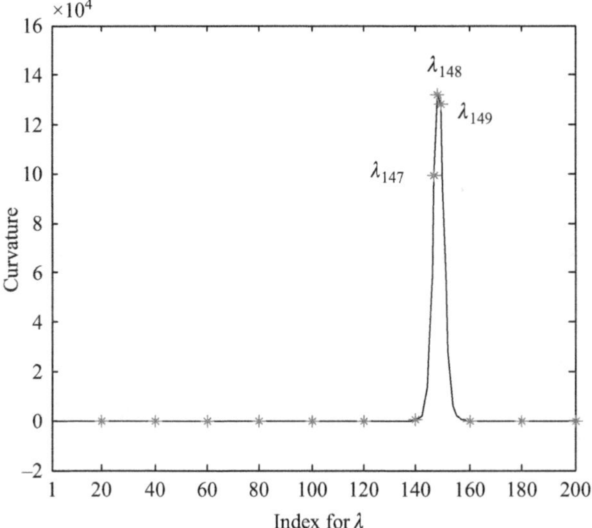

Figure 10.4 Curvature corresponding to Figure 10.6 for IEEE-14 bus system

Table 10.3 Regularization parameter for 550th Monte Carlo Sample of IEEE 14-bus system

Iteration	**Condition number**		λ
	E	**G**	
First	4.0512×10^{12}	1.711×10^{17}	0.0572
Second	4.4410×10^{12}	5.837×10^{8}	0.0485
Third	4.3716×10^{12}	2.097×10^{9}	0.1718
Fourth	4.3710×10^{12}	2.291×10^{9}	0.1768

Table 10.4 The SE performance for Case I of IEEE 14-bus system

Performance index	LU-SE	Regularized SE
Maximum absolute error in $\|V\|$ (pu, $\times 10^{-5}$)	5.0037	4.9287
Maximum absolute error in $\angle V$ (rad, $\times 10^{-5}$)	3.9713	3.9023
Average max absolute error in $\|V\|$ (pu, $\times 10^{-5}$)	2.1434	2.1259
Average absolute error in $\angle V$ (rad, $\times 10^{-5}$)	1.5626	1.6320
Average filter effects	0.15605	0.159193
TSEV ($\times 10^{-5}$)	3.103274	3.124207
Average objective function J	92.1093	90.1511
Average condition number of $\mathbf{E}$ ($\times 10^{12}$)	4.3822	4.3822
Average time (ms)	15.4185	53.7973
Average iteration counts	4.091	4.319

better as compared to the LU-SE, as is evident from Table 10.4. The average filter effect over the entire Monte-Carlo trials from the regularized method is comparable to the LU-SE. The small value of variance shows the accuracy of the solutions, whereas the filter effect indicates the fit of the estimated states for the measurements. The estimated values of the state variables through the Tikhonov regularization method by processing the erroneous measurements are accurate up to four decimal digits compared with the true values of the states. The average value of objective function represented by (10.3) is minimal for the regularized SE in comparison to LU-SE.

The accuracy of the SE is ensured by the proper selection of the regularization parameter. The condition number of the coefficient matrix is of the order of 10^{12} for IEEE 14-bus system, which indicates ill-conditioning. The regularization parameter selected to overcome the effect of ill-conditioning for providing the best estimate of the state change vector ranges from 0.0211 to 7.987×10^6 at the solution point for the Monte-Carlo trials.

10.7.2 Case II for IEEE 14-bus test system

In an effort to generate a case of SE divergence, a lot of measurement meter placement strategies are tested for IEEE 14-bus system, and one of those placements is listed for Case II in Table 10.5. The SE problem for this case was solvable by using the regularized method only.

Nonconvergence issues are reported for 757 samples of measurement sets out of 1,000 Monte-Carlo trials for LU-SE methods. The regularized SE provides the converged solution for all Monte-Carlo trials in Case II. The SE output for this case with regularized SE is listed in Table 10.6.

Regularized method works is able to solve the SE problem in case of numerically ill-conditioned problem, i.e., Case II. This divergence is due to the measurement meter placement. This issue is resolved by considering the regularized norm along with the residual norm using Tikhonov regularization. The trade-off between the both norms is controlled by proper selection of λ using L-curve method. For Case II, the voltage magnitude and angle estimated by Tikhonov

Table 10.5 Measurement meter placement of Case II for IEEE 14-bus system

Meter type	Reference	Location
Injection	Bus	1, 2, 6, 7, 9, and 14
Flow	Line	1–2, 2–4, 4–5, 5–6, 6–12, 7–8, 10–11, 13–14

Table 10.6 The regularized SE performance for Case II of IEEE 14-bus system

Average filter effect	0.9227
TSEV	0.0383
Average objective function J	0.9577
Average time (ms)	44
Average iteration counts	4.089
Average λ at solution point $\times 10^6$	6.6119

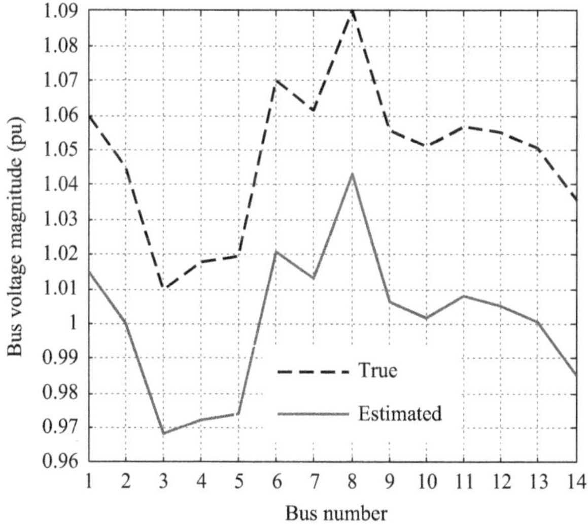

Figure 10.5 Bus voltage magnitude (pu) for Case II of IEEE 14-bus system by the regularized SE

regularization–based SE are plotted against the true values obtained from load flow studies and shown in Figures 10.5 and 10.6.

In case of nonconvergence, the value of regularization parameter plays a major role to provide the converged solution by trading-off between the solution norm and the residual norm. From Figures 10.5 and 10.6 for Case II, it is evident that the solution may not be very accurate, but a converged and reasonable one.

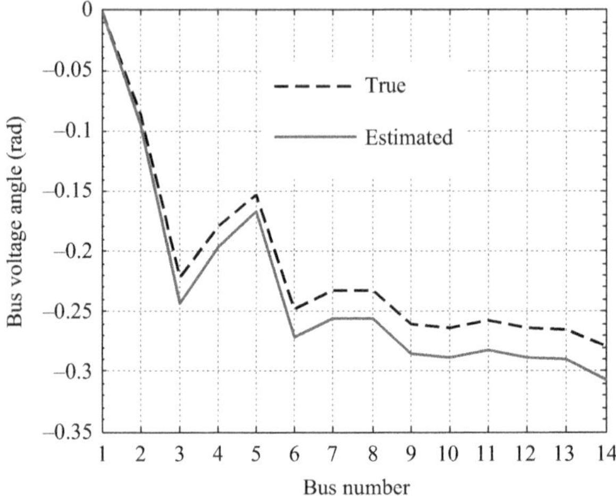

Figure 10.6 Bus voltage angle (rad) for Case II of IEEE 14-bus system by the regularized SE

The regularized method is also tested for an ill-conditioned 13-bus test system [39]. The injection meters are placed at buses 1, 6, 9, and 12; flow meters are placed at the branch connecting buses 10 and 11 near bus 10. Zero injection measurements are treated as equality constraints. With the given measurement placement, divergence issues are reported for the LU-SE. Out of 1,000 Monte-Carlo trials, 375 sets of measurement samples failed to converge for LU-SE method. The Tikhonov regularization–based SE converged successfully for the same 1,000 trials used for the LU-SE. One important advantage of the regularization method is its ability to solve ill-conditioned problems. The regularized method has focused mainly on the numerical stability aspect via Tikhonov regularization, while keeping the solution time low and maintaining high accuracy by minimizing the SE objective function.

10.8 Conclusion

In this chapter, a single phase, single-area, static SE is presented in presence of the PMUs along with the conventional SCADA measurements. In absence of the PMU, the conventional SCADA measurements are processed through conventional SE to estimate the states of the system. With sufficient number of PMUs being placed in the network, the linear estimator is a good choice ignoring the conventional measurements. In the presence of the limited number of PMUs in the network, the sequential SE is preferred due to reuse of the existing SCADA software running at the control center. The hybrid SE has been extensively used by the researchers to exploit the advantages of the PMUs. However, the presence of the PMUs may cause ill-conditioning, which results in inaccuracy in the estimates and in extreme cases, it may diverge also. This chapter presents an analysis for the inclusion of

PMU current phasor in different form with uniform distribution of measurement errors. From the results, it is concluded that the pseudo-branch-flow measurements derived from the PMU current measurements eliminates the high weights corresponding to the low values of current measurements, resulting in improvement of the convergence as well as the accuracy.

A number of methods are available to solve the numerical problem associated with the hybrid state estimation. The numerical problem mainly appears from an ill-conditioned information matrix, which may further escalate into divergence. A regularized SE is presented in this chapter to solve the ill-conditioned hybrid state estimation problems. The computational effort with the regularized SE is on the higher side, compared to the conventional methods. However, conventional methods often fail to converge due to numerical ill-conditioning. The SE algorithm in the control center may be equipped with this regularized method along with existing SE methods. Whenever some convergence issues are encountered, the SE program can switch to the regularized method to obtain a reasonable solution.

The SE algorithm may also be employed for transmission system as well as distribution system. In addition, for a large power system network, the computational effort for this type of single-area SE may be high. To resolve this issue, the control area is usually divided into a number of subareas, and each area has its own SE using local measurements. The estimates from these subareas are combined at the unified control center. This type of estimator is known as the multiarea SE (MASE). Many researchers have shown interest in analyzing different aspects of the MASE [40–42].

To simplify the static SE process, the different type of measurements are processed at different level. At the substation level, a local distributed linear estimation is performed, and then, the output of the first-level estimator, comprising of state variables and associated variances, are passed in a centralized nonlinear estimator at the control center [43]. This two-level SE has the advantage in computation. Also, all measurements need not be transmitted to the control center and thus, relieving the congestion in communication bandwidth.

References

[1] F. C. Schweppe and J. Wildes, "Power system static state estimation. Part I: exact model," *IEEE Transactions on Power Apparatus and Systems*, vol. PAS-89, no. 1, pp. 120–125, 1970.

[2] F. C. Schweppe and D. B. Rom, "Power system static state estimation. Part II: approximate model," *IEEE Transactions on Power* Apparatus and Systems, vol. PAS-89, no. 1, pp. 125–130, 1970.

[3] F. C. Schweppe, "Power system static state estimation. Part III: implementation," *IEEE Transactions on Power* Apparatus and *Systems*, vol. PAS-89, no. 1, pp. 130–135, 1970.

[4] D. Bailey and E. Wright, *Practical SCADA for Industry*. Oxford: Elsevier, 2003.

[5] A. G. Phadke and J. S. Thorp, *Synchronized Phasor Measurements and Their Applications*. New York: Springer, 2008.
[6] S. Chakrabarti, E. Kyriakides, T. Bi, D. Cai and V. Terzija, "Measurements get together," *IEEE Power and Energy Magazine,* vol. 7, no. 1 (January), pp. 42–49, 2009.
[7] A. G. Phadke, "Synchronized phasor measurements a historical Overview," *Proc. IEEE PES Asia Pacific Transmission and Distribution Conference and Exhibition*, vol. 1, pp. 476–479, October 2002.
[8] *IEEE Standard for Synchrophasors for Power Systems*. IEEE Std. C37.118-2005, 2005.
[9] *IEEE Standard for Synchrophasor Measurements for Power Systems*. IEEE Std. C37.118.2-2011, 2011.
[10] K. E. Martin, D. Hamai, M. G. Adamiak, *et al.* "Exploring the IEEE Standard C37.118 2005 Synchrophasors for Power Systems," *IEEE Transactions on Power Delivery*, vol. 23, no. 4 (October), pp. 1805–1811, 2008.
[11] A. Abur and A. G. Expósito, *Power System State Estimation: Theory and Implementation*. New York: Marcel Dekker, 2004.
[12] A. Monticelli, *State Estimation in Electric Power Systems: Generalized Approach*. Kluwer: London, 1999.
[13] G. R. Krumpholz, K. A. Clements and P. W. Davis, "Power system observability – a practical algorithm using network topology," *IEEE Transactions on Power Apparatus and Systems*, vol. PAS-99, no. 4, pp. 1534–1542, 1980.
[14] R. R. Nucera and M. L. Gilles, "Observability analysis: a topological algorithm," *IEEE Transactions on Power Systems*, vol. 6, no. 2 (May), pp. 466–475, 1991.
[15] B. Gao and A. Abur, "A direct numerical method for observability analysis," *IEEE Transactions on Power Systems*, vol. 15, no. 2 (May), pp. 625–630, 2000.
[16] B. Gou, "Jacobian matrix based observability analysis for state estimation," *IEEE Transactions on Power Systems*, vol. 21, no. 1 (February), pp. 348–356, 2006.
[17] Q. Ding and V. A. Emesih, "A simple factorization based observability analysis and meter placement method," *International Journal of Electrical Power & Energy Systems*, vol. 29, pp. 729–737, 2007.
[18] M. C. de Almeida, E. N. Asada and A. V. Garcia, "Power system observability analysis based on gram matrix and minimum norm solution," *IEEE Transactions on Power Systems*, vol. 23, no. 4 (November), pp. 1611–1618, 2008.
[19] C. Solares, A. J. Conejo, E. Castillo and R. E. Pruneda, "Binary arithmetic approach to observability checking in state estimation," *IET Generation, Transmission and Distribution*, vol. 3, no. 4, pp. 1336–1345, 2009.
[20] R. E. Pruneda, C. Solares, A. J. Conejo and E. Castillo, "An efficient algebraic approach to observability analysis in state estimation," *Electric Power Systems Research*, vol. 80, pp. 277–286, 2010.

[21] ISO-IEC-OIML-BIPM: Guide to the Expression of Uncertainty in the Measurement, 1992.

[22] T. S. Bi, X. H. Qin and Q. X. Yang, "A novel hybrid state estimator for including synchronized measurements," *Electric Power Systems Research*, vol. 78, no. 8, pp. 1343–1352, 2008.

[23] "Evaluation of measurement data – supplement 1 to the 'guide to the expression of uncertainty in measurement' – propagation of distributions using a Monte-Carlo method," Joint Committee for Guides in Metrology, JCGM 101:2008, 2008.

[24] J. W. Gu, K. A. Clements, G. R. Krumpholz and P. W. Davis, "The solution of ill-conditioned power system state estimation problems via the method of Peters and Wilkinson," *IEEE Transactions on Power Apparatus and Systems*, vol. PAS-102, no. 10, pp. 3473–3480, 1983.

[25] R. Ebrahimian and R. Baldick, "State estimator condition number analysis," *IEEE Transactions on Power Systems*, vol. 16, no. 2 (May), pp. 273–279, 2001.

[26] A. Simoes-Costa and V. H. Quintana, "A robust numerical technique for power system state estimation," *IEEE Transactions on Power Apparatus and Systems*, vol. PAS-100, no. 2, pp. 691–698, 1981.

[27] C. E. Lin and S. J. Huang, "Integral state estimation for well-conditioned and ill-conditioned power systems," *Electric Power Systems Research*, vol. 12, pp. 219–226, 1987.

[28] G. N. Korres, "A robust algorithm for power system state estimation with equality constraints," *IEEE Transactions on Power Systems*, vol. 25, no. 3 (August), pp. 1531–1541, 2010.

[29] S. K. Mallik, S. Chakrabarti and S. N. Singh, "A robust regularized hybrid state estimator for power systems," *Electric Power Components and Systems*, vol. 42, no. 7 (April), pp. 671–681, 2014.

[30] M. Zhou, V. Centeno, J. S. Thorp and A. G. Phadke, "An alternative for including phasor measurements in state estimators," *IEEE Transactions on Power Systems*, vol. 21, no. 4 (November), pp. 1930–1937, 2006.

[31] S. Chakrabarti, E. Kyriakides, G. Ledwich and A. Ghosh, "Inclusion of PMU current phasor measurements in a power system state estimator," *IET Generation, Transmission and Distribution*, vol. 4, no. 10, pp. 1104–1115, 2010.

[32] S. K. Mallik, S. Chakrabarti, and S. N. Singh, "Improving the convergence characteristic of hybrid state estimation using pseudo measurement," *17th Power Systems Computation Conference (PSCC)*, Sweden, August 2011.

[33] V. Agarwal, "Total variation regularization and *L*-curve method for the selection of regularization parameter," *ECE 599*, University of Tennessee, Knoxville, Summer 2003.

[34] P. C. Hansen, "The *L*-curve and its use in the numerical treatment of inverse problems," in *Computational Inverse Problems in Electro Cardiology*, P. Johnston Ed. 2001, pp. 119–142, WIT Press, Moscow.

[35] B. Lewis and L. Reichel, "Arnoldi–Tikhonov regularization methods," *Journal of Computational and Applied Mathematics*, vol. 226, pp. 92–102, 2009.

[36] P. C. Hansen and D. P. O'Leary, "The use of the *L*-curve in the regularization of discrete ill-posed problems," *SIAM Journal on Scientific Computing*, vol. 14 (November), no. 6, pp. 1487–1503, 1993.

[37] http://www.ee.washington.edu/research/pstca, IEEE Test Systems.

[38] S. K. Mallik, S. Handa, S. Chakrabarti, and S. N. Singh, "Performance study of a regularized method for solving non-converging power system state estimation problems," in Proceedings of *Fifth International Conference on Power and Energy Systems*, Kathmandu, Nepal, October 2013.

[39] N. D. Rao and S. C. Tripathy, "Power system static state estimation by the Levenberg–Marquardt algorithm," *IEEE Transactions on Power Apparatus and Systems*, vol. PAS-99, no. 2 (March), pp. 695–702, 1980.

[40] T. Van Cutsem, J. L. Horward and M. Ribbens-Pavella, "A two-level static state estimator for electric power systems," *IEEE Transactions on Power Apparatus and Systems*, vol. PAS-100, no. 8, pp. 3722–3732, 1981.

[41] L. Zhao and A. Abur, "Multiarea state estimation using synchronized phasor measurements," *IEEE Transactions on Power Systems*, vol. 20, no. 2 (May), pp. 611–617, 2005.

[42] G. N. Korres, "A distributed multiarea state estimation," *IEEE Transactions on Power Systems*, vol. 26, no. 1 (February), pp. 73–84, 2011.

[43] A. G. Exposito and A. de la Villa Jaen, "Two-level state estimation with local measurement pre-processing," *IEEE Transactions on Power Delivery*, vol. 24, no. 2 (May), pp. 676–684, 2009.

Chapter 11
PMU-based wide-area security assessment

S.R. Samantaray[1], Innocent Kamwa[2] and Geza Joos[3]

11.1 Introduction

Dynamic security assessment (DSA) of a power system usually refers to the degree to which a particular system condition can withstand all credible contingencies considering the detailed dynamic characteristic of the system. Due to ever-increasing power demand exceeding the transmission and generation facilities, power systems are forced to operate with narrower margins of security. Security is defined as the capability of guaranteeing the continuous operation of a power system under normal operation even following some significant perturbations. As security is a major concern in power system, a fast and reliable security assessment is necessary. DSA can deal with transient stability problems and/or voltage stability problems that respectively require transient stability assessment and voltage stability assessment. Thus, predicting early warning of the deteriorating system conditions is one of the important requirements of the security assessment which enables the system operator to take corrective actions. The early warning systems should provide more diagnostic tools than the existing one which can allow effective use of automatic controls for self-correction including automatic switching or power flow control.

The defense plan against these rare contingencies is based on event detection [1] using breaker status and fault signals from relays in combination. This is because the more appealing response-based approach [2] is not fast enough for effective remedial actions. However, with recent developments in phasor measurement unit (PMU)-based wide-area measurement, fast-response-based stability assessment of extreme contingencies finds better prospects. This includes ranking the contingency (severity/stability) using online dynamic information which is potentially generalized and robust compared to the existing event-detection-based schemes alone relying heavily on off-line simulations of system conditions. The lack of transmission system expansion means that, in any case, current defense

[1]School of Electrical Sciences, Indian Institute of Technology, Bhubaneswar, Odisha-751 013, India
[2]Hydro-Québec/IREQ, Power System and Mathematics, Varennes QC J3X 1S1, Canada
[3]Department of Electrical and Computer Engineering, McGill University, McConnell, 633, 3480 University Street, Montreal, Québec, H3A 2A7, Canada

plans will reach their limit at a time when it may be appropriate to supplement them with more refined and context-sensitive wide-area response-based remedial actions or special protection systems.

Recently, a comprehensive time-frequency-based approach for contingency severity ranking and rapid stability assessment [3] has been proposed, and the aim was to assist the classical off-line simulations-based DSA task [4–12] by correctly classifying all single or multiple contingencies those may result in a loss of stability in the first 20 s following the fault-clearing. Here, the number of strategic monitoring buses is selected where the PMUs are placed to capture the representative wide-area voltage magnitudes and angles during real-time operation [13,14]. The STFFT [15,16] is then dynamically applied to the responses for extracting selected decision features as the postdisturbance time-frame undergoes significant changes. It is observed that the frequency-domain features such as the peak spectral density of the angle, the frequency and their dot product evaluated over the grid areas (referred to the system center-of-inertia) are reliable time-varying stability indicators which can form the basis of an entirely reliable classifier.

In the wide-area framework, there is an abundance of historical and real-time quality data generally collected through the PMU infrastructure [17]. Thus, in a data-centric smart-grid, converting data to information and corresponding controls is a fundamental step, triggering the development of new predictive analytical tools to predict the status of catastrophes following disturbances, or to quantify the relative status of the system from the edge of a safe operating boundary [12]. However, developing predictive models from historical- or simulated-PMU records is no small task because power system responses are highly nonlinear, and therefore very context-dependent. Although data-mining technique is one of the emerging tool to perform the task of synthesizing such massive data bases into predictive models; however, they need thousands of configurations and instances to minimally capture the intrinsic characteristics of the underlying power grid dynamics [18].

Amongst the data-mining tools, decision tree (DT) has gained lots of momentum in power engineering applications. Most high-accuracy DTs are opaque (they contain a large number of deep terminal nodes), and it is not possible for a human to understand the logic behind the prediction which is mandatory for stability control and power system protection. This is to note that if simpler and transparent models are built, the accuracy is often sacrificed, resulting in the accuracy vs transparency trade-off. In this context, two competing data-mining trends are emerging: (1) the fuzzy-logic rules-based approach [19–22], which has the advantage of transparency and interpretability [23] and (2) the machine learning-based black-box approach which relies on several complementary tools from statistical learning such as neural network [3,10,24], support vector machine (SVM) [25,26], DT [2,12,27,28], and ensemble DTs (random forests (RFs)) [18,29]. All of the aforementioned data-mining tools have strengths and weaknesses in terms of computational burden, training time, accuracy, robustness to noise, etc. Normally, they are selected purely on the basis of their immediate availability to the user or prior familiarity with them.

Thus, trade-off between the choice of a powerful model with high performance (i.e. a model with high accuracy) and a transparent model (i.e. a model with high comprehensibility) needs to be discussed. To address this trade-off in the context of PMU data analysis for wide-area stability assessment and control, we will consider three black-box models [30] (neural net (NNET), SVM, RF) and two transparent (comprehensible models) [31] (DT, DT-induced Fuzzy Approach (DT_Fuzzy)) for catastrophe predictor. For the high performance black-box models, the RF has been found to be the most accurate model while SVM is the second [30]. Regarding the transparent model family, the DT_Fuzzy combines the strength of DT and fuzzy-logic to form a systematic fuzzy logic and is emerging as the most transparent predictive model.

11.2 System studied using wide area monitoring

The Hydro-Québec (HQ) 783-bus system using PMU base wide-area monitoring shown in Figure 11.1 is used for operations planning [32]. For the present study, winter and summer operation planning models are used with about 1,000 load flow patterns generated by the transfer limit search and critical clearing time search processes based on 30 carefully chosen 735-kV contingencies.

The HQ network is divided into a number of electrically coherent areas [33] associated with weak intertie lines or systematically identified stability boundaries [34]. In the shown representation, these intertie lines are the primary clusters of the system, defining the finite number of possible ways to split the system into unconnected islands with embedded generation. Further, selecting an optimum PMU set allocated for monitoring using a sequential addition algorithm to expand their number, while maximizing the amount of information added by each new PMU [13]. In this study, the busses with PMU are only considered for monitoring the wide area postdisturbance records. The system is configured in such a way that the simulation will be smoothly replaced by actual measurements in real time situation considering the same PMU-based wide-area monitoring system.

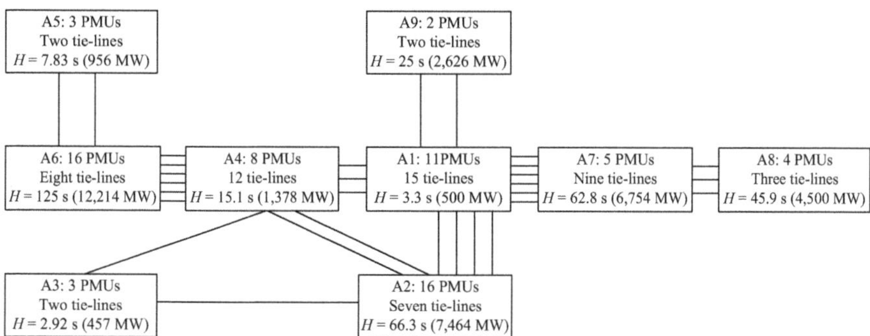

Figure 11.1 Hydro-Québec system (783-bus) with intraarea and intertie PMUs distributed over nine electrically coherent areas (H-inertia and MW-generation)

11.3 Wide-area severity indices

11.3.1 WASI features

The wide-area severity indices (WASI) are defined by an equivalent inertia associated with each area, representing the total inertia of the generation located in that area. It is reasonable to consider its behavior to that of a single large machine with the same inertia, and generation assuming each area is coherent following a disturbance. Even though this assumption is not perfect, it offers a straightforward means of deriving the center-of-inertia (COI), which is very useful information for tracking the stability of interconnected areas. In real time, a defense plan or an special protection scheme (SPS) [6,32,35] could readily derive these inertia constants through low-speed communication with the control center state-estimator, which holds the actual load and generation dispatch.

The voltage and COI angle deviation are the time-domain features considered in the present study and are the key descriptor of the power grid stability state. To determine the level of system stress induced by a contingency, the most obvious criterion is the transient stability after a single or multiple swings of the generator angles. The proposed study considers the system as unstable if there is a loss of synchronism between any of the areas monitored earlier in the prescribed observational time-frame, i.e. the angle shift exceeds 180°. When an early discovery of impeding instability is sought, this criterion is useless because it offers little pre-emption time. The COI angle deviation with respect to the prefault value is computed for each area, and the maximum value of these differences is found to be a good measure of the topological stress induced by the contingency.

It is observed that the frequency-domain features such as the peak spectral density of the product of angle and frequency evaluated over the grid areas and referred to the system COI are reliable time-varying stability indicators that can form the basis for assessing contingency severity [3]. Here, the WASI concept is extended to rapid stability assessment by using simultaneously a short (16-cycle) and long (192-cycle) short-time-Fourier-transform (STFT) analysis window [15,16], which is responded fast enough for the first swing instability as well as the postinertial and midterm dynamic instability [10]. Further, multisignal state-space-based modal analysis [36] of the output of the long filters is shown is highly useful for assessing the postcontingency damping condition of the grid [37].

Following a fault (which is automatically detected by the PMUs), the computation sequence follows is as follows:

1. Compute the pilot phasor and frequency of each area by averaging the within-area measurements: $\overline{V}_i$, $\overline{\omega}_i$
2. Compute the system COI of the angle and frequency variables using the available area inertias (M_i)
3. Project the pilot angle and frequency of step (1) into the COI reference of step (2): $\overline{\theta}_i^{COI}$, $\overline{\omega}_i^{COI}$
4. Compute the shift from the prefault to postfault COI-angle for each area: PostFltAngle
5. Compute the power spectrum of the dot-product of energy and angle to obtain a tracking severity index by area.

From the above algorithm, WASI in time and frequency domains are defined as follows [22]:

V min *Ts*	System-wise minimum voltage over the time span of $T = 1$ s or $T = 2$ s after fault-clearing
V min *Ts j*	Area-wise minimum voltage over the time span $T = 1$ s or $T = 2$ s following fault-clearing, considering only the buses in area *j*
V min *TsR*	Area-wise minimum voltage over the time span of $T = 1$ s or $T = 2$ s after fault-clearing, considering only the buses in the faulted area
FastWASI(*Ts*)	System-wise frequency-domain severity index defined over the time span of $T = 1$ s or $T = 2$ s after fault-clearing
FastWASI(*Ts*) *j*	Area-wide frequency-domain severity index for area *j*, defined over the time span of $T = 1$ s or $T = 2$ s after fault-clearing
FastWASITsR	Area-wise frequency-domain severity index for the faulted area, defined over the time span of $T = 1$ s or $T = 2$ s after fault-clearing
VLowPass2s j, *VCriterion2s j*	Filtered area-wise minimum voltage
VLowPass2s, *VCriterion2s*	Filtered system-wise minimum voltage
TDEF	Fault duration
PostFltAngle	System-wise maximum COI angle deviation from steady-state to fault-clearing time
PostFltAngle j	Area-wide maximum COI angle deviation from steady-state to fault-clearing time, for area *j*
*TsimSt*600	Duration of simulation up to normal end or loss of synchronism, whichever occurs first.

The schematic diagram for computing frequency-domain severity indices is shown in Figure 11.2. It is observed that for transient-stability assessment short window lengths are the most suitable. The 16-point Hamming window is used in the present study.

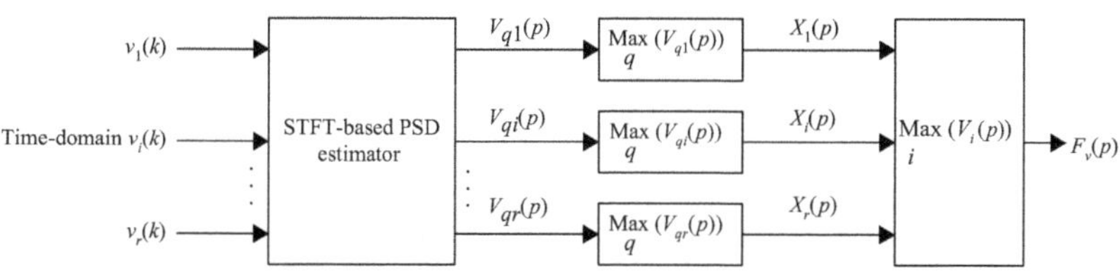

Figure 11.2 Schematic of computing frequency-domain-based WASI feature (number of areas $r = 1, \ldots, 9$)

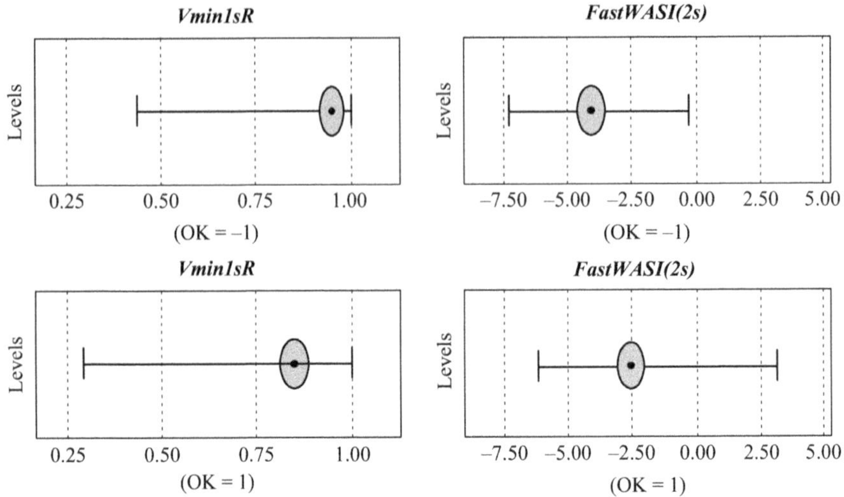

Figure 11.3 Vmins1R and FastWASI (2s) for stability limits

11.3.2 Wide-area severity indices and stability condition

The relationship between WASI features and stability conditions are depicted in Figure 11.3. It shows that the stability limits with respect to the voltage-based *Vmin1sR* (which is the 1 s voltage minimum over the faulted area) and the energy-based *FastWASI (2s)* (the system-wise maximum value of the dot-product power spectrum density (PSD)). These plots are categorized according to the stability condition of the case as: OK = –1 (stable) and OK = 1 (unstable) and defines 50% of the whole sample. The left limit of the box defines the first quartile of the data, whereas the right limit corresponds to the third quartile. It is observed that $Vmin1sR > 0.91$ includes 75% of stable cases, and $Vmins1R < 0.90$ includes 75% of unstable cases. Further, the relationship $FastWASI2s \leq 3.2$ includes 75% of stable cases, whereas $FastWASI2s \geq 3$ includes 75% unstable cases. Thus, *FastWASI2s* is able to split the stability domain into two largely disjointed sets.

11.4 Data-mining models

11.4.1 Decision tree

DT [38,39] is an effective supervised data-mining tool for solving the classification problem in high dimension. Each internal node in the tree tests the value of a predictor, while each branch of the tree represents the outcome of a test. The terminating nodes, also referred to as leaf nodes, represent a classification. The number of predictors used in the classification problem indicates the dimension of the problem. Associated with each decision (leaf) of the tree is the confidence of the decision. This is simply a measure of the ratio of the particular class to all the classes present in the dataset for that node.

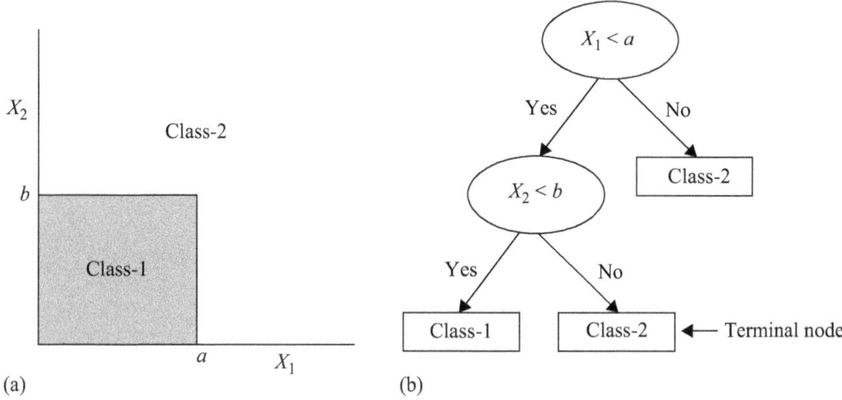

Figure 11.4 (a) Two-dimensional binary classification and (b) DT for binary classification

Any binary DT T is a collection of nested binary partitions, denoted here by a quintuple (u, v, s, p^L, q^R), which is represented in the following recursive form [40,41]:

$$T = \{(u, v, s, p^X, p^Y), T^X, T^Y\} \tag{11.1}$$

where u denotes a decision node label for the partition, v is the feature axis, s is the threshold value used for the partition, and p^X and p^Y are the node labels for the partition of the left and right sets, respectively. T^X and T^Y denote the subtrees defined on the left and right sets of a partition. In summary, (11.1) defines the DT T in terms of the pattern lattice L created by partitioning the features plane. The equation states that the lattice L can be binary-partitioned on the feature axis v into mutually exclusive left and right sets. The left set includes lattice elements with feature v values smaller than the threshold value, whereas the right set includes lattice elements with feature v values exceeding the threshold value. The example of two-dimensional binary classification is shown in Figure 11.4(a), and the corresponding DT is shown in Figure 11.4(b).

A common measure of impurity of *the predictive model T is the entropy, which is defined* for the subset T_k of all possible states of the prediction model T by the following equation [22,41]:

$$i(k) = -\sum_{S} p(S|T_k)\mathbf{log}_2 p(S|T_k) \tag{11.2}$$

The information carried by the prediction model T with K subset data models is then the weighted average of the entropies, given by (11.2). Thus, the impurity of the prediction model T is given by:

$$I(T) = \sum_{k=1}^{K} i(T_k)p(T_k) \tag{11.3}$$

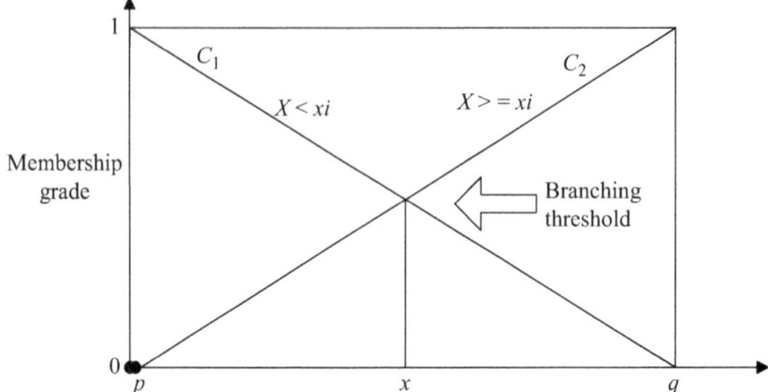

Figure 11.5 DT transformation to fuzzy MFs

Assessment of the entropy reduction ascribed to each subset model T_k provides the importance ranking of independent features toward the model total impurity.

11.4.2 DT-induced fuzzy approach

The sharp boundaries used by the crisp DT will cease to exist [42,43] when the fuzzy theory is applied to the individual tree branches, and the partitions become a series of fuzzy regions. The objective is to relax these sharp decision boundaries by introducing fuzzification onto the DT. This involves in introducing a pair of membership functions at each decision node, and the decision space will then become fragmented with a number of overlapping fuzzy regions [44,45]. In order to classify a particular case, all branches will fire to some degree. The degree of membership at each branch for a given attribute value depends upon the margin of the fuzzy set for that particular branch. A fuzzy region around any decision node in a crisp DT is defined using a pair of complementary membership functions C_1 and C_2 around a decision threshold x (Figure 11.5). For example, a 5-node tree would be represented by ten membership functions. Figure 11.5 illustrates two complementary membership functions over the domain $p \ldots . q$ of attribute i, where x_i represents the branching threshold of attribute i. A specific value m of attribute i, passing through the tree, would then be assigned a membership grade based on its proximity to x_i. Each membership function will have an associated domain (p_i, q_i) whose scope is determined by the attribute at each specific branch, which is defined as follows [46]:

$$p_i = x_i - n_j \sigma_i \tag{11.4}$$

$$q_i = x_i + n_j \sigma_i \tag{11.5}$$

where σ_i is the standard deviation of attribute i, n is a real number, $n \rightarrow [0.0, \infty]$ is used to determine the effect of σ_i on the membership function domain, and p and q are the lower and upper bounds of membership function, respectively. In practice,

"n" will remain typically small, $n \rightarrow [0.0, 5]$. This is because, the large value of n would introduce too much fuzzification onto the tree, and the process of making the decision would become too vague.

The DT is transformed into a fuzzy rule base by developing the fuzzy membership functions from the partition boundaries of the DT. After a pair of complementary fuzzy MFs has been developed at each node, the fuzzy IF-THEN rules are framed considering each node and leaf of the DT for final decision-making. The crisp output is obtained after a defuzzyfication process [23].

11.4.3 Ensemble decision trees

Ensemble DTs [38] (also known as RF) are a large combination of decorrelated tree predictors such that each tree depends on the values of a random vector sampled independently. Individual trees are noisy and unstable compared to the ensemble DTs as it grows to sufficiently deep level having relatively low bias. Thus, ensemble DTs are the ideal candidates for ensemble growing as they can capture complex interactions, while fully benefit from aggregation-based variance reduction. Using a random selection of features to split each node and resampling (with replacement) the training set to grow each tree yields error rates that are decorrelated and more robust with respect to noise. The generalization error for forests converges as to a limit as the number of trees in the forest becomes large.

The ensemble tree growing procedure [18] is that for the kth tree ($k \le n_{\text{tree}}$, the number of trees in the ensemble), a random vector $\mathbf{\Phi}_k$ is generated, independent of the past random vectors $\mathbf{\Phi}_k, \ldots, \mathbf{\Phi}_{k-1}$ but with the same distribution. A single tree is grown using the training set $\mathbb{S}$ and the set of attributes in $\mathbf{\Phi}_k$, resulting in a classifier $T_k(x, \mathbf{\Phi}_k)$, where x is an input vector. In random split selection, $\mathbf{\Phi}$ consists of a number n_{try} of independent random integers, where $n_{\text{try}} < n_a$, the number of attributes in $\mathbb{S}$. A RF consists of a collection of tree-structured classifiers $\{T_k(x, \mathbf{\Phi}_k), k = 1, \ldots, n_{\text{tree}}\}$, where $\{\mathbf{\Phi}_k\}$ are independent identically distributed random vectors, and each tree casts a unit vote for the most popular class at input x.

The algorithmic view the RF growing process is summarized below:

1. For $k = 1$ to n_{tree}:
   (i) Draw a boostrap sample $\mathbb{S}^*$ of size N from the training data $\mathbb{S}$ (which contains $M > N$ samples)
   (ii) Grow a RF tree $T_k(x, \mathbf{\Phi}_k)$ to the boostrapped data by recursively repeating the steps below for each terminal noted of the tree, until the no other split is possible (unpurned tree of maximal depth):
      (a) Select n_{try} variables from the n_a WASI features
      (b) Pick the best variable/split-point among the n_{try}
      (c) Split the node into two daughter nodes
2. Output the ensemble of trees $\{T_k(x, \mathbf{\Phi}_k), k = 1, \ldots, n_{\text{tree}}\}$.

Although RF is a relatively young data mining tool, scholars [31,47] have started recognizing its strengths: (1) it is simple and easy to use, (2) very high accuracy,

(3) its relatively robust to outliers and noise, (4) it gives useful internal estimates of error, strength, correlation, (5) not over fitting if selecting large number of trees, and (6) insensitive to choice of split.

11.5 Data-mining model–based catastrophe predictors

11.5.1 Scenarios and data count generation

The model studied in the proposed approach includes combined HQ and test as shown in the schematic diagram (Figure 11.6). The data sets studied with different scenarios are trained with extended data files and tested with performance data files. The extended data files are generated after extending the actual data sets triplicating the unstable cases in each scenario. The complete statistics of the scenarios studied are summarized in Table 11.1.

11.5.2 Data-mining models for catastrophe predictor

The catastrophe predictors proposed in this study use open-source software R [40] which includes the implementation of conventional DTs, RFs, NNET, and SVM. The training is carried out using the extended data files separately for each scenario, and the testing is performed using the performance files (flat files). The data-mining models were configured to randomly select 70% only of the assumed data file to build the model during training. DT_Fuzzy is developed from the decision nodes of the DT. The NNET-based predictor has a feed-forward structure with ten hidden layers. Similarly, the SVM is set with a Radial Basis Kernel. The DT is trained with the "R" package, which is called with the following settings: convergence error of 10^{-4}, with a minimum bucket, before and after a node split, equal to 600 and 300 samples, respectively.

11.5.3 Performance assessment

The performance assessment is carried out using various statistical indices defined as follows [3,8]:

1. *Reliability:* (Total number of unstable cases – total number of cases converted to stable cases)/Total number of unstable cases
2. *Security:* (Total number of stable cases – total number of cases converted to unstable cases)/Total number of stable cases
3. *Accuracy:* (Total numbers of cases – number of mis-classification)/Total number of cases

11.5.3.1 Decision tree

The study considers 70% of the data for training, the rest unseen data being reserved for testing. The simulation carried out on MATLAB®/SIMULINK® platform and the training took less than 22 s on a 2-GHz Centrino duo Laptop. The DT generated during the training process for 1 s decision time is shown in

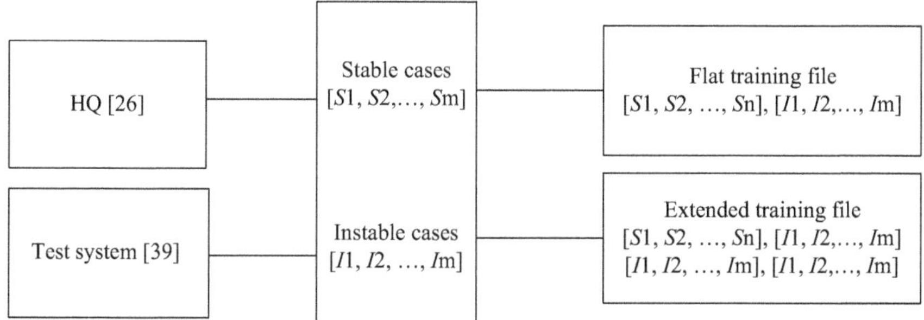

Figure 11.6 Schematic of the data-count generation process

Table 11.1 Summary count of the data sets used in the study

Model	Extended data file	Flat data file
System-wise		
HQ_1s	**93,425**	**55,196**
Stable	42,453	42,453
HQ_2s	**93,425**	**55,196**
Stable	42,453	42,453
Area_wise_1s		
BJN_1s	**16,434**	**10,647**
Stable	8,718	8,718
BJS_1s	**24,725**	**16,766**
Stable	14,113	14,113
CHU_1s	**12,440**	**8,684**
Stable	7,432	7,432
MQ_1s	**39,826**	**19,099**
Stable	12,190	12,190
Area_Wise_2s		
BJN_2s	**16,434**	**10,647**
Stable	4,680	8,718
BJS_2s	**24,725**	**16,766**
Stable	14,113	14,113
CHU_2s	**12,440**	**8,684**
Stable	7,432	7,432
MQ_2s	**39,826**	**19,099**
Stable	12,190	12,190

Figure 11.7. While observing the relative importance of the candidate features which played a role in the classification process, it is found that the *Vmin1sR* is the most important feature based on total entropy reduction. Similarly, for other features, the importance of candidate features follows the descending order. Similar observations are made for the DT generated for 2-s decision time with same training strategy, where *FastWASI2s* turned out to be the most relevant feature. An

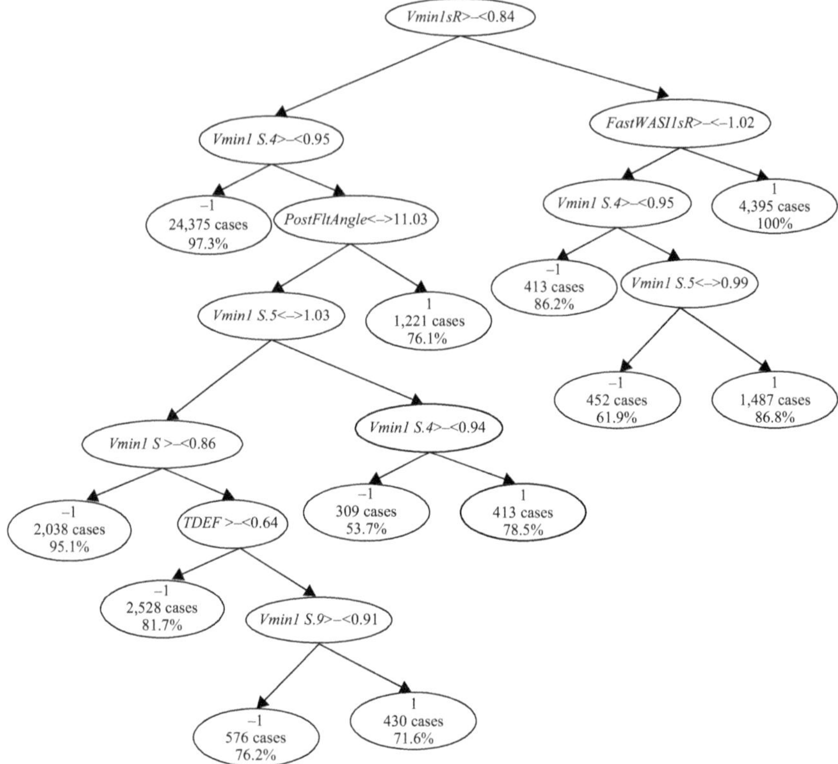

Figure 11.7 DT generated with training sample of 70% the complete dataset for 1-s decision time

attempt was made to improve these results by developing separate classifiers according to the geographic area where the fault occurred. This led to four different area-wise DTs as follows, each for $T = 1$ and $T = 2$ s:BJN: faults in the northern James Bay area (areas 5 and 6 in Figure 11.1), BJS: faults in the southern James Bay (area 4), CHU: faults in the Churchill Falls area (area 8), and MQ: faults in the Manic-Québec area (area 7).

The performance results for flat data file with single DT are shown in Figure 11.8. It is observed that the reliability for HQ_1s and HQ_2s are 78.46% and 76.70%, respectively, while the accuracy and security are above 92.0% and 96.0%, respectively, as shown in Figure 11.8(a). However, the reliability is more than 85% in BJN, BJS, and CHU in cases of 1- and 2-s early termination for area-wise assessment as shown in Figure 11.8(b) and (c), respectively. It is found that reliability for Manic-Québec (MQ) is less than 80% considering 1- and 2-s early termination. Thus, the reliability stays at low compared to security and accuracy for flat data sets. To observe the improvement further, the single DT has been trained and tested with extended data file. Figure 11.9(a) shows the results for extended data sets with single DT. It is observed that the reliability improves to 89.90% and

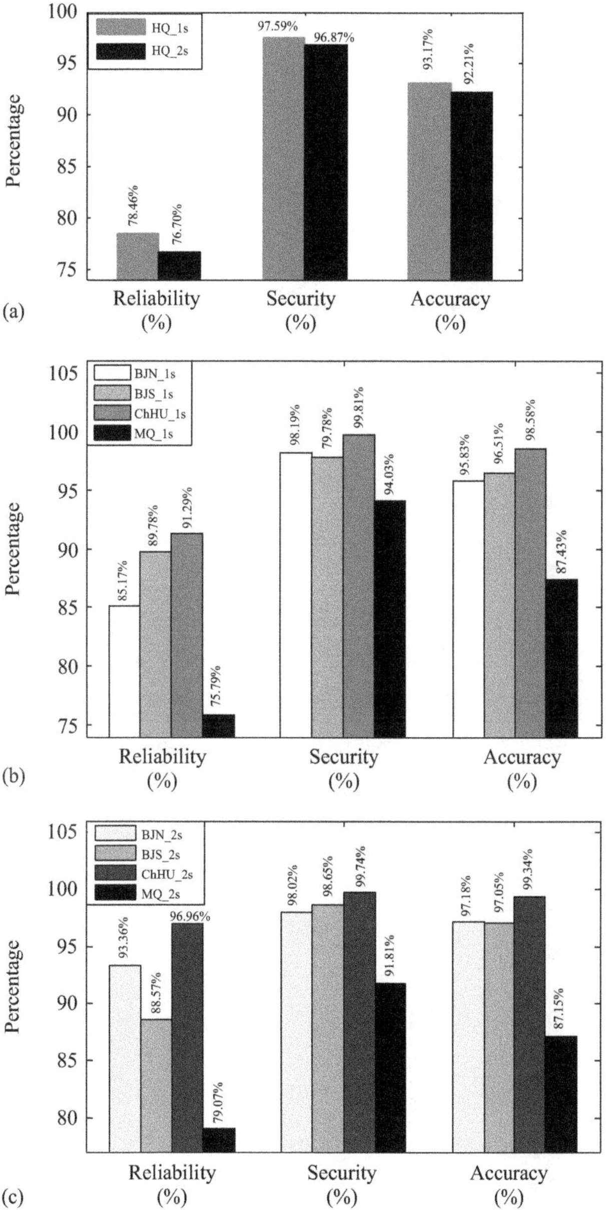

Figure 11.8 (a) Performance of DT for HQ_1s and HQ_scenarios, (b) performance of DT for area-wise scenarios for 1 s (flat data), and (c) performance of DT for area-wise scenarios for 2 s (flat data)

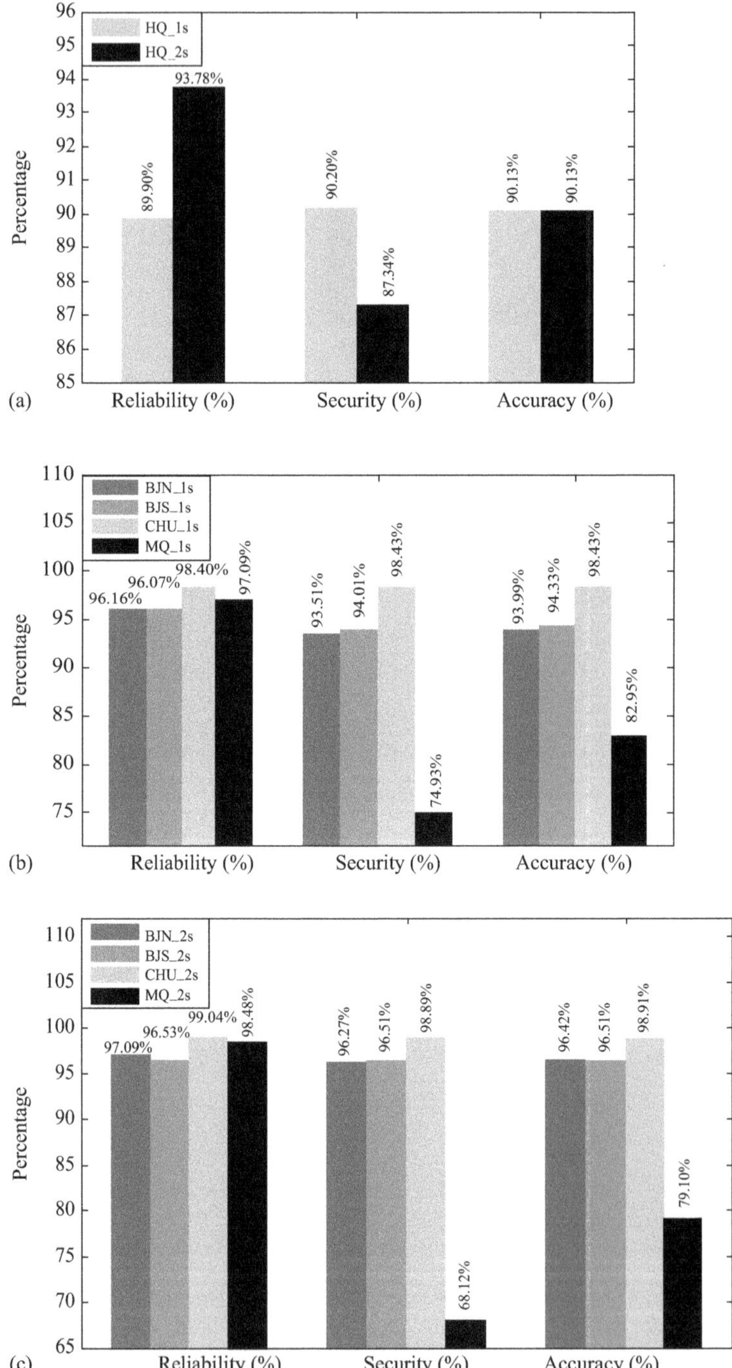

Figure 11.9 (a) Performance of DT for HQ_1s and HQ_scenarios, (b) performance of DT for area-wise scenarios for 1 s, and (c) performance of DT for area-wise scenarios for 2 s (extended data)

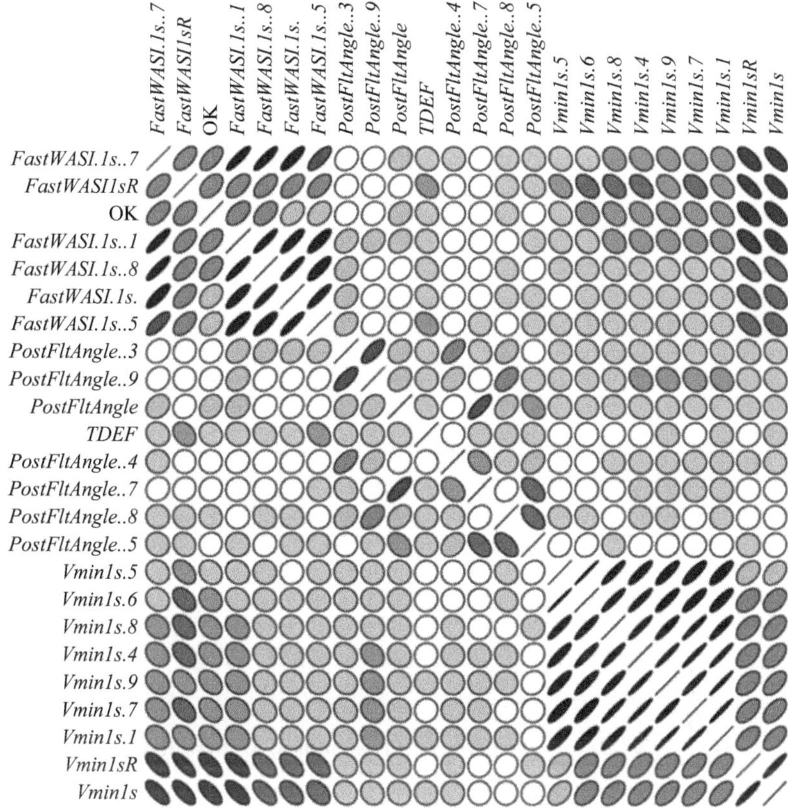

Figure 11.10 Correlation between candidate features. © 2010 IET. Reprinted, with permission, from Reference [50].

93.78% for HQ_1s and HQ_2s, respectively. Though there is a little sacrifice in security (90% from 97%) and accuracy (90% from 93%), but reliability is the important measure compared to security and accuracy in stability assessment. There is also a substantial jump in reliability in case of BJN, BJS, CHU, and MQ crossing 96% considering both 1- and 2-s early termination as shown in Figure 11.9(b) and (c).

11.5.3.2 Ensemble decision trees (random forests)

The classification accuracy mostly depends on the quality of attributes describing the security concept. The correlation between the decision (OK) and the attributes for HQ_1s resulted from RF during training is Figure 11.10. Highly correlated variables are close together and presented in the same intensity: frequency domain features on one side and voltage features on the other side, with *TDEF* in the middle. We interpret the degree of any correlation by both the shape and intensity of the graphic elements [48]. Any variable is perfectly correlated with itself, and

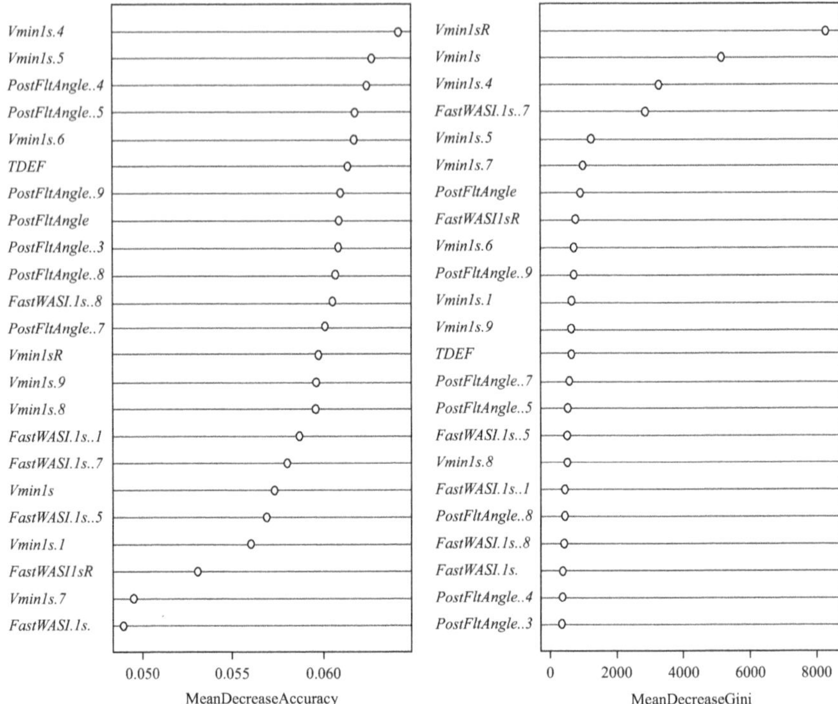

Figure 11.11 Top-down importance of the variables. © 2010 IET. Reprinted, with permission, from Reference [50].

this is reflected as the diagonal lies on the graphic. The graphic element is a perfect circle for the cases where there is no correlation between the variables (as is the case in the correlation between *PostFaultAngle.3, PostFaultAngle.9, and PostFaultAngle.7*).

The intensity is used to shade the circles give another clue to the strength of the correlation. The intensity is maximal for a perfect correlation and minimal (white) if there is no correlation. This provides the information on the high relevance of the energy-based features like WASI for explaining the security concept based on PMU measurements. *FastWasi1s.7* is the variable most positively correlated to OK, whereas *Vmin1s* is the most negatively correlated to OK (largest ellipses in both cases). This complementary behavior is highlighted by the variable importance analysis results from the RF learning in Figure 11.11.

The performance test results for both system-wise and area-wise for rapid stability assessment are shown in Table 11.2, which is really exciting [49,50]. Here, the extended data file is used to train the RF, where the training occurs for larger number of tree generations instead of single DT. It is observed that the reliability becomes 99.93% and 99.96% in case of HQ_1s and HQ_2s, respectively.

Table 11.2 Reliability, security, and accuracy for extended data sets with random forest

Early termination	Reliability (%)	Security (%)	Accuracy (%)
System-wise			
HQ_1s	99.93	98.92	99.16
HQ_2s	99.96	99.01	99.23
Area-wise_1s			
BJN_1s	99.84	99.32	99.47
BJS_1s	99.92	99.44	99.51
CHU_1s	100	99.91	99.93
MQ_1s	99.94	97.62	98.46
Area-wise_2s			
BJN_2s	99.89	99.57	99.63
BJS_2s	99.92	99.76	99.79
CHU_2s	100	99.94	99.95
MQ_2s	99.97	97.75	98.55

Area-wise study provides similar performance as depicted in Table 11.2. It is observed that the reliability is more than 99% (except BJS, it is 99.9% in all cases), considering both 1- and 2-s early termination. It is also found that there is substantial increase in security and accuracy in system-wise and area-wise study. The security and accuracy are around 99% for HQ_1s and HQ_2s system-wise as well as area-wise (BJN_1s and 2s , BJS_1s and 2s). Only in case of MQ, the security and accuracy are around 97.5% and 98%, respectively.

From the aforementioned result analysis, it is observed that the reliability achieved is 99.9% (3 nine) using RF compared to single DT. Our analysis of recent DT applications to DSA results 80%–85% reliability [51] on system-wise study, which is definitely unsatisfactory, giving that modern electric power systems are normally designed and operated to meet a "3 nines" reliability standard [52]. Even if a DT predictor is easily interpretable and transparent for the user compared to RFs, however, the reliability and security measures are substantially low to be accepted for the critical issues such as DSA. The RF not only provides high jump in reliability, but also the security and accuracies are around 99%, considering system-wise and area-wise security assessment for 1- and 2-s early termination.

11.5.3.3 DT_Fuzzy

The DT_Fuzzy is derived from the partitioning boundaries of the decision node of the DT. From the boundaries of the system-wise and area-wise classification, the fuzzy MFs of each and every feature involved in both the 1 and 2-s response time DTs are developed. The corresponding fuzzy rule base is also developed considering the respective fuzzy MFs and looking at the original DT structure. The Mamdani Model with Centroid defuzzification is used for implementing the developed DT-induced fuzzy rule base. The fuzzy rule base developed for 1-s early termination is as follows:

Rule base for 1-s early termination decision:

R-1: If X_1 is P_{11} and X_2 is Q_1, then "-1" (stable).

R-2: If X_1 is P_{11} and X_2 is Q_{11} and X_3 is R_1, then "1" (unstable).

R-3: If X_1 is P_{11} and X_2 is Q_{11} and X_3 is R_2 and X_4 is S_{11} and X_5 is T_2, then "-1" (stable).

R-4: If X_1 is P_1 and X_2 is Q_{11} and X_3 is R_2 and X_4 is S_{11} and X_5 is T_2 and X_6 is U_1, then "-1" (stable).

R-5: If X_1 is P_5 and X_2 is Q_{11} and X_3 is R_2 and X_4 is S_{11} and X_5 is T_2 and X_6 is U_1, then "-1" (stable).

R-6: If X_1 is P_1 and X_2 is Q_9 and X_3 is R_2 and X_4 is S_{11} and X_5 is T_2 and X_6 is U_2, then "-1" (stable).

R-7: If X_1 is P_1 and X_2 is Q_{11} and X_3 is R_2 and X_4 is S_{11} and X_5 is T_2 and X_6 is U_2, then "-1" (stable).

R-8: If X_1 is P_1 and X_2 is Q_{11} and X_3 is R_2 and X_4 is S_7 and X_5 is T_2 and X_6 is U_2 and X_7 is V_1, then "-1" (stable).

R-9: If X_1 is P_1 and X_2 is Q_{11} and X_3 is R_2 and X_4 is S_7 and X_5 is T_2 and X_6 is U_2 and X_7 is V_2, then "1" (unstable).

R-10: If X_1 is P_1 and X_2 is Q_{11} and X_3 is R_2 and X_4 is S_4, then "1" (unstable).

R-11: If X_1 is P_1 and X_2 is Q_{10} and X_3 is R_2 and X_4 is S_4, then "-1" (stable).

R-12: If X_1 is P_4 and X_8 is W_2, then "1" (unstable).

R-13: If X_1 is P_4 and X_2 is Q_1 and X_8 is W_1, then "-1" (stable).

R-14: If X_1 is P_4 and X_2 is Q_{11} and X_4 is S_{11} and X_8 is W_1, then "-1" (stable).

R-15: If X_1 is P_4 and X_2 is Q_{11} and X_4 is S_6 and X_8 is W_1, then "1" (unstable).

where $X_1 = Vmin1sR$, $X_2 = Vmin1s.4$, $X_3 = PostFaultAngle$, $X_4 = Vmin1s.5$, $X_5 = Vmin1s$, $X_6 = TDEF$, $X_7 = PostltAngle.8$, $X_8 = FastWASI1sR$. Similar framework for building fuzzy rule base for 2-s early termination is also carried out for classifying stable cases from unstable cases.

The performance of the proposed DT_Fuzzy predictor is shown in Figure 11.12(a) for 1- and 2-s early termination. It is observed that the reliability improves with fuzzy-rule base formulation, which reflects the robustness of the proposed fuzzy rule base. Similar performance results for the area-wise study are shown in Figure 11.12(b) and (c) for 1- and 2-s early termination time decision-making, respectively. As expected, the overall accuracy of the area-wise classifiers is better compared to the system-wise classifiers in all scenarios except for the MQ area, which connects directly to the major load center of Québec City. Therefore, most contingencies with a line outage under heavy load resulted in voltage collapse or in an inadmissible transient voltage profile. While the latter situation is well covered by the present severity indices-based method, the former is definitely not in its scope and should be managed by other complementary means. Further, it is

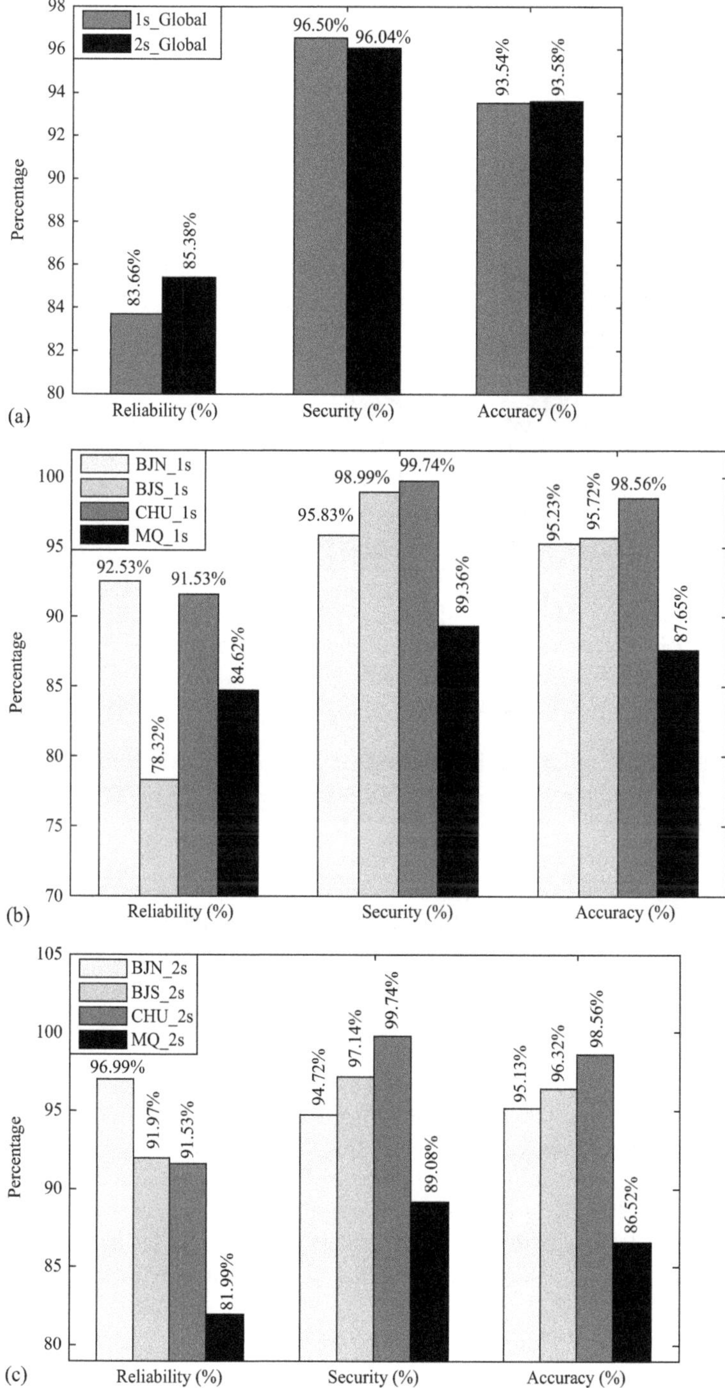

Figure 11.12 (a) Performance of DT_Fuzzy for 1s_Global and 2s_Global scenarios, (b) performance of DT_Fuzzy for area-wise scenarios (1 s), and (c) performance of DT_Fuzzy for area-wise scenarios (2 s)

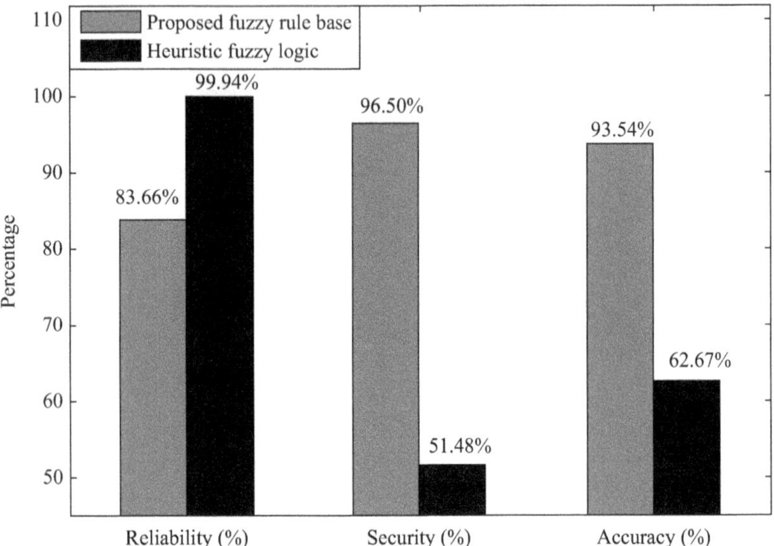

Figure 11.13 Performance comparisons between proposed DT_Fuzzy and heuristic fuzzy for 1s_Global

observed that the reliability improves even after fuzzy-rule base simplification reflecting the robustness of the proposed fuzzy-rule base approach.

The performance of the proposed DT_Fuzzy is compared with heuristic fuzzy logic classifiers for two system-wise scenarios which are hand-developed after a lengthy and tedious process [3] as shown in Figure 11.13. The heuristic fuzzy logic classifiers share the same 15 inputs selected from the pool of best candidate features. They are combined in 16 and 10 rules for 1 and 2 s response-time decision-making, respectively. It is found that the fuzzy-rule classifiers developed automatically from the DTs provide a substantially better security and overall accuracy compared to the heuristic fuzzy classifiers. It is found that the yield of the proposed automated fuzzy rule bases exceeds 93% compared to 51.48% and 64.21%, respectively, with the heuristic fuzzy logic for 1- and 2-s response times decision-making. However, a key point is that the reliability of the heuristic method is close to 100%. The responsible factor for low-performance indices in case of heuristic fuzzy classifier is the classifier design process which is heuristic in nature. This example demonstrates that when reliability is a prime factor, the automated DT-based solution may need additional fine-tuning to meet requirements.

11.5.4 Accuracy vs transparency trade-off

The performance of the different data-mining tools has been discussed extensively in previous sections. The complete process for building the data-mining models for catastrophe predictor is summarized in Figure 11.14. Data mining for predictive analytics starts invariably with the adoption of a process of data collection and

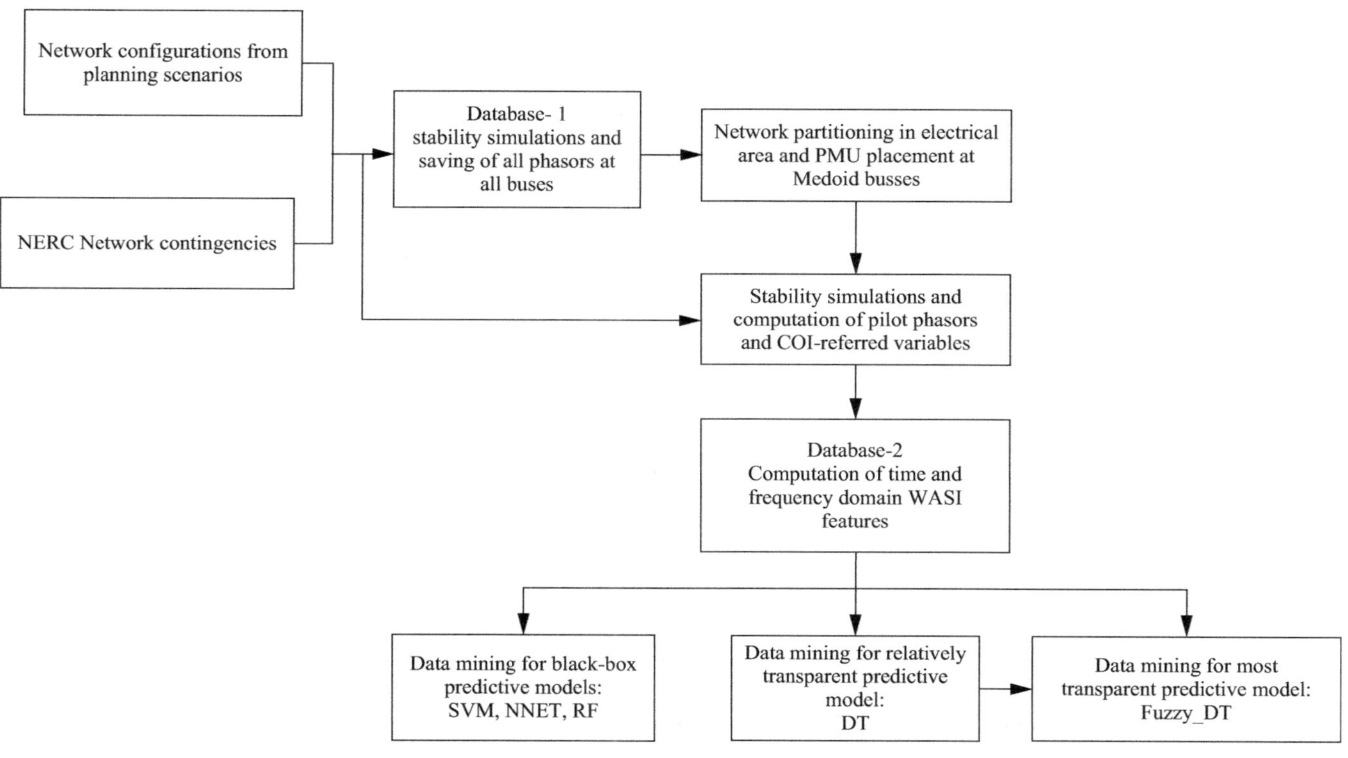

Figure 11.14 The complete framework for building data-mining model for catastrophe predictors

Table 11.3 Performance comparisons between black-box and transparent predictive models

Data-mining tools	**Performance (%)**		
	Reliability	**Security**	**Accuracy**
NNET	82.0	87.9	85.7
SVM	83.9	92.5	87.7
RF	99.9	99.0	99.2
DT	80.1	86.0	84.3
DT_Fuzzy	86.4	82.9	83.7

preparation. Summarizing further, the first stage includes: (1) collecting information about network configurations and contingencies, (2) obtaining a reduced-size database of stability results through a high-performance computation center, and (3) partitioning in to medoid busses, which are the most suitable locations for PMUs. The second stage includes extended set of simulations and collection of the bus phasor signals aggregated by area and refereed to the COI reference frame. The data-mining features in the time and frequency domains (WASI) are finally derived according to [3,22] and saved in a large database for subsequent data-mining investigations. However, while moving from black-box data mining tools to transparent one, there is a trade-off [53] exists between accuracy and transparency in classification which needs to be addressed.

Table 11.3 provides the complete statistics of the catastrophe predictors developed different data-mining tools including NNET, SVM, RF, DT, and DT_Fuzzy. It is observed that NNET provides a classification accuracy of 87.0% for the combined HQ and test system considered in the proposed study. SVM stays at 87.7%, whereas DT and DT_Fuzzy stay below NNET and SVM for accuracy measure. However, RF provides most accurate results with more than 99.0% accuracy. While comparing the reliability and security measures, it is observed that the reliability is more than 99.9% with the RF-based predictor compared to other data-mining models for the same system configuration. The reliability is above 80% for NNET, SVM, and DT. On the other hand, the reliability index of DT_Fuzzy stands at 86.0% which is improved compared to NNET, SVM, and DT. In fact, the security performance of all data-mining models is better than 86% except for the DT_Fuzzy.

While analyzing the complete statistics of accuracy, reliability, and security indices, the RF-based predictor emerges as the only tool with a more than 99% level of performance on all three indices. It is observed that, while switching from black box solutions such as NNET, SVM, and RF to semitransparent solutions such as DT, and transparent solutions such as DT_Fuzzy, there is a trade-off between accuracy, reliability, and security for the combined HQ and test, HQ and test systems.

Table 11.4 Robustness testing under noisy condition (SNR: 30 dB)

Data mining tools	Performance (%)		
	Performance accuracy	Reliability	Security
RF	98.18	95.33	99.02
DT	81.49	84.36	80.66
DT_Fuzzy	84.46	64.62	91.07

Table 11.5 Performance loss of data-mining models

Data-mining tools	Performance loss (%)		
	Reliability	Security	Accuracy
RF	4.50	−0.25	0.75
DT	−4.25	5.25	2.75
DT_Fuzzy	−0.5	1.00	1.25

Further validation is carried out to test the robustness of the algorithms. The black-box RF, semitransparent algorithm (DT) and transparent algorithm (DT_Fuzzy) are further tested under noisy conditions with signal-to-noise ratio (SNR) 30 dB, and the performance is depicted in Table 11.4. It is observed that the performance of the DT is more degraded compared to that of RF and DT_Fuzzy with noise. Thus, RF and DT_Fuzzy are found to be more robust algorithms than DT for the proposed application. Table 11.5 provides a more synthetic view of the same findings. It is observed that in presence of noise, crisp models (DT and RF) tend to suffer a more significant loss of reliability than the fuzzy logic-based model. This observation is in line with the previous investigations [54] and represents another benefit of fuzzy logic in addition to transparency and comprehensibility [23,54]. Indeed, a fuzzy rule-based system which includes symbolic rules enables ease of understanding and transfer of high-level knowledge. Further, the fuzzy sets along with fuzzy logic and approximate reasoning provide the ability to model the fine knowledge base which can handle uncertainties in application such as power systems which is highly uncertain. Thus, fuzzy representation is becoming increasingly popular which is more transparent, leading to easy implementation and ability to deal with problems of uncertainty, noise, and inexact data [31].

11.6 Conclusion

This chapter discusses data-mining-based catastrophe predictors using PMU-based WASI features. The study validates the performance of the black-box models such as SVM and RF through semitransparent data-mining models such as DT and transparent models such as DT_Fuzzy. It is observed from the extensive studies of the developed data-mining-based catastrophe predictors that while switching from the black box solutions to transparent and interpretable solutions, there is an unavoidable trade-off between accuracy, reliability, and security measures. The more transparent the predictor, the easier the implementation and maintenance by human actors. Overall, the fuzzy logic-based transparent solutions are preferred over black-box solutions to ease the implementation with improved robustness and enhance their suitability for auditing process, even sacrificing the predictive performance indices.

References

[1] G. Trudel, S. Bernard, G. Scott, "Hydro-Québec's Defense Plan Against Extreme Contingencies," *IEEE Trans. Power Syst.*, **PWRS-8**(2), May 1993, pp. 445–451.

[2] K. Mei, S. M. Rovnyak, "Response-based Decision Trees to Trigger One-Shot Stabilizing Control," *IEEE Trans. Power Syst.*, **PWRS-19**(1), Feb. 2004, pp. 531–537.

[3] I. Kamwa, R. Grondin, L. Loud, "Time-varying Contingency Screening for Dynamic Security Assessment Using Intelligent-Systems Techniques," *IEEE Trans. Power Syst.*, **PWRS-16**(3), August 2001, pp. 526–536.

[4] D. Ernst, D. Ruiz-Vega, M. Pavella, P. Hirsch, D. Sobajic, "A Unified Approach to Transient Stability Contingency Filtering, Ranking and Assessment," *IEEE Trans. Power Syst.*, **PWRS-16**(3), August 2001, pp. 435–443.

[5] Y. Mansour, E. Vaahedi, A. El-Sharkawi, "Dynamic security contingency screening and ranking using neural networks," *IEEE Trans. Neural Netw.*, **8** (15), July 1997, pp. 942–950.

[6] CIGRE Technical Brochure on "*Review of On-line Power System Security Assessment Tools & Techniques*," January 2007 (K. Morison, Convener).

[7] G. C. Ejebe, C. Jing, J. G. Waight, *et al.*, "On-Line Dynamic Security Assessment: Transient Energy Based Screening and Monitoring for Stability Limits," Presented at the *1997 IEEE/PES Summer Meeting*, Berlin, Germany.

[8] H. D. Chiang, C. S. Wang, H. Li, "Development of BCU Classifiers for On-line Dynamic Contingency Screening of Electric Power Systems," *IEEE Trans. Power Syst.*, **PWRS-14**(2), May 1999, pp. 660–666.

[9] C. Fu, A. Bose, "Contingency Ranking Based on Severity Indices in Dynamic Security Analysis," *IEEE Trans. Power Syst.*, **PWRS-14**(3), August 1999, pp. 980–986.

[10] V. Brandwajn, A. B. R. Kumar, A. Ipakchi, A. Bose, S. D. Kuo, "Severity Indices for Contingency Screening in Dynamic Security Assessment," *IEEE Trans. Power Syst.*, **PWRS-12**(3), August 1997, pp. 1136–1142.

[11] R. Schainker, P. Miller, W. Doubleday, P. Hirsch, G. Zhang, "Real-time Dynamic Security Assessment: Fast Simulation and Modeling Applied to Emergency Outage Security of the Electric Grid," *IEEE Power Energy Mag.*, **4**(2), March–April 2006, pp. 51–58.

[12] K. Sun, S. Likhate, V. Vittal, V. S. Kolluri, S. Mandal, "An Online Dynamic Security Assessment Scheme Using Phasor Measurements and Decision Trees," *IEEE Trans. Power Syst.*, **PWRS-22**(4), November 2007, pp. 1935–1943.

[13] I. Kamwa, R. Grondin, "PMU Configuration for System Dynamic Performance Measurement in Large Multi-Area Power Systems," *IEEE Trans. Power Syst.*, **PWRS-17**(2), May 2002, pp. 285–394.

[14] I. Kamwa, J. Béland, G. Trudel, R. Grondin, C. Lafond, D. McNabb, "Wide-Area Monitoring and Control at Hydro-Québec: Past, Present and Future," *Panel Session on PMU Prospective Applications, 2006 IEEE/PES General Meeting*, Montreal, QC, Canada, June 18–22, 2006.

[15] S. H. Nawab, T. E. Quatieri, "Short-Time Fourier Transform," in: *Advanced Topics in Signal Processing* (J. S. Lim, A. V. Oppenheim, eds.), Prentice Hall, Englewoods Cliff, NJ, 1988.

[16] D. R. Ostojic, "Spectral Monitoring of Power System Dynamic Performances," *IEEE Trans. Power Syst.*, **PWRS-8**(2), May 1993, pp. 445–451.

[17] C. W. Taylor, D. C. Erickson, K. E. Martin, R. E. Wilson, V. V. Venkatasubramanian, "WACS—Wide-Area Stability and Voltage Control System: R&D and On-Line Demonstration," *Proc. IEEE*, **93**(5), May 2005, pp. 892–906.

[18] I. Kamwa, S. R. Samantaray, G. Joos, "Catastrophe Predictors from Ensemble Decision-Tree Learning of Wide-Area Severity Indices," *IEEE Trans. Smart Grid*, **1**(2), September 2010, pp. 144–158.

[19] B. Kirby, S. Richards, J. Attas, R. Grondin, I. Kamwa, M. Rousseau, "Fast Topology Detection Based on Open Line Detection," *Eighth IEE International Conference on Developments in Power System Protection*, Amsterdam, Netherlands, 5–8 April 2004, pp. 1–4.

[20] M.-C. Shin, C.-W. Park, J.-H. Kim, "Fuzzy Logic-Based Relaying for Large Power Transformer Protection," *IEEE Trans. Power Deliv.*, **18**(3), February 2003, pp. 718–724.

[21] W. Rebizant, K. Feser, "Fuzzy Logic Application to Out-of-Step Protection Of Generators," *Proc. 2001 IEEE/PES Summer Meeting*, 2001, vol. 2, pp. 927–932, Vancouver, BC, Canada, 15–19 July 2001.

[22] I. Kamwa, S. R. Samantaray, G. Joos, "Development of Rule-Based Classifiers for Rapid Stability Assessment of Wide-Area Post-Disturbance Records," *IEEE Trans. Power Syst.*, **24**(1), Feb. 2009, pp. 258–270

[23] S. Guillaume, "Designing Fuzzy Inference Systems from Data: An Interpretability-oriented Review," *IEEE Trans. Fuzzy Syst.*, 9, June 2001, pp. 426–443.

[24] N. Amjady, S. F. Majedi, "Transient Stability Prediction by a Hybrid Intelligent System," *IEEE Trans. Power Syst.*, **22**(3), Aug. 2007, pp. 1275–1283.

[25] L. S. Moulin, A. P. Alves da Silva, M. A. El-Sharkawi, R. J. Marks II, “Support Vector Machines for Transient Stability Analysis of Large Scale Power Systems,” *IEEE Trans. Power Syst.*, **19**(2), May 2004, pp. 818–825.
[26] F. R. Gomez, A. D. Rajapakse, U. D. Annakkage, I. T. Fernando, “Support Vector Machine Based Algorithm for Post-Fault Transient Stability Status Prediction Using Synchronized Measurements,” *IEEE Trans. Power Syst.*, **26**(3), August 2011, pp. 1474–1483.
[27] Z. Li, W. Wu, “Phasor Measurements-Aided Decision Trees for Power System Security Assessment,” *2009 IEEE Computer Society 2nd Int. Conf. on Information and Computing Science*, pp. 358–361.
[28] J. A. Huang, S. Harrison, L. Wehenkel, G. Vanier, F. Lévesque, A. Valette, “Operation Rules Determined by Risk Analysis for Special Protection Systems at Hydro-Quebec,” *CIGRÉ Session 2004*, Paper B5-205, Paris, pp. 1–8.
[29] M. Sadeghi, M. A. Sadeghi, S. Nourizadeh, A. M. Ranjbar, S. Azizi, “Power System Security Assessment Using AdaBoost Algorithm,” *Proceedings of the North American Power Symposium (NAPS 2009)*, Starkville, Mississippi, October 2009.
[30] T. Hastie, R. Tibshirani, J. Friedman, *The Elements of Statistical Learning*, 2nd Edition, Springer-Verlag: New York, 2009, 745p.
[31] J. Abonyi, B. Feil, *Cluster Analysis for Data Mining and System Identification*, Basel-London-Berlin: Birkhauser, 2007, 319p.
[32] G. Trudel, J. -P. Gingras, J.-R. Pierre, “Designing a Reliable Power System: The Hydro-Québec’s Integrated Approach,” *IEEE Proc. Spec. Issue Energy Infrastruct. Def. Syst.*, **93**(5), May 2005, pp. 907–917.
[33] I. Kamwa, A. K. Pradhan, G. Joos, S. R. Samantaray, “Fuzzy Partitioning of a Real Power System for Dynamic Vulnerability Assessment,” *IEEE Trans. Power Syst.*, 24(3), August 2009, pp. 1–10.
[34] V. Kolluri, S. Mandal, M. Y. Vaiman, M. M. Vaiman, S. Lee, P. Hirsch, “Fast Fault Screening Approach to Assessing Transient Stability in Entergy’s Power System,” 2007 *IEEE/PES General Meeting*, Tampa, FL, USA, June 24–27, 2007
[35] C. W. Liu, M. C. Su, S. S. Tsay, Y. J. Wang, “Application of a Novel Fuzzy Neural Network to Real-Time Transient Stability Swings Prediction Based on Synchronized Phasor Measurements,” *IEEE Trans. Power Syst.*, **PWRS-14**(2), May 1999, pp. 685–692.
[36] I. Kamwa, R. Grondin, E. J. Dickinson, S. Fortin, “A Minimal Realization Approach to Reduced-Order Modelling and Modal Analysis for Power System Response Signals,” *IEEE Trans. Power Syst.*, **PWRS-8**(3), August 1993, pp. 1020–1029.
[37] J. F. Hauer, “Application of Prony Analysis to the Determination of Modal Content and Equivalent Models for Measured Power System Response,” *IEEE Trans. Power Syst.*, **6**(3), August 1991, pp. 1062–1068.
[38] L. Breiman, J. H. Friedman, R. A. Olshen, C. J. *Stone: Classification and Regression Trees*. Chapman & Hall: New York, 1984.

[39] L. Rokach, O. Maimon, "Top-Down Induction of Decision Trees Classifiers—A Survey," *IEEE Tran. Syst. Man Cybern.—Part C*, **35**(4), November 2005, pp. 476–487.

[40] D. Williams "Rattle (the R Analytical Tool to Learn Easily)," ver. 2.3, May 2008: http://rattle.togaware.com/.

[41] S. R. Safavian, D. Landgrebe, "A Survey of Decision Tree Classifier Methodology," *IEEE Tran. Syst. Man Cybern.*, 21, May/June 1991, pp. 660–674.

[42] Y. Yuan, M. J. Shaw "Induction of Fuzzy Decision Trees," *Fuzzy Sets Syst.*, **69**, 1995, 125–139.

[43] J. Zeidler, M. Schlosser "Continuous-Valued Attributes in Fuzzy Decision Trees," *Proceedings of Information Processing and Management of Uncertainty in Knowledge-Based Systems*, Granada, 1996, pp. 395–400.

[44] E. C. C. Tsang, X. Z. Wang, Y. S. Yeung "Improving Learning Accuracy of Fuzzy Decision Trees by Hybrid Neural Networks," *IEEE Trans. Fuzzy Syst.*, **8**(5), 2000, 601–614.

[45] X. Z. Wang, B. Chen, G. Qian, F. Ye "On the Optimization of Fuzzy Decision Trees," *Fuzzy Sets Syst.*, 112, 2000, 117–125.

[46] K. Crockett, Z. Bandar, D. Mclean, J. O'Shea "On Constructing a Fuzzy Inference Framework Using Crisp Decision Trees," *Fuzzy Sets Syst.*, **157**(1), 2006, 2809–2832.

[47] D. S. Siroky, "Navigating Random Forests and Related Advances in Algorithmic Modeling," *Stat. Rev.*, **3**, 2009, pp. 147–163. [Online]: http://projecteuclid.org/DPubS?service=UI&version=1.0&verb=Display&handle=euclid.ssu.

[48] K. Alcheikh-Hamoud, N. Hadjsaid, Y. Bésanger, J. P. Rognon, "Decision Tree Based Filter for Control Area External Contingencies Screening," *2009 Bucharest PowerTech*, Bucarest, Romania, 2009.

[49] S. R. Samantaray, "Ensemble Decision Trees for High Impedance Fault Detection in Power Distribution Network," *Int. J. Electr. Power Energy Syst.*, **43**(1), December 2012, pp. 1048–1055.

[50] S. R. Samantaray, I. Kamwa, G. Joos, "Ensemble Decision Trees for PMU based Wide-Area Security Assessment in the Operations Time-Frame," *IET Gen. Transm. Distrib.*, **4**(12), December 2010, pp. 1334–1348.

[51] Smart Grid: "Enabler of the New Energy Economy," U. S. Dept. Energy, Dec. 2008 [Online]. Available: http://www.oe.energy.gov/Doc-umentsand-Media/final-smart-grid-report.pdf, Electric Advisory Committee.

[52] R. E. Banfield, L. O. Hall, K. W. Bowyer, W. P. Kegelmeyer, "A Comparison of Decision Tree Ensemble Creation Techniques," *IEEE Trans. Pattern Anal. Mach. Intell.*, **29**(1), November 2007, pp. 173–180.

[53] I. Kamwa, S. R. Samantaray, G. Joos, "On the Accuracy vs Transparency Trade-off of Data-mining Models for Fast-response PMU-based Catastrophe Predictors," *IEEE Trans. Smart Grid*, **3**(1), pp. 152–161, 2012.

[54] J. Abonyi, J. A. Roubos, F. Szeifert, "Data Driven Generation of compact. Accurate, and linguistically sound fuzzy classifiers based on a decision tree initialization," *Int. J. Approximate Reasoning*, **32**, 2003, pp. 1–21.

Index

accuracy index (AI) 140–1
accuracy vs transparency trade-off 306, 324–7
adaptive search–based algorithm 255–8, 270
adaptive simulated annealing (ASA) algorithm 251
Advanced Modeling and Programming Language (AMPL) 43–4
alternating current (AC) system 2–3
Americas, wide area measurement system (WAMS) in 11
angular difference curves for different clearing time 117–18
artificial neural network (ANN) algorithms 233, 251
Asia, wide area measurement system (WAMS) in 12–13
asymmetric faults 178

Backward Euler's Method 259
Bonneville Power Administration (BPA) 11
bus voltage deviation 64–5

catastrophe predictor, data-mining models for 314
center-of-inertia (COI) angle deviation 308, 326
center of inertia (COI)-based rotor angle 63–4
centralized real-time applications 240
China Southern Power Grid 12
China's power transmission grid 12
chi-square test–based bad data identification method 131
circuit breakers (CBs) 60
clustering methods 147, 155, 161, 167, 169
  application of density-based spatial clustering of applications with noise clustering 152–4
  application of *k*-means clustering 148–52
  multitier *k*-means clustering method 154
  openPDC setup and data extraction process 147–8
combined uncertainty, calculation of 282
composite load model (CLOD) approach 247–9
  in PSSE 248–9
  WECC 248
computational run time (CPU) 52
confidence intervals (CIs) 140–1
constant current load 138, 244
constant impedance load 244–5
constant power load 244, 252
controlling unstable equilibrium point (CUEP) 213, 261–2
conventional state estimation 282–5
current transformer (CT) 1, 112, 183, 250

damping torque 17
data-count generation process 315
data mining and visualization providing basis for decision-making 149
data-mining model–based catastrophe predictors 314

accuracy vs transparency trade-off 324–7
data-mining models for catastrophe predictor 314
performance assessment 314
decision tree 314–19
DT_Fuzzy 321–4
ensemble decision trees (random forests) 319–21
scenarios and data count generation 314
data-mining models 310
decision tree 310–12
decision tree-induced fuzzy approach 312–13
ensemble decision trees 313–14
decentralized voltage stability (DVS) assessment 240
distributed coordination platform 240–1
initialization of decentralized application 241
monitoring method 242–3
decision tree (DT) 233, 306–7, 310–28
-induced fuzzy approach 312–13
density-based spatial clustering 152–4
density-based spatial clustering of applications with noise (DBSCAN) 33, 147, 152–6, 161–5, 167, 169–74
differential and algebraic equations (DAEs) 258–60, 262
direct current (DC) 3, 180, 218–19
direct underreaching transfer trip (DUTT) scheme 60, 85–9, 91–3
discrete Fourier transform (DFT) 31, 62, 91, 104–5, 108, 278
discrete wavelet transform (DWT) 104–5, 108
Distributed Coordination Computational Algorithms (DCBlocks) 240
distributed generations (DGs) 29
disturbance detection index 71
disturbance records (DR)/event logger (EL) outputs 15
DOE (Department of Energy) 11
double line-to-ground (LLG) faults 178
Dshapelet 105
DT_Fuzzy 307, 314, 321–4, 326–8
Dunn's index (DI) 161, 174–5
dynamic load models 137, 246–7
dynamic security assessment (DSA) 214, 305–6, 321
dynamic state estimator (DSE) tool 116–17

eastern India test system
block diagram of 67
case study for 66–9
Edison, Thomas 3
EHC (electro hydraulic controller) 20
Eigenvalue analysis 17
Electric Power Group (EPG) 233
electric power system 2, 231
electromagnetic transient design and control (EMTDC) 62
empirical mode decomposition (EMD) 65–6
energy function-based direct methods 213
energy management system (EMS) 4, 8, 123–4, 128, 262, 279
ensemble decision trees 313–14, 319–21
equivalent current phasor angle (ECPA) 179
fault detection using 182–91
fault localization using 195–208
oscillations in 183
real-time estimation of 179–82
variation 182, 184, 196, 204–9
equivalent voltage phasor angle (EVPA) 179
deviation 195–204
fault detection using 182–91
oscillations in 183
real-time estimation of 179–82

Europe, wide area measurement system (WAMS) in 12
Explicit Euler Method 259
Explicit Runge–Kutta Methods 259
exponential model 245–6

failure index 78, 81
fast Fourier transform (FFT) 182–4, 190–2, 195–202, 204–9
fault detection using EVPA and ECPA 182–91
fault detector index 71–4
fault localization 191, 194
  using ECPA in faulty branch 195–208
  using SVM classifier 196
fault location algorithm 129, 131, 143–5, 195
fault monitoring, detection, and correction
  clustering methods 147
    application of density-based spatial clustering of applications with noise clustering 152–4
    k-means clustering, application of 148–52
    multitier *k*-means clustering method 154
    openPDC setup and data extraction process 147–8
  hybrid method, need for 155
    advantages of using proposed multitier *k*-means method 160
    development of multitier *k*-means method 155, 157–9
  results and discussions 161
    computational time 170–3
    Dunn's index 174–5
    quantification of the visual data representation 169–70
    results obtained from data clustering algorithms 161–75
    system under heavy load condition (test case) 163–5
    system under light load conditions (test case) 165–7
    system under single line to ground fault condition (test case) 167–9
    system under steady state (normal) conditions (test case) 161–3
    test results with current data 169
    visual fidelity 173–4
fault monitoring systems 178, 209
faulty branch discrimination methodology 195
filter effect (FE) of state estimator algorithm 295
FLS output unit 69
frequency-based power imbalance index 64
frequency domain–based methodologies 179

Gaussian error distribution 280–1
generalized cross validation (GCV) method 291
global positioning system (GPS) 2, 7, 100, 232
global positioning system (GPS) clock 30–1, 232, 278
Gram–Schmidt reorthonormalization (GSR) 217
GTNET card 266
guide to the uncertainty in measurements (GUM) 280

heuristic techniques, algorithms using 251
Hilbert Huang transform (HHT), damping estimation using 65–6
HVDC (high voltage direct current) line 15, 27
hybrid method, need for 155
  multitier *k*-means method
    advantages of using 160
    development of 155, 157–9

hybrid state estimation, preprocessing 287–90
Hydro-Québec (HQ) 783-bus system 307

IEEE-14 bus system 36–7, 39–43, 184–93, 196–209, 252, 297
  case study for 296–300
  observability index for various placement solutions in 41
  with zero-injection bus 40
IEEE 39 bus system 70
IEEE 118 bus system 49–50, 52
  with contingency cases 53
IEEE 300-bus system 45
IEEE synchrophasor standard 76
Independent System Operator New England (ISONE) 239
independent system operators (ISOs) 6–7, 30, 33
India, synchrophasors in 12–15, 19–20
Indian power system 5
input–output dynamic models 246
instrument variable (IV)–based technique 251
integral mode functions (IMFs) 65
intelligent electronic devices (IEDs) 124, 278

KEPCO system, WAVI in 236–7
Kirchhoff's current law 38
k-means clustering 147–52, 165
  advantages of 160
  application of 148
  multitier *k*-means method 155–9
  pseudocode for 151
k-nearest neighbor (kNN) algorithm 106
K-step slope value 108
K-WAMS (Korean Wide Area Measurement and Monitoring System) project 236

L-curve method 291
  selection of regularization parameter by 292–3

least-square based parameter estimation 251
least-square method 253
Lie derivative 220
linear programming (LP), time computation using 52
linear regression (LR) method 52
  forecasting the scalability of OPP using 52–5
linear state estimation 284–6
line-to-line (LL) faults 178
line-to-line-to-line (LLL) faults 178
line-to-line-to-line-to-ground (LLLG) faults 178
load encroachment 113
load estimation and stability analysis, synchrophasor applications for 231
  PMU-based load modeling: *see* phasor measurement unit (PMU)-based load modeling
  transient stability assessment using synchrophasor data 257
    energy function–based approach 260–2
    equal area criteria–based approach 259–60
    MLE-based transient stability assessment 262–9
    time domain approach 258–9
  voltage stability assessment 232
    decentralized voltage stability assessment 240–3
    industrial voltage stability application 233–40
    review of existing PMU-based voltage stability 232–3
load modeling tool 231
load model validation procedure 140–2
low-frequency electromechanical oscillations (LFO) 17–18
Lyapunov exponents (LEs) 214, 232
Lyapunov method 260–1

malfunction index 78
MATLAB® 148, 170, 252
maximum likelihood estimation 125
maximum Lyapunov exponent 214–15, 231
 -based transient stability assessment 262
  estimation of MLE from PMU data 263–4
  test study 264–9
 calculation algorithm 215
  computation of MLE 217
  prediction of system stability 217–18
  spectral analysis 215
  state calculator 215–17
metal oxide varistor (MOV) 60, 84–5
model (.mod) files 44
modern grid, challenges of 3–4
multiarea state estimator (MASE) 301
multi-machine power system 259, 266
multitier *k*-means method 154–9
 advantages of using 160
 categories 157
 development of 155, 157–9
 line-to-ground fault 168

N_bus 69
New England (NE) 39 bus systems 263, 266
New England 10-generator system 117–18
Newton–Raphson technique 144
Newton–Raphson's method 217
nonlinear system identifications 250
 analytical-based approach 250
 optimization-based approach 250–2
normalized voltage coefficients (NVCs) 191–2
North American Reliability Council (NERC) 11, 102

observability matrix 221, 225–6
Ohm's law 3, 32, 34
one machine infinite bus (OMIB) system 259–60
one-step SVM classifier 192
online load model validation 138–40
 data collection 138
 event identification 138–9
 event replication and results comparison 139–40
 generalization capability test 139
online transmission line parameter estimation 129–36
Open Multistep Formulas 259
openPDC setup and data extraction process 147–8
optimal placement problem (OPP) 43
 Advanced Modeling and Programming Language (AMPL) 43–4
 enforcing optimal redundancy criterion (ORC) 46
 forecasting the scalability of OPP using linear regression 52–5
 justification of ORC index 49–52
 PMU placement costs 48–9
 SORI 46–8
 for synchrophasors 34
 time computation using LP 52
 zero-injection buses 44–6
optimal redundancy criterion (ORC) 29, 41–2
 enforcing 46
 justification of 49–52
 optimal redundancy criterion (ORC) 41–2
 phasor measurement units (PMU), optimal placement of 33
  need for OPP for synchrophasors 34
 PMU placement problem formulation using LP 34
  problem formulation 36
  system with no zero-injection buses 36–8
  system with zero-injection measurements 38–9

on standard test systems 43
Advanced Modeling and Programming Language (AMPL) 43–4
enforcing optimal redundancy criterion (ORC) 46
forecasting the scalability of OPP using linear regression 52–5
justification of ORC index 49–52
PMU placement costs 48–9
system observability redundancy index (SORI) 46–8
time computation using LP 52
zero-injection buses 44–6
synchrophasors 30
advantages of synchrophasors over SCADA measurements 32
challenges with synchrophasor measurements 32–3
system observability redundancy index (SORI) 39–41
oscillations monitoring system (OMS) 19
out-of-step (OOS) detecting relay 115–17
out-of-step condition 63, 66–9

parent bus identification methodology 195
Park's transformation 180–1, 183, 191
particle filter (PF) 116–17
P_bus 69–70
performance assessment 314
decision tree 314–19
DT_Fuzzy 321–4
ensemble decision trees 319–21
phasor data concentrator (PDC) 7, 9, 12–13, 27, 31, 59–61, 100, 114, 147
phasor measurement 7
phasor measurement unit (PMU) 2, 7–8, 11, 13, 21–2, 30–2, 59, 100, 102, 104, 115, 147, 211, 231, 277–9, 284–8, 295
-based monitoring and assessment of rotor-angle stability 214
Maximum Lyapunov exponent (MLE) calculation algorithm 215–18
Maximum Lyapunov exponent (MLE) method 214–15
-based monitoring of power systems 213–14
-based monitoring of system dynamics 218
dynamic observability 220–1
power system dynamics 221–2
static observability 218–19
topology observability 219–20
cascaded events and vulnerability assessment 211–13
example 222–5
measurement accuracy of 278
optimal placement of 33
need for optimal placement problem (OPP) for synchrophasors 34
placement costs 48–9
placement problem formulation, using LP 34
problem formulation 36
system with no zero-injection buses 36–8
system with zero-injection measurements 38–9
phasor measurement unit (PMU)-based load modeling 243
classification 244
composite load 247–9
dynamic load 246–7
static load 244–6
definition 243–4
load model parameters, estimation of 249–50
measurement-data based approach 249–50
physical-component based approach 249

parameter identification techniques 250–2
static load parameters, estimation of 252–7
synchronized phasor measurements, advantages of 250
phasor measurement unit (PMU)-based wide-area security assessment 305
data-mining model–based catastrophe predictors 314
accuracy vs transparency trade-off 324–7
data-mining models for catastrophe predictor 314
performance assessment 314–24
scenarios and data count generation 314
data-mining models 310
decision tree 310–12
decision tree (DT)-induced fuzzy approach 312–13
ensemble decision trees 313–14
system studied using wide area monitoring 307
wide-area severity indices (WASI) 308
features 308–10
and stability condition 310
phasor measurement unit (PMU) measurements 123, 177
power system model, monitoring of 127
load model monitoring 136–42
online transmission line parameter estimation 129–36
system-level model monitoring 142
power system operation, monitoring of 123
background 123–4
state estimation solution 124–6
use of PMUs for state estimation 126–7
wide-area fault location based on PMU measurements 143
fault location algorithm 143–5
physical-model based dynamic models 246–7
polynomial model 244–5
postprocessing sequential state estimation 286–7
potential energy boundary surface (PEBS) 261
power grid 1, 6, 9–10, 123
defined 2
evolution of 2–3
real-time transmission line fault monitoring in 177–9
power swing, protection support during 63
eastern India test system, case study for 66–9
method 66
simulated test system 66
wide-area severity indices 63
bus voltage deviation 64–5
center of inertia (COI)-based rotor angle 63–4
damping estimation using Hilbert Huang transform (HHT) 65–6
frequency-based power imbalance index 64
power system computer aided design (PSCAD) 62
power system dynamics, PMU-based monitoring of 221–2
power system model, monitoring of 127
load model monitoring 136–42
online transmission line parameter estimation 129–36
system-level model monitoring 142
power system operation, monitoring of 123
background 123–4
phasor measurement units (PMUs), for state estimation 126–7
state estimation solution 124–6

Power System Operation Corporation Ltd. 14
power system stabiliser (PSS) 24–6
Predictor–Corrector Methods 259
principal component analysis (PCA) 105
principal components (PCs) 105
programmable logic controllers (PLCs) 278
protection element failure, online identification of 75
  component failure indices 75–6
  faulted line identification 75
  method 77–8
  relay underreaching (case study) 78–81
  signal communication 76–7
protection support scheme, block diagram of 68
protective relaying 59
pseudo-gradient evolutionary programming 240
PSLF^{TM} dynamic simulation tool 107

Quanta Technology 237

random forests (RFs): *see* ensemble decision trees
R-bus relay setting 90–1
reactive power margin method 237
Real Time Dynamics Monitoring System (RTDMS) 233–5
real-time power system (RTDS) 265–6
real-time transmission line fault monitoring in power grids 177–9
recursive least-square estimation 252–5
region of stability existence (ROSE)
  by VR energy 239
regularization method 290
  by *L*-curve method 292–3
  regularized state estimation problem 293–4
regularized state estimation problem 293–4
remote terminal units (RTUs) 232, 278
rotor angle–based prediction of loss of synchronism between two areas 63
rotor-angle dynamics of power system 216
rotor-angle instability 212
rotor angle stability 17
  PMU-based monitoring and assessment of 214–18

S difference indicator (SDI) 235–6
search algorithms 250–1
security constrained optimal power flow (SCOPF) 100
series compensated line 60, 82
  adaptive method 83–5
  result 85
shapelets 105, 108
short-time-Fourier-transform (STFT) analysis window 308
simulated test system 66
single line-to-ground (LG) faults 178
singular value decomposition (SVD) 221, 292
slope sequence of DS (SSD) 108, 111
small signal stability 17
smart power grids (SPG) 177
  transmission line fault localization in 194
special protection scheme (SPS) 16–17, 308
  validation of performance of 26
stable equilibrium point (SEP) 261
Stanley, William, Jr. 3
state estimation 8, 277
  assigning weights to the measurements 280
    calculation of combined uncertainty 282
    Gaussian error distribution 280–1
    uniform error distribution 281–2

case study 294
for IEEE 14-bus test system 296–300
conventional state estimation 282–4
flowchart 285
functions 279
linear state estimation 284–6
measurement model for static state estimation 279–80
phasor measurement unit 278–9
postprocessing sequential state estimation 286–7
preprocessing hybrid state estimation 287–90
regularization method 290
regularized state estimation problem 293–4
selection of regularization parameter by *L*-curve method 292–3
SCADA measurements 278
sequential state estimation flowchart 288
state estimator (SE) algorithms 8, 232, 277, 282
State Grid Corporation of China 12
static load model 137–8, 244–6
static load parameters, estimation of 252
adaptive search–based algorithm 255–7
recursive least-square estimation 252–5
static state estimation 213, 218–19, 221
measurement model for 279–80
statistical techniques, algorithms using 251
Steinmetz, Charles Proteus 7
sum of squared error (SSE) 150
supervisory control and data acquisition (SCADA) system 1, 4–6, 30, 99–100, 124, 127, 277–8, 286, 288
architecture of 5
-based energy management systems (SCADA/EMS) 232
data acquisition process in 5
hierarchical SCADA system for large grid 6
phase angle measurement in 10
vs PMU measurements 32
synchrophasors vs 32
vs wide area measurement system (WAMS) 10–11
supervisory protection 115
support vector machine (SVM) 106–7, 191–2, 194, 196, 209
symmetric faults 178
synchronising torque 17
synchrophasor 30, 59–61, 102
challenges with synchrophasor measurements 32
economics 32
reliability 33
utilization 33
classification 105
k-nearest neighbor (kNN) algorithm 106
support vector machine (SVM) classifier 106–7
disturbance identification using PMU-generated disturbance files (case study) 107
data 107–8
results 109–11
selection of features and classifiers 108
feature extraction 104
discrete Fourier transform (DFT) 104
discrete wavelet transform (DWT) 104
principal component analysis (PCA) 105
shapelets 105
measurements 2
for power system protection 111

potential improvements in existing protection applications 113–14
real-time constraints 111
suitability of PMU measurements for protection 111–13
supervisory protection—potential use of synchrophasors 114–15
synchrophasor-assisted out-of-step protection for generators (case study) 115–20
vs SCADA measurements 32
state monitoring with 100–1
synchrophasor technology 1–2, 6, 284
applications of synchrophasors data 13
fault detection, classification and analysis 14–17
insight to coherency of generators 21
low-frequency oscillations 17–20
power plant performance, improving 24
remote detection of system separation 21–3
synchronising two grids from remote 23
challenges of modern grid 3–4
evolution of power grid 2–3
phasor measurement 7
supervisory control and data acquisition (SCADA) system 4–6
vs wide area measurement system (WAMS) 10–11
time synchronised measurements 7–9
way-forward 27
wide area measurement system (WAMS) 9, 11
in Americas 11
in Asia 12–13
in Europe 12
system integrity protection scheme (SIPS) 114
system-level model monitoring 142
system observability redundancy index (SORI) 39–41, 46–9
system protection center (SPC) 177

Tehri Hydro Electric Power Plant 24
Thevenin-based voltage stability monitoring algorithm 242
Thevenin Equivalent parameters 233
3-bus power system 222
three-terminal line 60, 85
adaptive method 86–91
results 91
system condition, test for variation in 92–4
Tikhonov regularization 290–2, 298, 300
time computation using linear programming (LP) 52
time domain–based methodologies 179
time-domain simulation methods 213
time synchronised measurements 7–9
total statistical error variance (TSEV) 295
total vector error (TVE) 278
transducer 4
transient stability 17
transient stability assessment using synchrophasor data 257
energy function–based approach 260–2
equal area criteria–based approach 259–60
MLE-based transient stability assessment 262–9
time domain approach 258–9
transmission line fault monitoring 177
EVPA- and ECPA-based transmission line fault classification 191–4
EVPA- and ECPA-based transmission line fault detection 179

fault detection using EVPA and ECPA 182–91
real-time estimation of EVPA and ECPA 179–82
EVPA- and ECPA-based transmission line fault localization 194
fault localization using ECPA in faulty branch 195–208
faulty branch discrimination methodology 195
parent bus identification methodology 195
in power grids 177
emerging methodologies for fault monitoring 178–9
two-source power system
double-voltage source model for 234
with synchrophasor measurement 234

UMPP (Ultra Mega Power Project) 15
Unified Real-Time Dynamic State Measurement (URTDSM) project 27
uniform error distribution 281–2
Universal Time Coordinate (UTC) time information 266
Universal Time Frame (UTC) 31
unstable equilibrium points (UEPs) 213, 261
USP of Synchrophasor technology 10

Visual Studio 148
voltage stability assessment, using synchrophasor data 232
decentralized voltage stability assessment 240–3
industrial voltage stability application 233–40
review of existing PMU-based voltage stability 232–3
voltage stability index (VSI) 232, 236
voltage stability using SDI 235–6
voltage transformer (PT) 1

WAVI (wide-area voltage stability index)
in KEPCO system 236–7
weighted least square (WLS) estimation 125–6
weighted least squares (WLS)-SE 280
Western Electricity Coordinating Council (WECC) 239, 248
wide-area fault location based on PMU measurements 143
fault location algorithm 143–5
wide-area measurement system (WAMS) 2, 9, 11, 21, 30–1, 59
adaptive protection 82
series compensated line 82–5
three terminal line 85–94
adaptive technique, flow diagram for 92
in Americas 11
in Asia 12–13
architecture 60–1
communication 61
delay causes 62
in Europe 12
hierarchy 61
improved network protection during stressed conditions 62
improved zone 3 operation 69–74
protection support during power swing 63–9
online identification of protection element failure 75
component failure indices 75–6
faulted line identification 75
method 77–8
relay underreaching (case study) 78–81
signal communication 76–7
supervisory control and data acquisition (SCADA) system vs 10–11

wide area monitoring (WAM) system 2, 8–9
wide-area severity indices (WASI) 63, 308
  bus voltage deviation 64–5
  center of inertia (COI)-based rotor angle 63–4
  damping estimation using Hilbert Huang transform (HHT) 65–6
  features 308–10
  frequency-based power imbalance index 64
  and stability condition 310

zero-injection buses 34, 38, 44–6
  IEEE 14 bus system with 40
ZIP load model 244–5, 252
zone 3 operation 69
  case study 72–4
  method 70–2
  synchrophasor placement 69–70